W9-ADG-394

INSECT–FUNGUS SYMBIOSIS

Insect–Fungus Symbiosis

NUTRITION, MUTUALISM, AND COMMENSALISM

Proceedings of a symposium organized and sponsored by the
SECOND INTERNATIONAL MYCOLOGICAL CONGRESS
August 27–September 3, 1977 at the
University of South Florida, Tampa, Florida

Edited by LEKH R. BATRA

ALLANHELD, OSMUN & CO.
MONTCLAIR

A HALSTED PRESS BOOK
JOHN WILEY & SONS
NEW YORK CHICHESTER BRISBANE TORONTO

ALLANHELD, OSMUN & CO. PUBLISHERS, INC.

19 Brunswick Road, Montclair, N.J. 07042

Published in the United States of America in 1979
by Allanheld, Osmun & Co.
Distribution: Halsted Press,
a division of John Wiley & Sons, Publishers
605 Third Avenue, New York, New York 10016

Library of Congress Cataloging in Publication Data

Main entry under title:

Insect-fungus symbiosis.

"A Halsted Press book."
Includes index.
1. Insect-plant relationships—Congresses.
2. Fungi—Ecology—Congresses. 3. Symbiosis—
Congresses. I. Batra, Lekh R. II. International
Mycological Congress, 2d, University of South
Florida, 1977.
QL461.1495 595.7'05'24 78-20640
ISBN 0-470-26671-6

Printed in the United States of America

Contents

Contributors and Their Affiliations

Lekh R. Batra
U.S. Department of Agriculture
Agricultural Research Service
Plant Protection Institute
Mycology Laboratory, BARC
Beltsville, Maryland 20705

Suzanne W. T. Batra
U.S. Department of Agriculture
Beneficial Insect Introduction Laboratory
Insect Identification and Beneficial Insect Introduction Institute, BARC
Beltsville, Maryland 20705

M. P. Coutts
Forestry Commission
Northern Research Station
Roslin, Midlothian
Scotland

Gerhard Jurzitza
Botanisches Institut und Botanischer Garten der Universität (TH)
Karlsruhe, Germany

L. T. Kok
Department of Entomology
Virginia Polytechnic Institute and State University
Blacksburg, Virginia 24061

J. L. Madden
Faculty of Agricultural Science
University of Tasmania
Hobart, Tasmania

Stephen T. Moss
Department of Biological Sciences
Portsmouth Polytechnic
Portsmouth, U.K.

Dale M. Norris
Department of Entomology
University of Wisconsin
Madison, Wisconsin 53706

Isabelle I. Tavares
Herbarium, Department of Botany
University of California
Berkeley, California 94720

Neal A. Weber
Department of Biological Science
The Florida State University
Tallahassee, Florida 32306

Howard C. Whisler
Department of Botany
University of Washington
Seattle, Washington 98195

Preface

MARCHA DA SAÚVA*

Either Brazil puts an end to the saúva,
Or the saúva will put an end to Brazil.

[repeat]

There are saúva in the field,
There are saúva in the yard,
But where there are the most saúva
Is in the Federal district.
These are the worst
Saúva, Mr. Cabral,†
Those who don't work
Get their hands into the capital.**

The interactions and interdependence at various levels between unrelated organisms have fascinated professional biologists and others throughout history. The above song tells about the dilemma of the common man and the economic importance of *saúva,* the attine ants, in Brazil. Similar folklore abounds in Africa and India about the termites and attests to the nonesoteric nature of the subject. Also, the number of monographs on mutualistic symbioses, commensalism and related topics appearing in recent years further testifies to rapidly growing interest in fungi associated

*Popular during Brazilian Carnival of 1955. Original Portuguese follows: "Ou o Brasil acaba com a saúva,/ Ou a saúva acaba com o Brasil./ [repeat] Tem saúva na lavoura,/ Tem saúva no quintal,/ Mas onde tem mais saúva/ E no Distrito Federal./ Essa é a pior/ Saúva, Seu Cabral,/ Quem nao trabalha/ Mete a nao no capital." By Arlindo Marques Jr. and Roberto Roberti, Alvarenga Ranchibo & Orchestra, Odeon Record 13740-B. Text and translation prepared by R. W. Lichtwardt.
†Reference to Pedro Alvares Cabral, who "discovered" Brazil and claimed it for Portugal.
**—This is a pun in Portuguese, just as in English.

with the Arthropoda. I was therefore pleased when R. W. Lichtwardt, a member of the Program Committee, asked me to organize a symposium on fungus-arthropod mutualism and commensalism for the Second International Mycological Congress in Tampa, Florida, August 27–September 3, 1977. This was in keeping with the tradition begun at the Exeter Congress in 1971, where several symposia were arranged under the collective title "Symbiosis and Pathogenicity."

In consultation with some of the prospective contributors, I soon realized that publication of the proceedings was desirable. During November 1975, I wrote to the contributors that the general theme of the symposium on mutualistic and commensalistic fungi would be the synecology of the symbionts, i.e. the interrelationships of the two symbionts and their morphological or biochemical adaptive modifications that enable them to perpetuate such subtle and complex relationships. The taxonomy of the fungal symbionts, as it may relate to the general theme, would also be discussed.

In this symposium, 'symbiosis' and symbiont (=symbiote), respectively, refer to mutualistic symbiosis or organism only. However, in Chapter 8, Moss aptly points out in conjunction with the discussion of other related terms and concepts, that ". . . commensalism, mutualism and parasitism should be considered as special cases of symbiosis, although it is often difficult and may not be desirable to separate one type of relationship from the other."

Mutualism is an obligatory, mutually beneficial interdependence between unrelated symbionts. In commensalism a measurable benefit accrues to one, but the other symbiont is not affected. The symbionts have evolved complex mechanisms to perpetuate such relationships. Once established, these relationships may determine further co-evolution of both partners. This symposium gives a comprehensive account of current research on the subject. It discusses unresolved, complex and daunting problems of the nutrition of symbiotic mycetophagous or xylomycetophagous insects, with opportunities and suggestions to solve them. The symposium offers clues to those engaged in developing pest control technology for dealing with these economically important insects and their associated fungi to best advantage.

Mutualism and commensalism must be ecological necessities, for the myco-symbionts do not live apart from their arthropod partners. If necessary, what mechanisms or modifications ensure their association from generation to generation? How are the quiescent or reproductive periods of the two coordinated? How do the symbionts cope with each other's metabolic products? How are the foreign organisms, with nearly similar attributes, discouraged or eliminated from the habitat, with preference for the specific fungus? For the endosymbionts, how does the fungal symbiont overcome rejection by the recipient protoplast and to be accepted as its "own"?

The first chapter of the book is devoted to a general view of arthropod-fungus associations (H. C. Whisler, Seattle). Several following chapters answer many questions with respect to mutualistic fungi and ants (N. A. Weber, Tallahassee), ambrosia beetles (D. M. Norris, Madison and L. T. Kok, Blacksburg), siricid wood wasps (J. L. Madden, Hobart, Tasmania, and M. P. Coutts, Roslin, Scotland), Anobiidae (G. Jurzitza, Karlsruhe), and Macrotermitinae (L. R. and S. W. T. Batra, Beltsville). The remaining part of the book is devoted to the commensalism of Trichomycetes (S. T. Moss, Portsmouth, England), Laboulbeniales (I. I. Tavares, Berkeley), and the final chapter by me deals with the conclusions and a discussion of aposymbiosis—a condition where the mutualistic or commensalistic symbionts may be deprived of their partners by chemical or surgical means. This has important implications for the control of major pests such as attine ants, termites and ambrosia beetles. One of the challenges that emerges from the symposium is this: is it possible to use aposymbiosis to our advantage in pest management in preference to, or in combination with, other means that may be uneconomical or otherwise undesirable?

Fungi are a highly versatile, plastic, and adaptable pleomorphic group. They often abstract and transport nutrients from diverse substrata, sometimes those with very low levels of nitrogen such as wood and plant sap. These nutrients thus can be used by the insects. Insect-fungus mutualism and commensalism have seemingly co-evolved several times since they occur in diverse orders and families of the Arthropoda and the fungi. The contributors to the symposium hope that by including up-to-date, comparative and comprehensive information in this volume it will: (1) benefit biologists of diverse interests; (2) enhance potential for specific future research in the nutritional requirements of insects and fungi that symbiotically exploit plants or their residues; and (3) explain symbiosis in biochemical terms and elucidate the spectacular evolutionary convergence, previously known only from the point of view of anecdotal natural history.

I express my appreciation to the following contributors and/or reviewers of one or more chapters: S. W. T. Batra, L. V. Knutson, P. L. Lentz, P. B. Marsh, P. D. Millner, and F. G. Pollack, all of U.S. Department of Agriculture, Beltsville; E. J. Moore, Glassboro State College, New Jersey; J. C. M. Jonkman, the Royal Netherlands Embassy, Washington; and D. M. Norris, University of Wisconsin; to several authors, editors and publishers for the permission to reproduce illustrations which are acknowledged in the text; to J. A. Romberger and S. W. T. Batra for editorial guidance; to the officers of the congress for their financial and administrative support, particularly to the members of the Congress Program Committee chaired by H. Aldrich; and to W. Klassen, National Program Staff, Science and Education Adminstration, U. S. Department of Agriculture, for counsel on the organization of this symposium.

Lekh Raj Batra
May 1978
Beltsville, Maryland.

INSECT–FUNGUS SYMBIOSIS

chapter 1

The Fungi Versus the Arthropods

by HOWARD C. WHISLER*

ABSTRACT

The diversity of interrelations between fungi and arthropods presents unlimited opportunities and challenges for today's biologists. Ranging from laboratory curiosities to participants in major world crises, entomogenous fungi are beginning to receive some of the attention they deserve. Study of these fungi necessarily leads us into foreign territories, where an entomologist may find himself raising mushrooms or a botanist talking about malaria. The rewards, however, are well worth these interdisciplinary hazards and this chapter presents one man's view of the interplay between bugs and molds.

This opening chapter† is designed as a biased, partial and untechnical ode to the field of entomological mycology. My sales pitch comes in two parts. First, there are so many different associations between bugs and fungi, that anyone can be easily hooked by them—if they'll just sit down and take a close look. Secondly, there is a real need for this kind of research—the chemists are losing the battle against the insects. There is a critical, global requirement for studies in biological control of pest arthropods. If one is

*Department of Botany, University of Washington, Seattle, Washington 98195.

†This chapter has been adapted from a lecture, entitled "Bugging the Molds," which was presented to a general session of the Second International Mycological Congress, 1977. In contrast to the subsequent scholarly chapters presented in this volume, this paper was designed for a general audience and stresses the whims and interests of the author. It is hoped that the reader will extend some degree of "poetic license" to an article that has been transformed from a minilight show.

INSECT-FUNGUS SYMBIOSIS /Batra (ed.) / Allanheld, Osmun, Montclair, NJ

concerned about man's abuse of the environment, or about maintenance of existing world health standards, then a career devoted to entomological mycology would be rewarding indeed.

The field is not overly crowded and tends to be overshadowed by its larger cousin discipline, plant pathology. Based on a recent poll of the 1400 members of the Mycological Society of America, one estimates that there are approximately 336 members expressing professional interest in plant pathology, while there are only 36 mycologists who are working with entomogenous fungi.

Although small in number of investigators, the field has a distinguished and important history. As early as 1836, Agostino Bassi established that the muscardine disease of silk worms, which was causing havoc in Northern Italy and France, was due to a pathogenic fungus (1). This well-documented study was completed a full 10 years before Berkeley convinced the world that a disease could be caused by a microorganism (late blight of potatoes) (29). Metchnikoff's description in 1884 of a yeast with needle-shaped ascospores that penetrated the gut of *Daphnia* is considered to be a milestone in microbiology (8)—yet it doesn't receive much notice in our introductory mycology and pathology texts.

In recent years one notes more activity in the field of entomological mycology. This is exemplified by the number of symposia, workshops and contributed papers presented at the Second International Mycological Congress (5), as well as the recent birth and rapid growth of the Society for Invertebrate Pathology.

It is interesting to note that there are a number of similarities between fungi and arthropods. Both have a fuselage-like body form with chitin as the main structural component. They are small, numerous, and specialized in both sexual and asexual activity. Tough and hardy, they are found in almost all ecological niches. With these similarities and potentials in mind, it is not surprising that "sparks fly" when they take each other on. Frequently it is difficult to decide who is the "winner" or "loser" in these confrontations. Nevertheless, one can attempt to list a few of these combinations in a format that suggests that either kingdom may be the exploiter (Table 1.1). In the middle of this spectrum of associations, one encounters a number of highly adapted and successful fungal symbionts that live in conjugal peace with their companions. The morphological and physiological specializations that underlie these continuing associations make these little-known fungi a source of continuing interest for the biologist who can find the opportunity and time to study them. Details of many of these relationships are presented by experts in the following chapters of this book. Here, I would like briefly to highlight a few of these associations, with emphasis on those fungi that I have had some personal experience with, or those that may have received relatively light coverage elsewhere in this text.

Any consideration of insects and fungi must include mention of the

Table 1.1 Spectrum of Fungal-Invertebrate Associations. (partial list)

Fungus	Invertebrate	
Fungal gardens	Ants, termites	Fungus consumed
Ambrosia fungi	Beetles	
Stinkhorns	Flies, beetles	
Ascomycetes	Mites	
Laboulbeniales	Flies, roaches, lice	
Eccrinales	Crabs, millipeds	
Harpellales	Aquatic insects	
Amoebidiales	Crustacea	
Septobasidium	Scale insects	
Cordyceps	Lepidoptera, etc.	
Monosporella	Zooplankton, beetle larvae	
Zoophagus	Rotifers	
Lagenidium	Crabs, mosquitoes	
Thalassomyces	Zooplankton	
Entomophthora	Mosquitoes, etc.	Animal consumed
Culicinomyces	Mosquitoes	
Metarrhizium	Mosquitoes, etc.	
Coelomomyces	Mosquitoes, etc.	

mushroom farming ants and termites (Fig. 1.1). N. A. Weber and S. W. T. Batra in following chapters discuss how these insects harvest a variety of substrates and compost them with care. The insects developed pure culture techniques long before Brefeld admonished mycologists "that without pure cultures one can only expect nonsense and *Penicillium glaucum.*" One wonders how the ants maintain the purity of their stocks, and there is a certain vengeful pleasure in speculating how these efficient fellows might treat any mites that invade their cultures! Clearly these animals are serious about keeping a good table, since they have chosen very edible species of fungi to grow in their mushroom houses. But they must share the title of "gourmets of the animal world" with the mushroom flies. Here are two specialized families of flies whose larvae frequently anger the eager mushroom collector. Their larvae feed in the choicest mushrooms, and one wonders what factors entice these highly efficient insects to their dinner. Swarms of truffle flies can tell an experienced observer where to dig for these treasures (40).

Ambrosia beetles, as their name suggests, are also fastidious farmers. Carrying their starter in a back pocket, the mycetangium, the adults burrow into a tree, inoculating their tunnels as they mine. The fungus acknowledges its debt to the beetles by producing an ambrosial meal which is the primary delight of the developing grubs (12; see Chapters 2, 3, this volume).

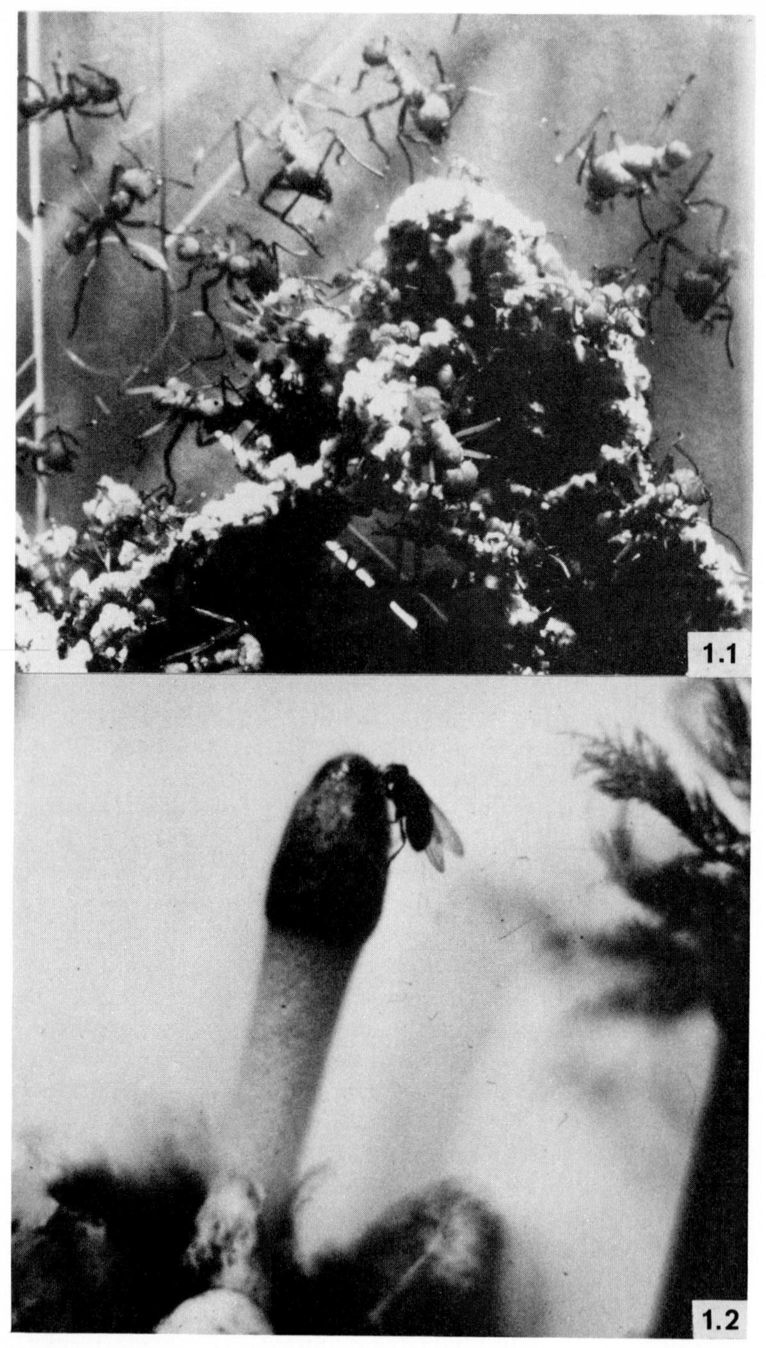

Figure 1.1 (top). Ants tending their mushroom house, as seen through a laboratory window. (Photo: Susan L. -Barnes.)
Figure 1.2 (bottom). Chez *Mutinus caninus.* One customer checking the menu. (Photo: Benjamin Woo.)

Before leaving the list of fungi on the menu, it is appropriate to consider the Phallales. We tend to think of this group of highly adapted puffballs as anthropomorphic curiosities, almost as if they were created for our own amusement. The bugs and the fungus have other ideas. After rapidly growing out of an egg-like case, the elevated spore mass releases a scent that smells like dinner to a variety of insects. The smell varies with the fungus involved, but frequently it is a distinctive blend of urine and rotting flesh. The fly seen in Figure 1.2 couldn't resist this come-hither, and in field situations the spore mass may be consumed in a few hours. Examination of their specks reveals large masses of apparently healthy spores. Obviously here is a group of fungi that knows how to get its spores around. The rapid extension of the fruiting body from the egg should present many interesting opportunities for the developmental mycologist.

Not all fungal arthropod associations involve exploitation by one or the other partner. This seems to be true of the Trichomycetes, the tapeworms of the fungal world. We are indebted to a number of French invertebrate zoologists, particularly J. F. Manier (*31*) of the University of Montpellier, for bringing these unique molds to the attention of the mycological world. These fungi grow in the gut of such diverse arthropods as crabs, millipeds and may-fly larvae. This environment is quite hospitable to the fungus. All it needs to do is hold on and leave the work to the animal. The Trichomycetes have evolved effective hold-fasts that secure the fungus to the cuticular lining of the gut. All is well until the animal molts or dies. The gut lining is shed with the molt, and the fungus abruptly finds itself in a very difficult world. It is not surprising to find that these fungi have evolved special spores for the trip outside of the animal. S. T. Moss (see Chapter 8, this volume) considers this group in more detail, but here I would like briefly to stress some examples of host-dependent development that are well marked in these fungi.

The Eccrinales, with their unbranched thalli, are commonly found in the guts of crabs and millipedes (Fig. 1.3). A recent study by Hibbits (25) of an eccrinid in shore crabs establishes that their thin-walled spores are responsible for colonizing the gut. These spores predominate during the winter season when the crab is not molting. As the spring-summer molting season approaches, thick-walled resistant sporangia make their appearance (Fig. 1.4). These spores carry the infection to the new host. Monthly examination of crab populations suggests that the thick-walled spores are initiated some weeks prior to actual ecdysis, and one wonders how the fungus learns about the impending change. Like many other entomogenous fungi, the eccrinids have never been grown in axenic culture. Once this problem is solved, we should be able to ask some interesting questions on host-controlled development in fungi.

The Harpellales frequent the gut of aquatic larvae such as black flies, stone flies, may flies, etc. These insects are common in fast-moving waters, and, in response to this habitat, their fungal commensals have evolved long appendages on their sporangia (Fig. 1.5). When the sporangia are released

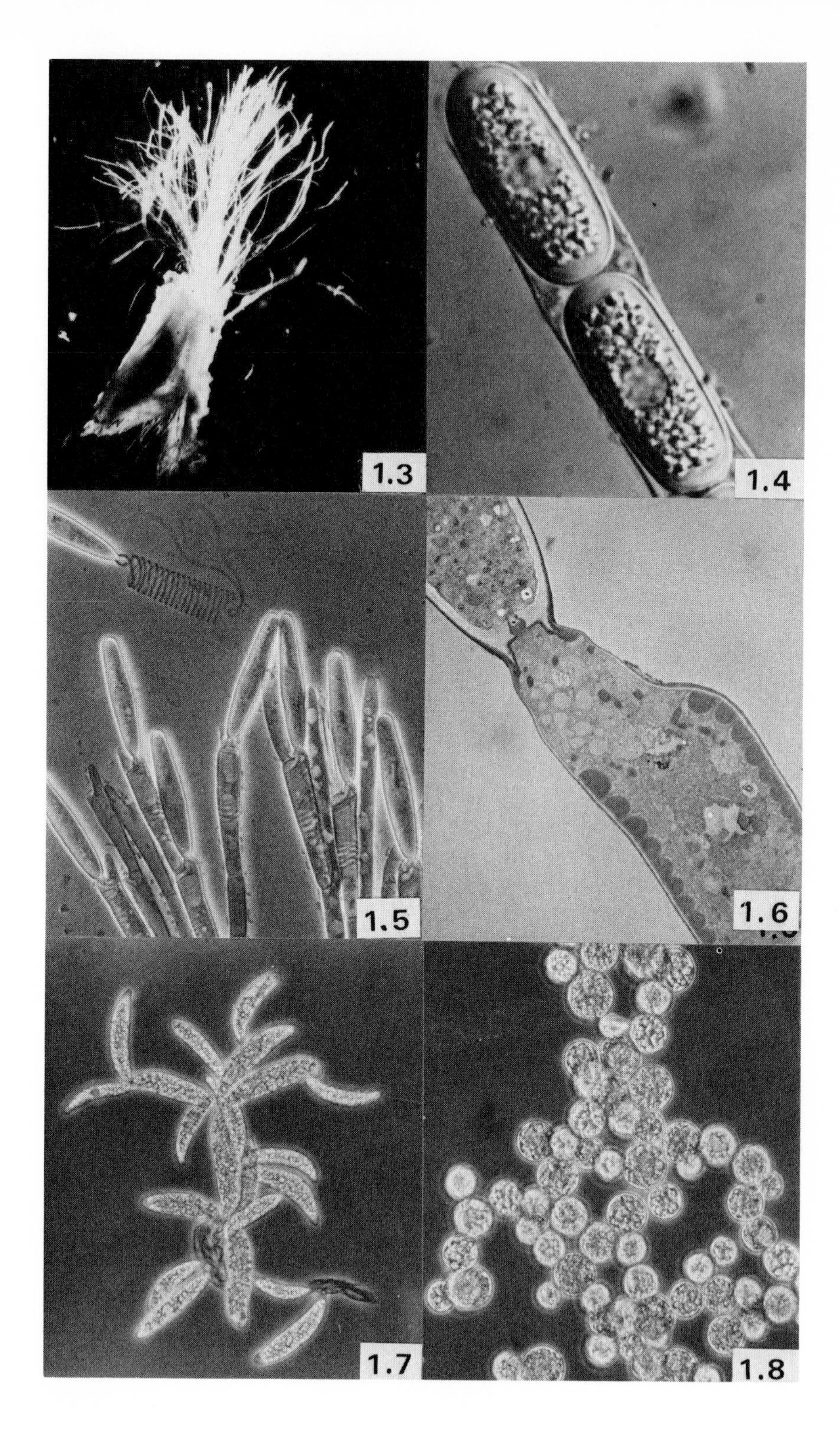
1.3
1.4
1.5
1.6
1.7
1.8

to the outside current, these "life-lines" tend to tangle in the surrounding vegetation and permit their reingestion by new host animals. The appendages are produced in special cells adjacent to the sporangia in a manner quite unique in the fungal world (Fig. 1.6). Moss and Lichtwardt (34,35) have illustrated the remarkable diversity and precision with which these sporangial filaments are differentiated. One genus of the order can now be cultured (13), and considerable information on the biology, systematics and host-relatedness of these fungi has appeared in the past few years.

The Trichomycete genus *Amoebidium* offers special advantages for the experimental study of host-dependent development. The small, unbranched thallus is typically found attached to the external cuticle of freshwater crustaceans and insects. The fungus is spread from host to host by small sporangiospores that are cleaved out of the mother thallus (Fig. 1.7). Death of the host, however, may lead to the release of amoebae which migrate over the host-cadaver and subsequently encyst (Fig. 1.8). When growth conditions improve, the cysts produce new spores for reinfection. Both developmental pathways may be followed in defined conditions, and the way is now open for biochemical studies of host-dependent morphogenesis in these very unique organisms (52).

In contrast to the endocommensalate Trichomycetes, there is no question as to the parasitic nature of the Laboulbeniales. These Ascomycetes, with their uniquely structured body, are found as ectoparasites on insects, millipedes and mites. This group is the largest and most highly specialized of entomogenous fungi, with approximately 1500 known species (4). But only a tiny fraction of the potential host arthropods has been examined, and this largely by a few people (see Chapter 9, this volume). Most notable is Roland Thaxter (Fig. 1.9), whose studies (47) of these peculiar Ascomycetes are a standard for excellence in biological description. Figure 1.10 is a drawing of *Zodiomyces,* a parasite of aquatic beetles, illustrating both Thaxter's artistic talent and how complex the thallus of these fungi may be. The mycelial habit expected in most fungi is exchanged for a near cellular organization. In addition to their highly specialized structure, the Laboulbeniales display amazing degrees of specificity. Host specificity is common, but certain species may even be restricted to certain locations on the

---

Figures 1.3–1.8. 1.3. Hind gut dissected from a shore crab (*Hemigrapsus nudus*) packed with the thalli of the eccrinid, *Taeniella* sp. (X17.) *1.4.* Thick-walled resistant sporangia of the eccrinid *Taeniella* sp. from "molt" of a shore crab. (X1000.) *1.5.* Thalli and appendaged sporangia of an undescribed member of the Harpellales. Before detachment the doubly helical appendages are coiled in the subterminal cells. (X400.) *1.6.* Electron micrograph of a sporangium and generative cell of an undescribed harpellid. The coiled appendages are seen as hemispheres against the generative cell wall. (X2500.) *1.7.* Thalli of *Amoebidium parasiticum,* epizooite of freshwater arthropods. Sporangiospore phase. (X300.) *1.8.* Mass of cysts derived from amoebae released from thalli of *Amoebidium parasiticum.* Amoeboid phase. (X500.)

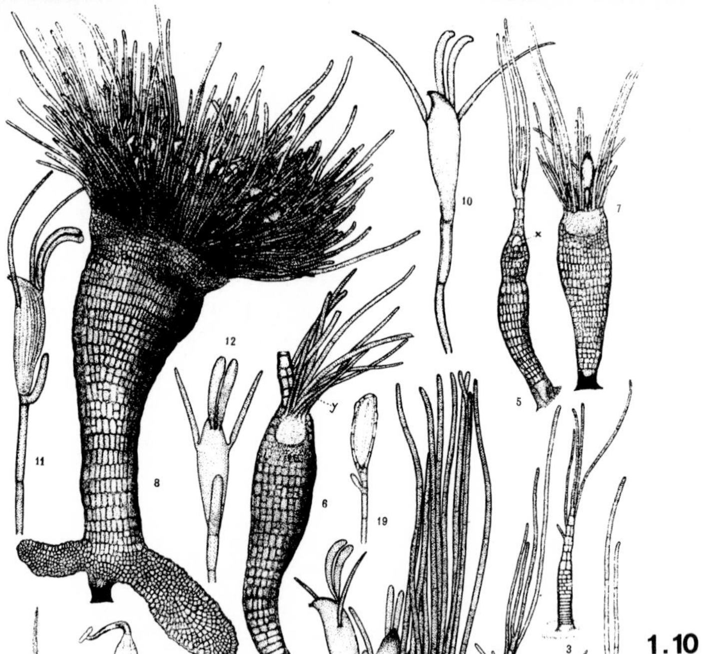

Figure 1.9 (top). Professor Roland Thaxter in his laboratory at Harvard University. Note illustration on wall. He is not, as some wags have suggested, infected with one of his *Cordyceps*! (Photo courtesy of R. Benjamin.)

Figure 1.10 (bottom). Zodiomyces vorticellus as illustrated in Vol. I of Thaxter's monograph on the Laboulbeniales. This ectoparasitic ascomycete prefers water beetles (hydrophilids). Large "cellular" body bears male and female sex organs and perithecia (10–12).

insect—and then on only one sex! The Laboulbeniales also emphasize their own sexual diversity. Asexual reproduction is ignored, and the sex organs are highly developed (Figs. 1.12–1.16). Some genera even have independent male and female plants (Fig. 1.11), a relatively rare situation in the fungal kingdom (11). If we could grow these fungi in culture (53), they would be ideal organisms for the study of host specificity, sex differentiation and thallus development.

Leaving the relatively innocuous Laboulbeniales, we encounter the really lethal Ascomycete, *Cordyceps* (32). Here the host is consumed and its body contents replaced by the fungal mycelium. When mature, the fungus produces a large fruiting structure, which will help disperse the ascospores to new host animals. The grub in Figure 1.17 had anticipated metamorphosing into an attractive June bug (Fig. 1.18) before *Cordyceps* got it.

Lethal fungi are also found in ponds. Consider, for example, the two water molds *Zoophagus* and *Lagenidium*.

Zoophagus insidians, as its name suggests, makes you think twice before you go near the water—particularly if you are a rotifer. Professor Weston has assigned this fungus to the lethal lollipop group of predacious molds (Fig. 1.19). One lick and you've had it. When browsing rotifers brush against the trapping pegs of the fungus, they trigger a glue-release mechanism which sequentially stimulates separation of the wall layers on the peg (Fig. 1.20), rapid secretion of a glue, and subsequent growth of the now unarrested peg into the host animal as a haustorium (Fig. 1.21) (39,54).

Perhaps less dramatic, but more worrisome for the shellfish fancier, is *Lagenidium callinectes.* This marine water mold enjoys one of America's seafood delicacies, the Dungeness crab (Fig. 1.22). Recent efforts by Armstrong (3) to rear the crabs from eggs in the marine laboratory at Newport, Oregon were abruptly terminated by infections of this fungus. The young planktonic larvae were killed by the mold, which then produced a number of discharge tubes to the external environment and finally released its zoospores (Fig. 1.23) to find a new meal. One wonders how active this parasite is in the open ocean. Crab lovers needn't panic, for the supply is currently adequate, and a number of significant studies on the biology of *Lagenidium* have already been accomplished by Gotelli (24) and Bland (6).

Parasites of marine zooplankton have received relatively little attention from mycologists. Undoubtedly, the most unusual parasite is the genus *Thalassomyces.* First described as a fungus, it has also been allied with the dinoflagellates and various protozoa. It can be found on a variety of zooplankton including euphausids, mysids, shrimp, copepods, and amphipods. In Figure 1.24 an immature thallus is growing out of the brood pouch of an amphipod, and where there should be eggs one finds a mass of parasite. The organism is so strange and poorly studied that one doesn't even know how to appropriately label the various parts. The swollen heads were thought to have something to do with reproduction; so, they were

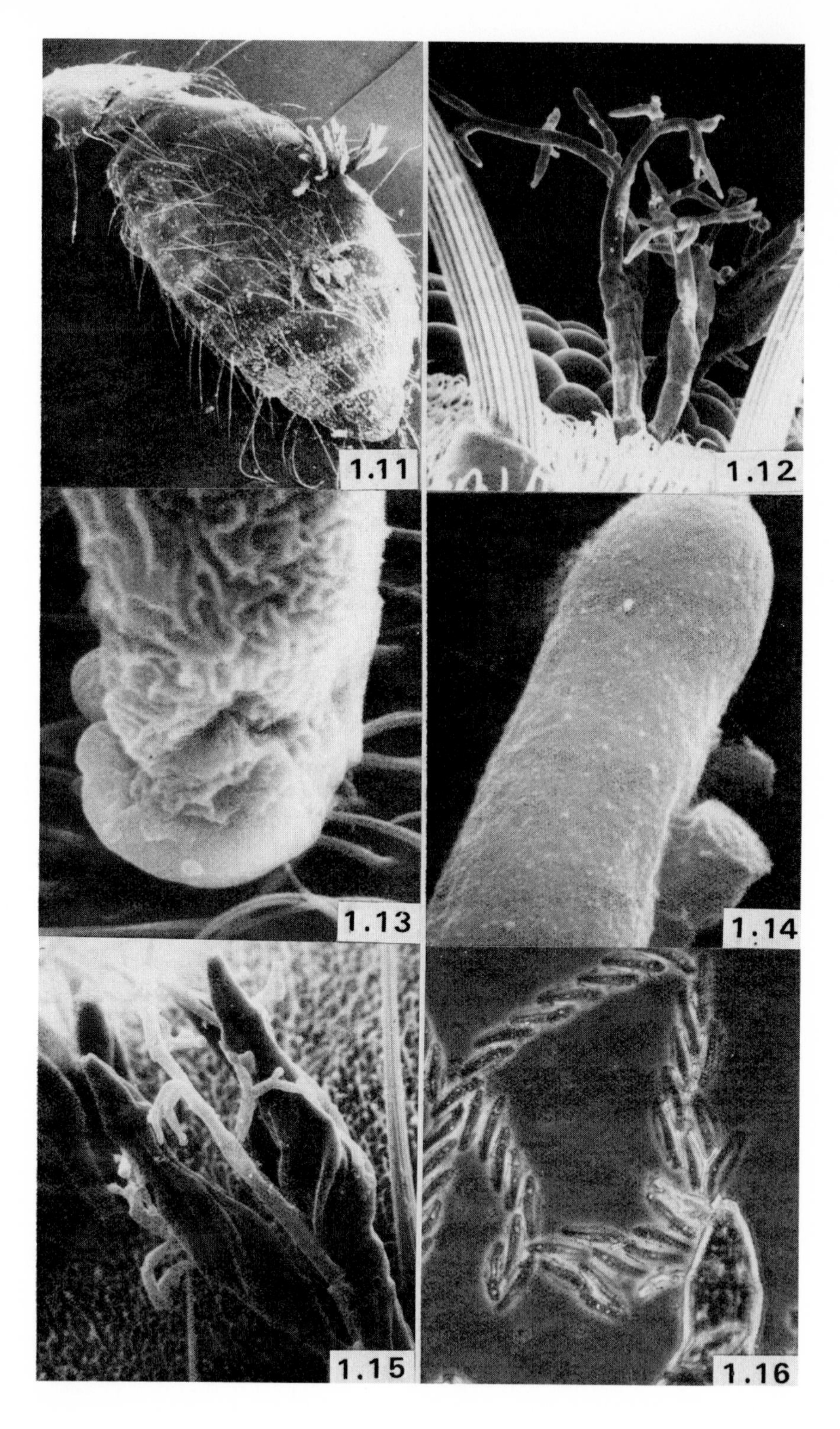
1.11
1.12
1.13
1.14
1.15
1.16

called gonomeres, the supporting structures trophomeres. The whole structure is plugged into the nervous system of the animal. Occasionally one can see some pebbling or infolding of the surface of the gonomere, but the reproductive mechanism of these major marine parasites was totally unknown until 1968 when Hibbits (22) discovered detached masses of the parasite covering the bottom of a dish that held a parasitized animal. Each mass separated into individual spores which then produced 2 flagella, one encircling the middle, the other one extending in a posterior direction, quite like a gymnodinoid flagellate. This plus other criteria led us to suggest that *Thalassomyces* was a colorless dinoflagellate just as Chatton (10) had guessed 50 years earlier. This diagnosis was immediately attacked by dinoflagellate specialists, and it would appear that *Thalassomyces* must remain as a protist without a taxon until some bright student finally gives this important organism the attention it deserves.

By this point I hope I have established the fact that there are a number of interesting organisms to study. Indeed, one of the major problems in working with entomogenous fungi is deciding on which combination to specialize. Perhaps an even more difficult problem is deciding how much weight one should assign to societal needs in selecting a research problem. Obviously, some of these fungi kill insects that are important to man and have potential as biological control agents. Consider, for example, Canada's effort to develop a dairy industry in northern Alberta. This new land is rich in oil and agricultural promise. The cows are willing to produce if it weren't for the black fly problem (21). Clouds of these biting dipterans harass the cattle to a point of depressing milk production and even killing animals that have become sensitized to their bites. The larvae of these flies develop in dense masses in the local rivers and streams. This concentration of the insects is attractive from a control viewpoint. If one dumps quantities of insecticides into the river system, black fly larvae are killed for hundreds of miles. Non-target insects are also killed, and the environmental costs are so high that the Canadians are looking to other means of protecting the cattle, such as dosing the cows with insect repellents, etc.

Figures 1.11–1.16. 1.11. Chicken louse parasitized by *Trenomyces.* This member of the Laboulbeniales has separate, compound, male and female thalli. The most obvious parasite is a single female with multiple perithecia. (SEM X60.) *1.12.* Young thalli of *Stigmatomyces ceratophorus* growing on head of a lesser-house fly. This member of the Laboulbeniales has both sex organs on same body. (SEM X450.) *1.13.* Base of *Stigmatomyces* thallus where it penetrates the fly's cuticle. (SEM X681.) *1.14.* Female receptive cell (trichogyne) with attached spermatia (*Stigmatomyces ceratophorus).* (SEM X11,000.) *1.15.* Mature thalli of *Stigmatomyces ceratophorus.* Branching male appendage and spirally developed perithecia. (SEM X400.) *1.16.* Perithecia (squashed) releasing ascospores. Note all the two-celled ascospores coming out "feet-first." The longer cell is the attaching cell. (Phase X400.)

Figure 1.17. A Junebug grub infected with *Cordyceps* sp. (Photo: R. Humber.)
Figure 1.18. A Junebug that *Cordyceps* missed. (Photo: B. Meeuse.)

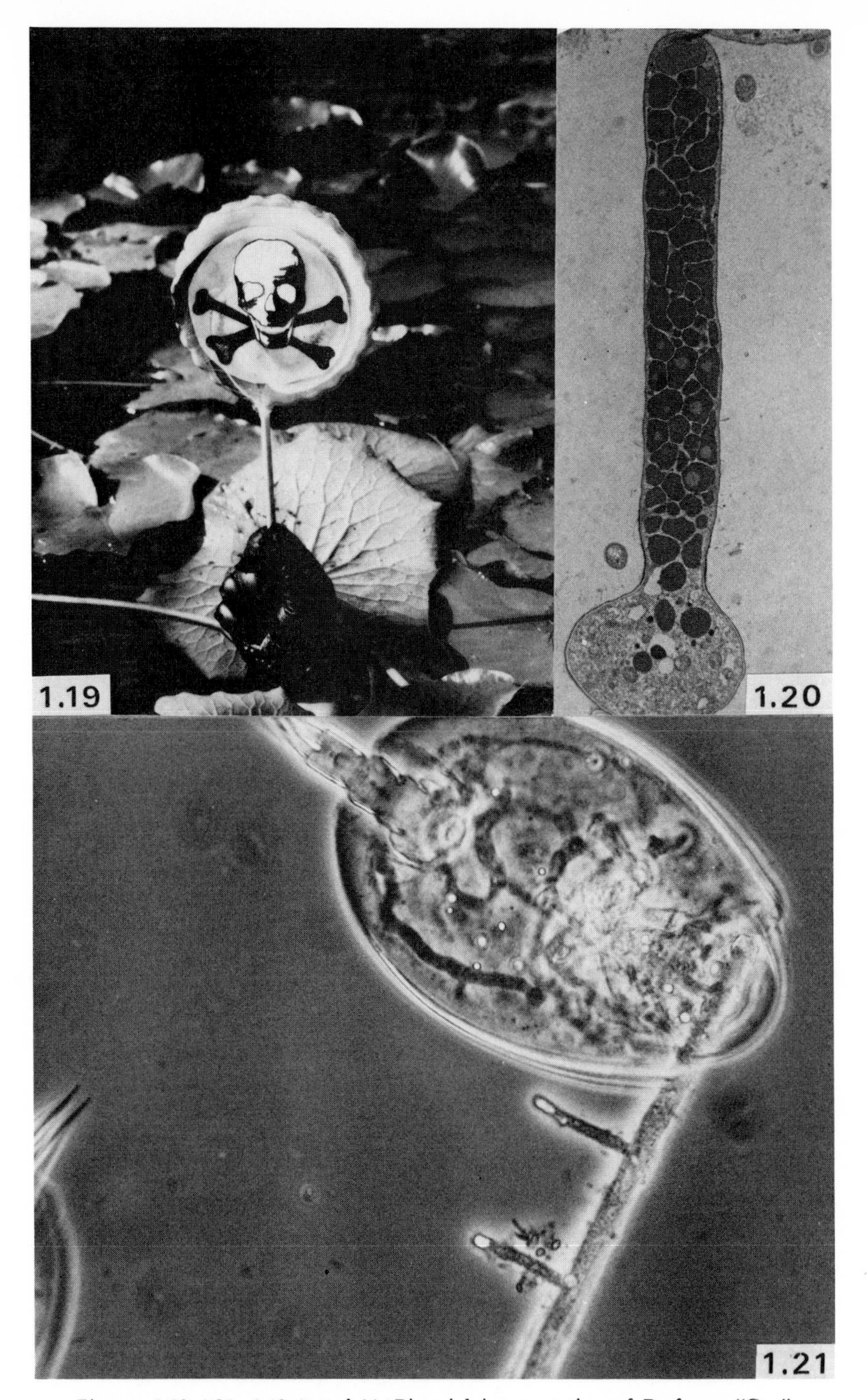

Figures 1.19–1.21. 1.19 (top left). Pictorial interpretation of Professor "Cap" Weston's lethal lollipop taxon. *1.20 (top right).* Longitudinal section of a trapping peg and main hyphal axis of *Zoophagus insidians.* Dense granules in peg will be secreted at the tip and stick to the oral cilia of browsing rotifers. (X4000.) *1.21 (bottom).* Trapped rotifer (*Lecane* sp.) on hypha of *Zoophagus.* Two unsprung traps showing typical apical refrectility induced by glue granules. The body of the rotifer is being digested by mycelium from the original trapping peg. (X1000.)

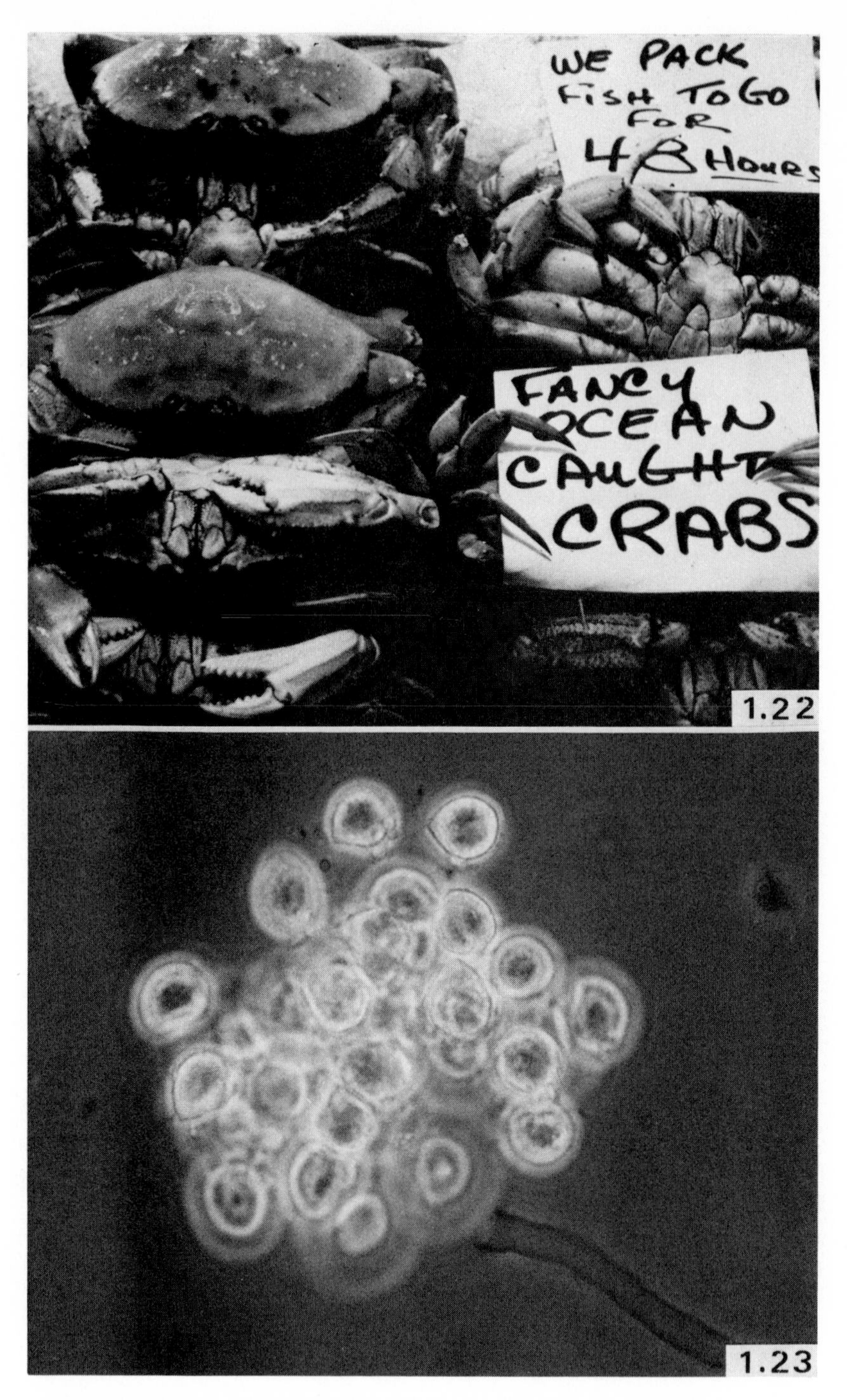

Figure 1.22 (top). Some of Seattle's finest. Dungeness crabs in the Pike Place Market.
Figure 1.23 (bottom). Zoospore release from a vesicle of *Lagenidium,* a crab pathogen. (Photo: D. Gotelli.) (X1700.)

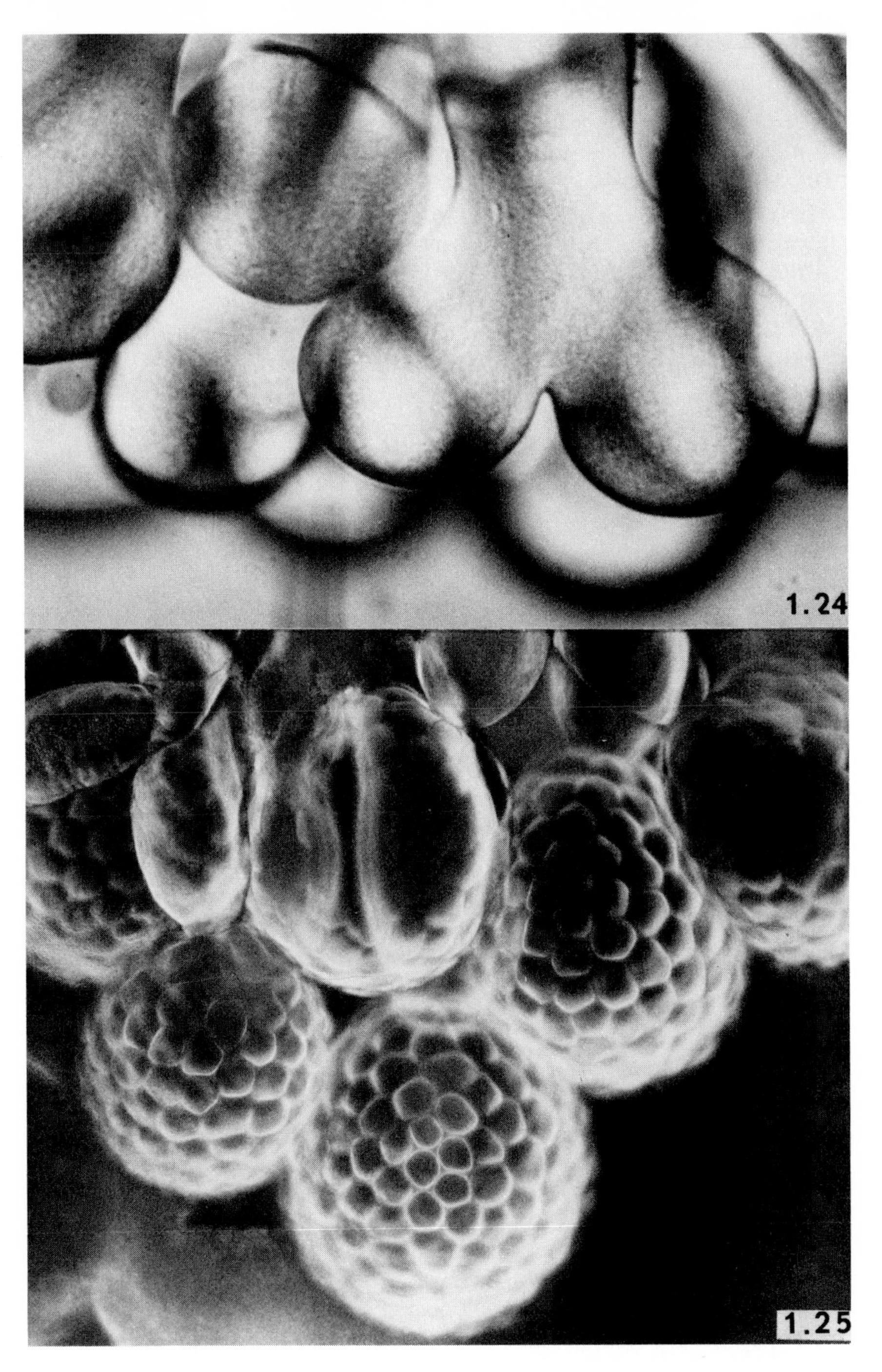

Figure 1.24 (top). Young trophomeres and gonomeres of *Thalassomyces* growing from the brood pouch of a marine amphipod. (X400.)
Figure 1.25 (bottom). Mature thallus of *Thalassomyces*. Gonomeres with pebbling which occurs before zoospore production. (X400.)

Such easy solutions are less handy in the Volta River Basin of Western Africa. Here, the effect on non-target insects must be measured against human suffering, for the black flies transmit a small worm that makes life miserable for 30 million people (28). Blindness is the common result of untreated infection. The river valleys are rich agricultural basins, yet full of flies, and the World Health Organization has been adding large quantities of insecticides to the rivers in an effort to alleviate the problem.

The development of biological control agents that select black flies is obviously needed. Some possibilities do exist. While working on the black fly problem in northern Alberta, Shemanchuk (45) and Humber reported epizootics among recently emerged flies caused by the fungus *Entomopthora* (Fig. 1.26). This genus is under intensive studies in a number of laboratories (46,30,26) and already shows great promise in the control of a number of different insects.

If one is to focus, however, on medically significant insects, then there should be no hesitation in naming the main culprit. The mosquito is still man's worst enemy. Malaria and other mosquito-borne diseases persist as the main cause of death and misery from infectious disease. Despite the miracles rendered by modern insecticides, there are still 1 million deaths from malaria in Africa alone each year (37). If this were not bad enough, specialists tell us that insecticide resistance is increasing at an alarming rate. A recent report of the World Health Organization indicates that approximately one-third of the area involved in malarial control has significant populations of mosquitoes that are resistant to the practical or currently available insecticides. The report concludes with a call to biologists to examine all organisms that might be useful in the battle with this persistent killer. The possibilities include viruses, bacteria, protozoa, nematodes, fish and fungi. Within the mycological world, one recognizes *Entomophthora* (2), *Metarrhizium* (41), *Culicinomyces* (17), *Lagenidium* (23) and *Coelomomyces* (16) as active pathogens of mosquitoes (9). With a problem so immense, we must examine all of these molds and determine if they do—or do not—have a role to play in stemming what is surely a developing crisis. Research on the biology of all these molds is underway. Available space and my own expertise make it appropriate to restrict further remarks to *Coelomomyces*.

This water mold is a pathogen of the aquatic mosquito larva (Fig. 1.27). First discovered by Keilin in 1921 (27), it has more recently undergone intense study by Dr. John N. Couch (Fig. 1.28) of the University of North Carolina (14,33). The fungus grows in the hemocoel of the aquatic larva as a weakly branched mycelium that appears to lack the typical chitinous wall expected of most fungi. The fungus utilizes the hemolymph nutrients at the expense of the competing host tissues. At maturity it differentiates into thick-walled, resistant sporangia (RS) which are typically ovoid and yellow-brown. The details of surface structure of the RS are of basic importance in taxonomic determinations. The scanning electron microscope has proven to be a very useful tool in these species diagnoses, as

Figure 1.26 (top). Black flies killed by *Entomophthora*. The white muffs are masses of sporangiophores that have broken through the cuticle of the host prior to their explosive release of terminal sporangia. (Photo: J. A. Shemanchuk.)
Figure 1.27 (bottom). Culiseta inornata infected with *Coelomomyces psorophorae*. Numerous ovoid resistant sporangia are obvious in the thorax. (Photo: J. A. Shemanchuk.)

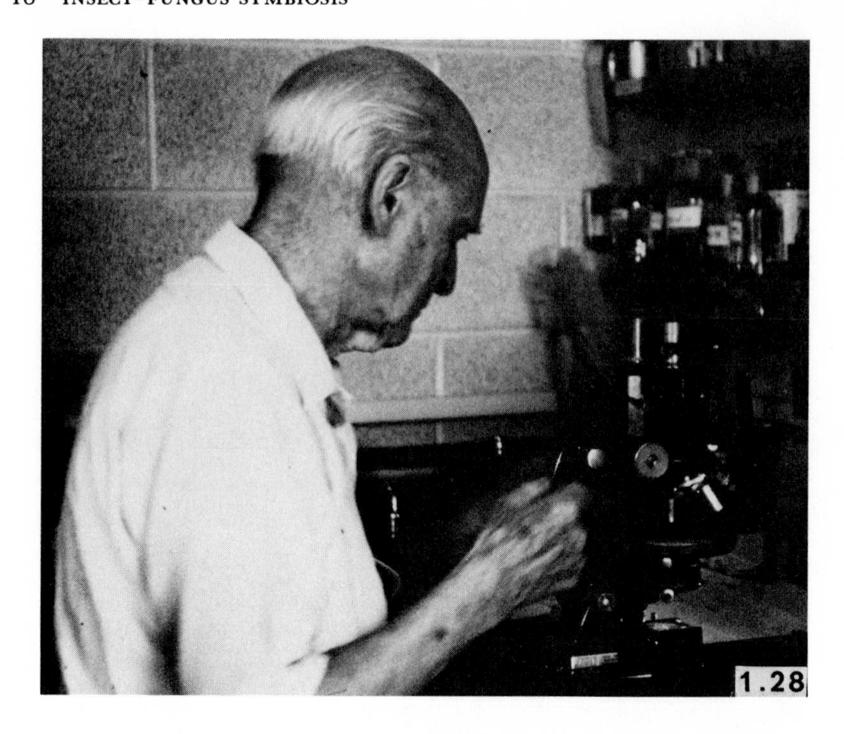

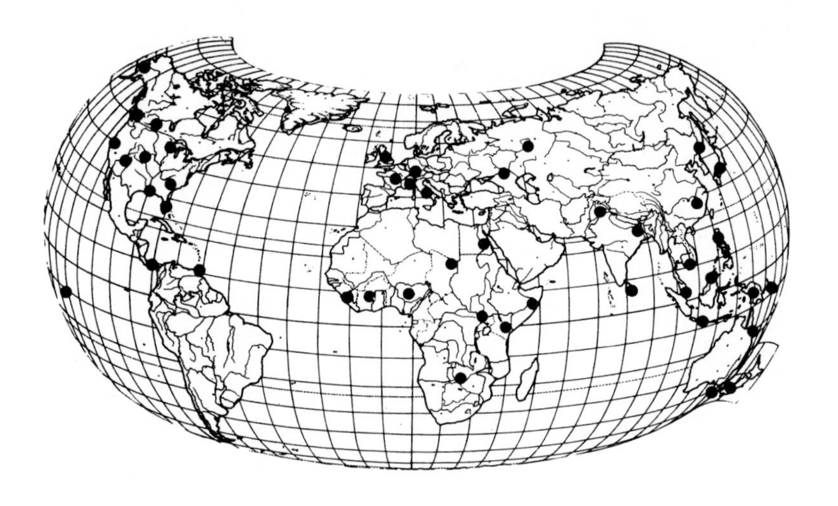

1.29

Figure 1.28 (top). Professor John N. Couch at work in his laboratory in the Botany Department, University of North Carolina. Summer 1977. (Photo by author.)
Figure 1.29 (bottom). World distribution of *Coelomomyces.* For details and more precise information see McNitt and Couch, 1978. *(33)*

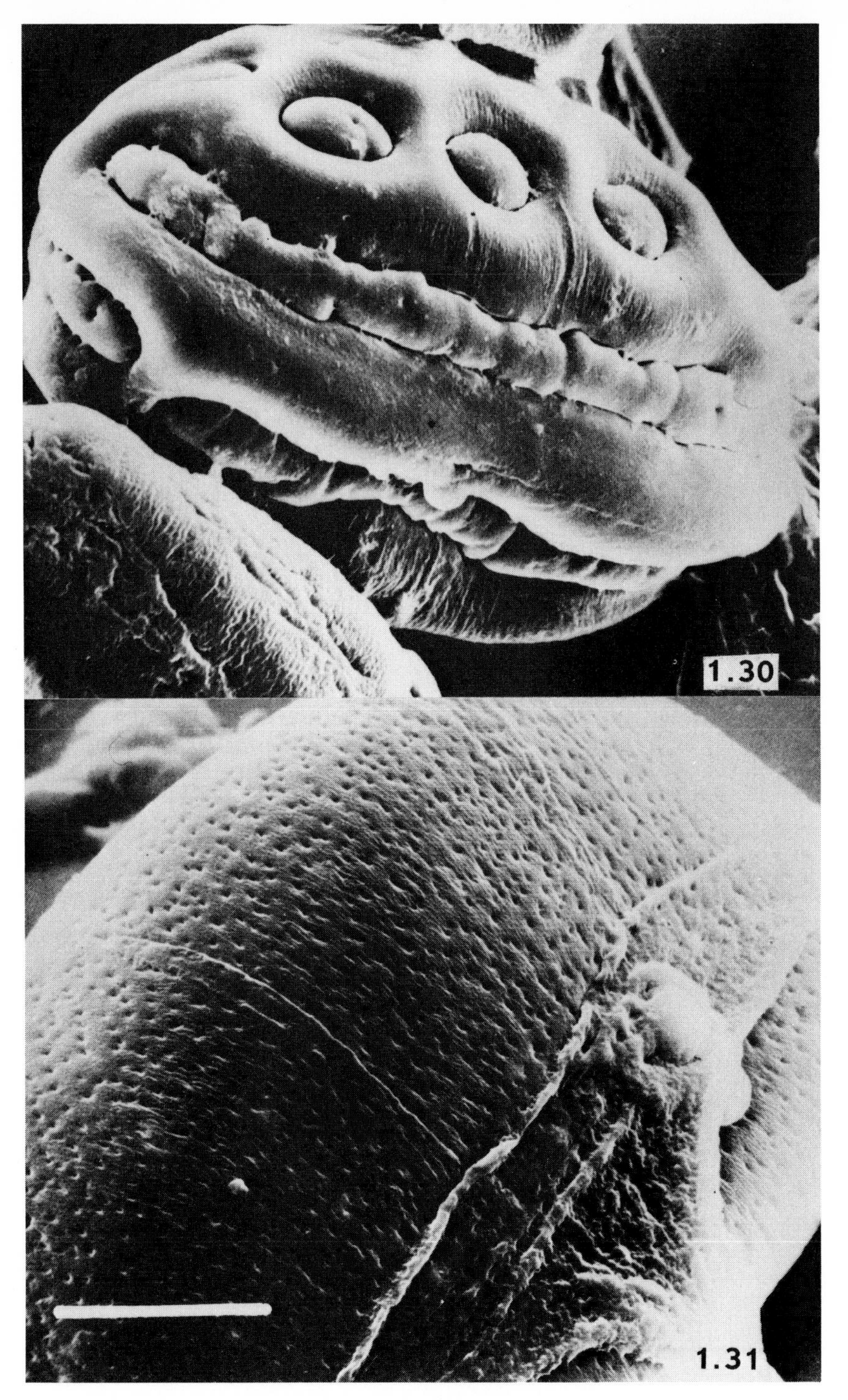

Figure 1.30 (top). Scanning electron micrograph of the resistant sporangium of *Coelomomyces sculptosporus,* a parasite of anopheline mosquitoes. (Photo: C. Bland.) (X3500.)

Figure 1.31 (bottom). Scanning electron micrograph of *Coelomomyces psorophorae.* Note characteristic surface pitting, which connects to the more complex internal labyrinth seen in Fig. 1.34. (Photo: C. Bland.) (Bar = 10μ.)

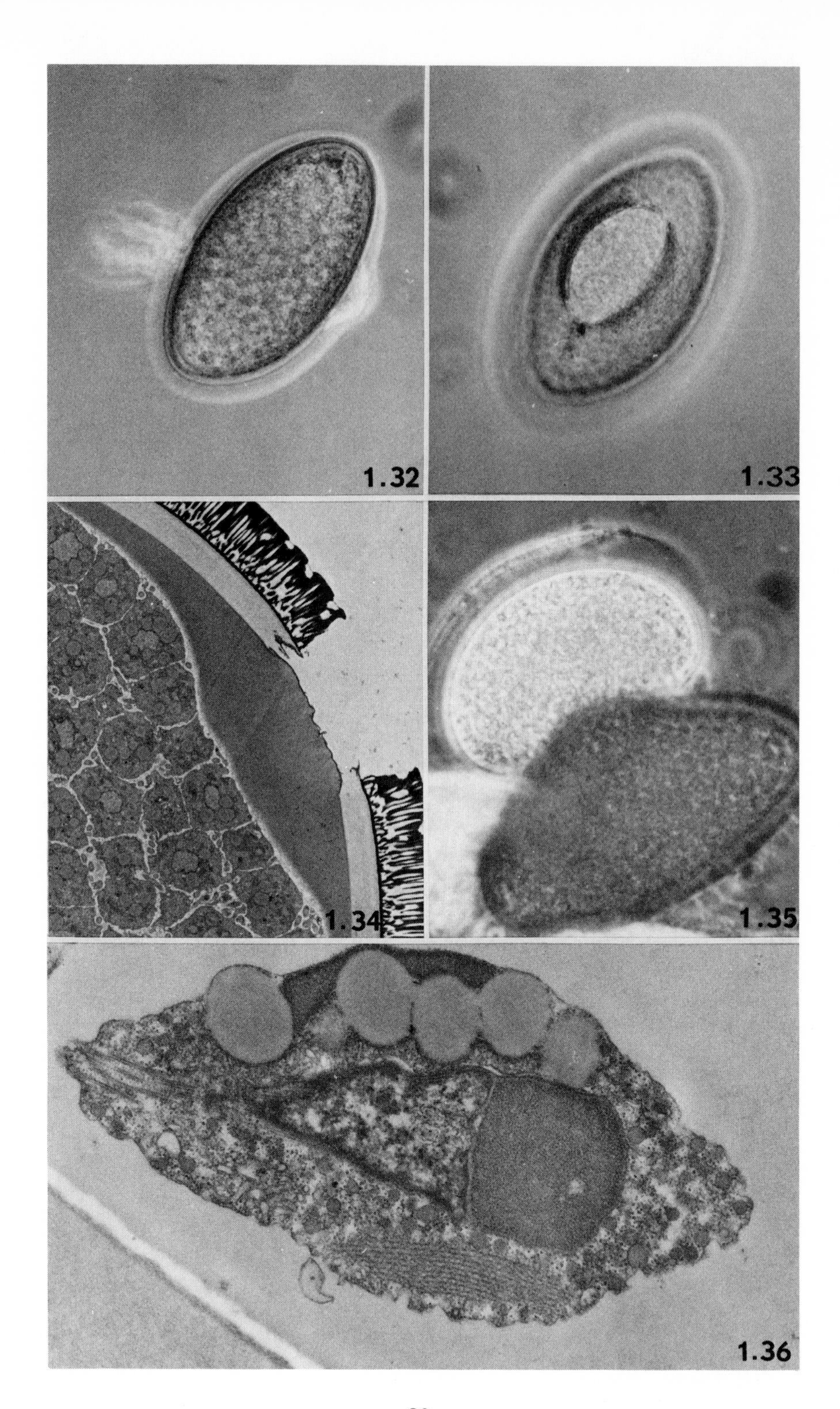
1.32
1.33
1.34
1.35
1.36

exemplified by the dramatic pictures of Bland and Couch (7) (Figs. 1.30, 1.31). There are approximately 40 different species and varieties of *Coelomomyces* that have been reported from numerous localities around the world (Fig. 1.29). They parasitize all the major mosquito taxa, most significantly, the chief disease transmitters *Anopheles* and *Aedes*. Muspratt's (36) observations in Africa are particularly interesting in that he reported significant kills by *Coelomomyces indicus* in *Anopheles gambiae*. This mosquito is one of the main malarial vectors in Africa, and these early observations established *Coelomomyces* as a potentially important biological control agent. Unfortunately, this potential has yet to be tested. Early laboratory efforts to infect larvae with zoospores from the resistant sporangia consistently failed, unless they were carried out in very complex systems that approached the character of the natural habitat (15).

Stephen Zebold and I experienced many of these failures when we tried to develop a laboratory colony of *Culiseta inornata* infected with *Coelomomyces psorophorae*. This particular combination had been discovered by J. A. Shemanchuk in southern Alberta in 1956 (44). He has informally monitored the disease since that time and believes the fungus is exerting a significant and persistent control pressure on that species of mosquito. In our early infection efforts we exposed mosquito larvae to zoospores (Fig. 1.36) from the resistant sporangia (RS) (Figs. 1.32–1.35) under varying conditions and densities. These efforts consumed large quantities of RS but failed to produce a single infected larva. Minor infection rates were obtained, however, after we created a laboratory "pond" which contained many components from an Albertan ditch. Subsequent studies with "mini-ponds" suggested there was a correlation between the presence of *Cyclops*, a common freshwater copepod, and successful infection trials.

Cursory examination of copepods from one of the infected ponds showed no signs of fungal infection. A number of these animals were then transferred to individual dishes and reexamined the following day. One dish was swarming with large numbers of motile spores. They resembled the zoospores from the resistant sporangia, except that some had two,

---

Figures 1.32–1.36. 1.32. Resistant sporangium of *Coelomomyces psorophorae*. (X500.) *1.33.* Resistant sporangium of *C. psorophorae* at "go" phase (Sensu Couch). Induction in light has led to meiosis, opening of a predetermined release-crack and differentiation of zoospores. (X500.) *1.34.* Electron micrograph of a resistant sporangium of *C. psorophorae* at "go" phase. Partially cleaved meiospores are contained by a discharge "shield," inner wall and the black, alveolated outer wall. (X3000.) *1.35.* Meiospore release. Anaerobiosis triggers zoospore escape from a "go" sporangium. The discharge "shield" swells as the zoospore mass moves out of the sporangium. (X550.) *1.36.* Meiospore of *Coelomomyces psorophorae*. Central cone-shaped nucleus carries scoop of ribosomes (= nuclear cap). Note basal flagellum, lipid-side body complex and paracristalline body. (X20,000.)

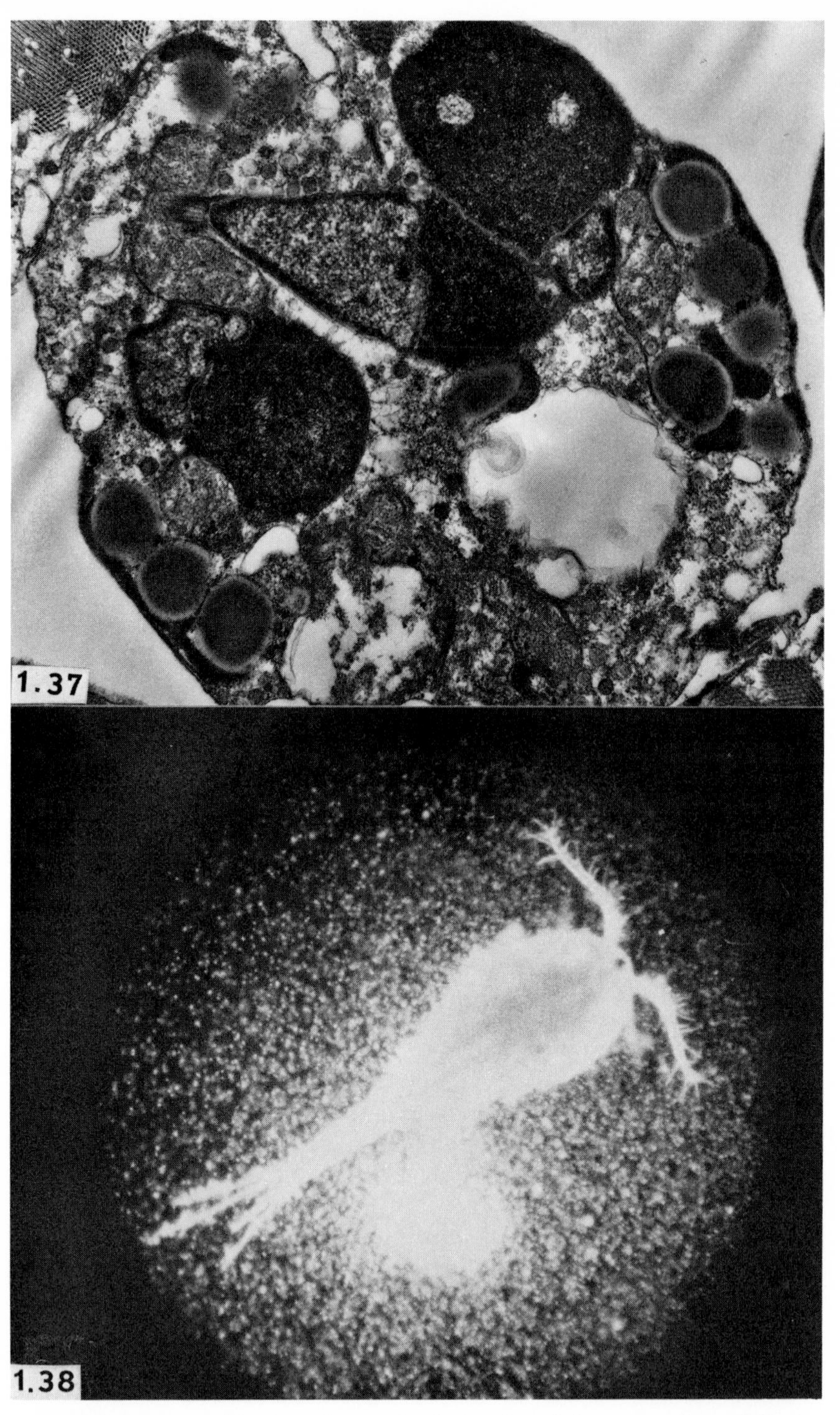

Figure 1.37 (top). Electron micrograph of a portion of the gametophyte of *C. psorophorae* just prior to gamete formation. Obvious cell wall is missing. (X18,000.)
Figure 1.38 (bottom). A cloud of gametes and zygotes escaping from the copepod, *Cyclops vernalis.* (Photo: D. McCabe.)

rather than one, flagella (Fig. 1.39). The copepod in the dish was dead. Further observations on copepods from the infection pans revealed the following scenario: Infected copepods appear to be swimming "happily" with no obvious sign of disease. They then abruptly slow down, stop swimming, fall to the bottom of the pond and die. The tail characteristically arches up and within minutes the whole body is filled with a scintillating mass of motile spores. After swarming within the corpse for some minutes, the cells break out to the exterior environment (Fig. 1.38), typically escaping from a rupture in the cuticle near the tail. Dissection of infected copepods revealed that the motile cells were derived from a weak, wall-less mycelium (Fig. 1.37) similar to that found in the mosquito larva. This totally unexpected development implied that copepods might serve as obligate, alternate hosts for *Coelomomyces*. This was confirmed when the mosquito larvae were exposed to the cells escaping from the copepod. Within a week, mycelia and RS typical of *C. psorophorae* appeared within the tested larvae. Not only had a laboratory infection of mosquitoes in a simple system finally succeeded, but heteroecism, or obligate alternation of hosts, was established for the first time outside of the rust fungi (55). These were exciting times in the laboratory, and, once in the midst of watching spores escape from the copepod, I accidently crushed the animal with the microscope objective. This precocious release of the motile cells explained their true character. Two uniflagellate cells (Fig. 1.40) were seen to fuse, reorient their flagella, and promptly swim away as a zygote (Fig. 1.41). Thus, we were obviously dealing with an isogamous sexual system. Further study implied that *Coelomomyces* had a life cycle in which alternation of generations coincided with alternation of hosts, with the diploid sporophyte in the mosquito and the haploid gametophyte in the copepod (56) (Fig. 1.46). This hypothetical life history assumed that meiosis occurred in the resistant sporangium, as is the case in the related water mold *Allomyces*. Subsequent studies have proven that this, indeed, is the situation. Tests with the three motile cells in the cycle, the meiospores from the RS, the gametes, and the zygotes established that the spore from the RS can only infect the copepod, and the zygote can only infect the mosquito. Pure populations of gametes were occasionally obtained from a copepod. They failed to produce zygotes and failed to infect larvae. Reciprocal crosses of these pure gamete populations did yield zygotes, and this indicated that a mating system was also involved in the life history of *Coelomomyces*.

With the fungus finally behaving properly in the laboratory, it was now possible to answer some other important questions about the biology of this pathogen. The fungus was generally thought to be highly specific, but was it restricted to a single host species? Broad spectrum insecticides create a number of environmental problems, but a fungus that is restricted to a single host species might be too much of a good thing. Tests with 10 different mosquito species indicated that *C. psorophorae* could infect

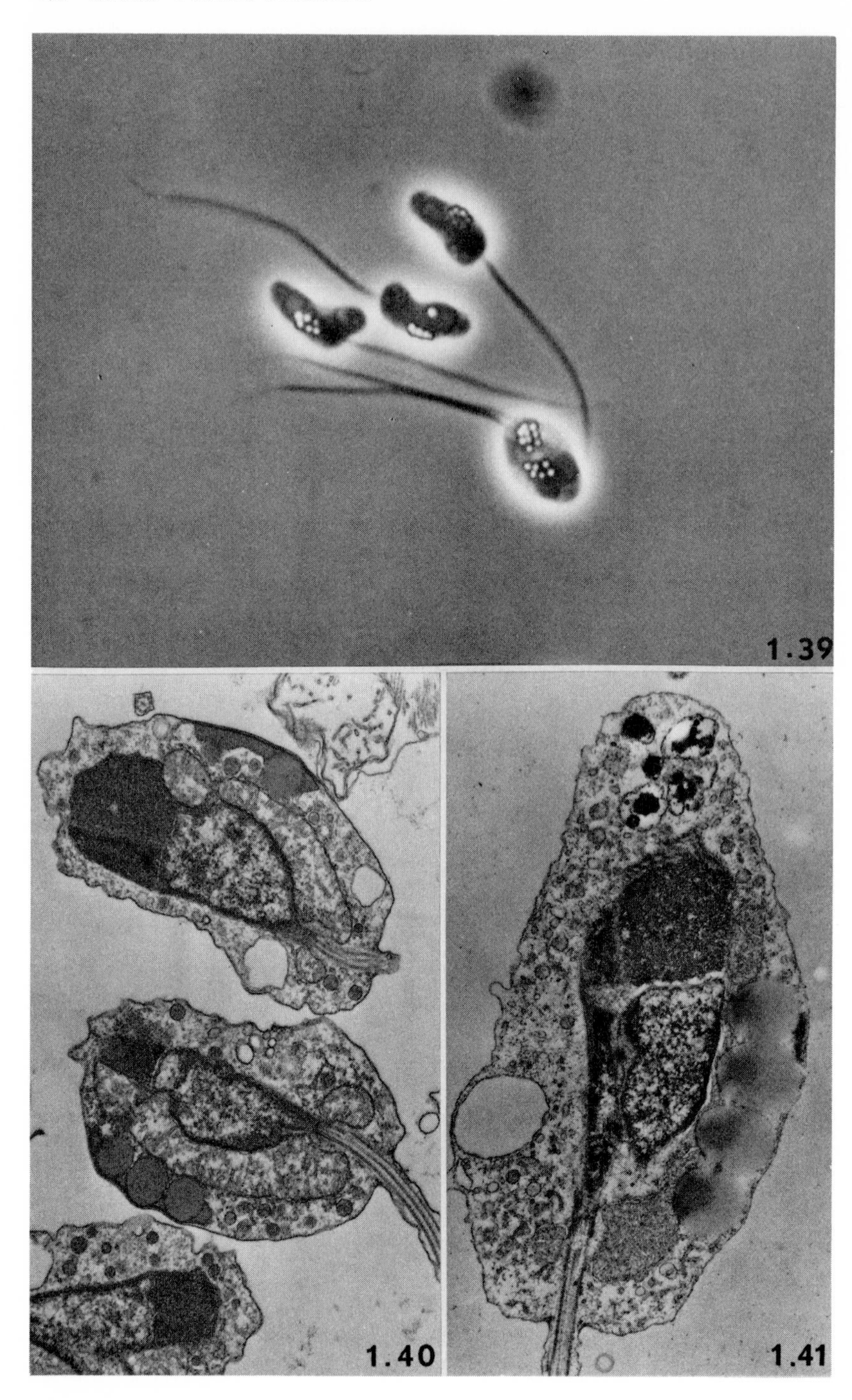
1.39
1.40
1.41

approximately half of them. Some species (e.g. *Aedes aegypti*) appear to be more susceptible than others, but maximum RS production is consistently best in the original host species, *Culiseta inornata*.

One significant life cycle question remained to be answered—how does the fungus enter the host animal? The two major routes employed by insect pathogens are via the mouth or by direct penetration of the cuticle. Young larvae that had been exposed to heavy concentrations of zygotes were treated with a variety of stains and examined both internally and externally. A dilute solution of methylene blue promptly answered the question. Heavy concentrations of encysted zygotes had selectively acquired the stain (Fig. 1.42). The cysts were particularly numerous on the intersegmental areas of the abdomen. No zygotes were found attached to the gut lining. Similar attachment of RS meiospores was found on the copepod host (57). Dr. Linda Travland has been able to follow the route of this infection with both transmission (Fig. 1.44) and scanning electron microscopy (48,49) (Fig. 1.43). The zygote appears to attach to the cuticle using a glue released by specialized cytoplasmic vesicles. They then encyst, differentiate an appressorium, and penetrate the cuticle with a narrow tube which grows out of the appressorium (Fig. 1.45). A vacuole in the main cyst body then appears to expand and inject the protoplast of the cyst into a subcuticular epidermal cell. Eventually the fungus passes into the hemocoel for its subsequent development.

Other workers studying other combinations have subsequently reported that copepods are also involved in their systems. M. Pillai, working in New Zealand (38), has reported a harpacticoid copepod to be essential in his combination of *C. opifexi* in salt marsh mosquitoes. Federici, Roberts (20), and Couch have found that *Cyclops vernalis* is an alternate host for two species of *Coelomomyces* in *Anopheles quadramaculatus*. The gametophyte in this system is particularly striking since it accumulates a red pigment. And Federici has evidence that the color preferentially accumulates in one mating type (19). This phenomenon would make crosses between different species of *Coelomomyces* much easier. Carotenoid accumulation in gametangia is well known in *Allomyces*, a relative of

---

Figures 1.39–1.41. 1.39 (top). Three gametes and one biflagellate zygote of *C. psorophorae*. Two groups of lipid granules and two nuclear caps may be seen in the zygote. (X2000.) *1.40 (bottom left).* Electron micrograph of sections of two gametes. Nucleus with nuclear-cap, adjacent mitochondrion, lipid body and flagellum may be seen. The paracristalline body found in the meiospore (Fig. 1.36) is absent in the mature gamete. (Photo: L. B. Travland.) (X11,500.) *1.41 (bottom right).* Electron micrograph of a section of a zygote. Nuclei do not completely fuse until encystment occurs. Apical vesicles with dense contents may be involved in cyst wall formation. Smaller vesicles (e.g. those at level of nuclear cap) may be involved in attachment to the host. (Photo: L. B. Travland.) (X12,000.)

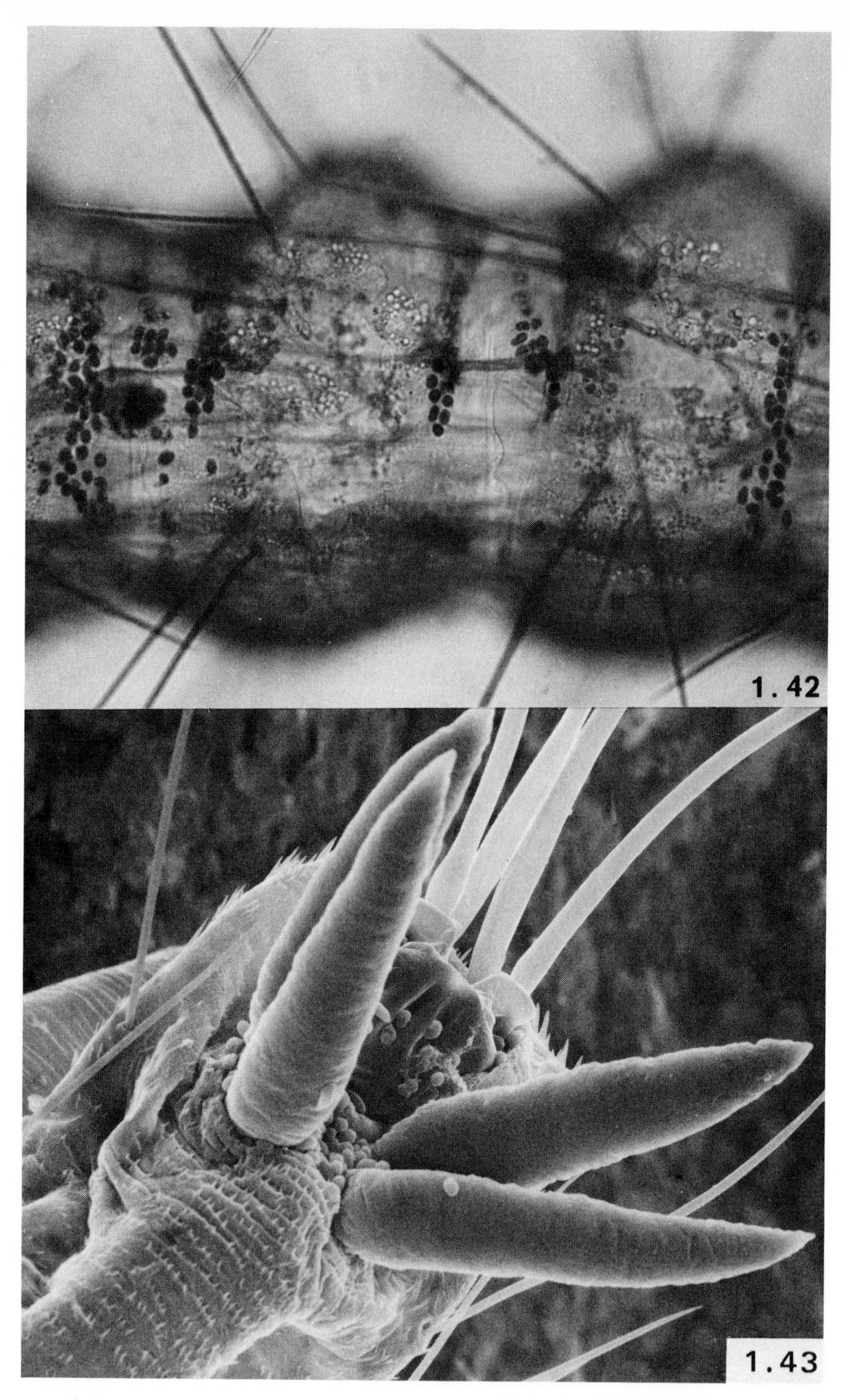

Figure 1.42 (top). Abdomen of a young mosquito larva with dark zygote-attacked intersegmented areas. Selective staining of zygote cysts with methylene blue. *Figure 1.43 (bottom).* Scanning electron micrograph of the anal "gills" and anal area with attached zygotic cysts. (Photo: L. B. Travland.) (X600.)

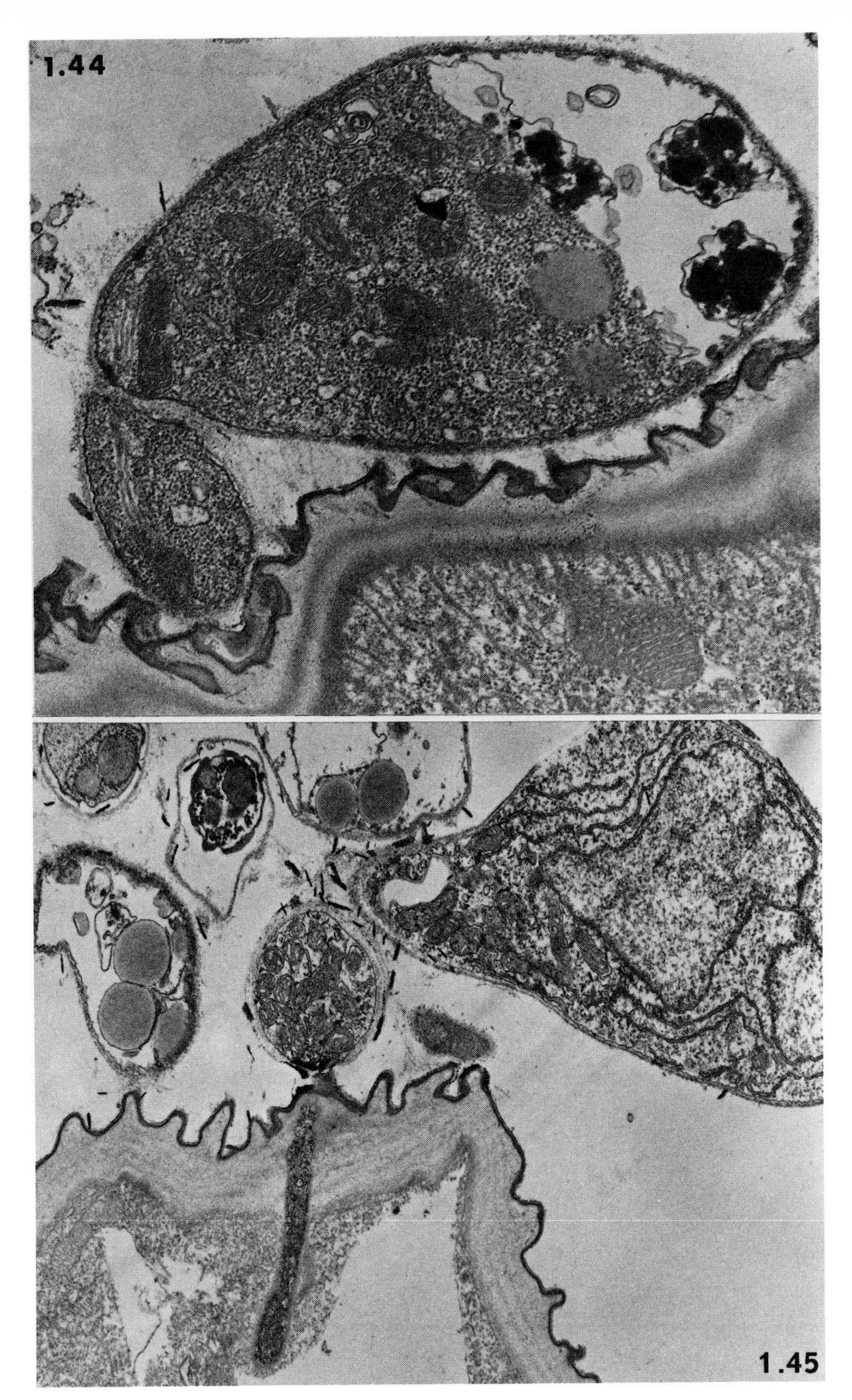

Figure 1.44 (top). Electron micrograph of an attached cyst. Smaller appressorium is attached to the cuticle of the host. Developing vacuole at distal end of cyst body will assist in moving cyst contents into the host cell. (Photo: L. B. Travland.) (X24,000.)
Figure 1.45 (bottom). Penetration tube from appressorium has passed through the cuticle to an epidermal cell. The nucleus and other cytoplasmic components will pass through the tube and leave an essentially empty cyst wall. (Photo: L. B. Travland.) (X15,500.)

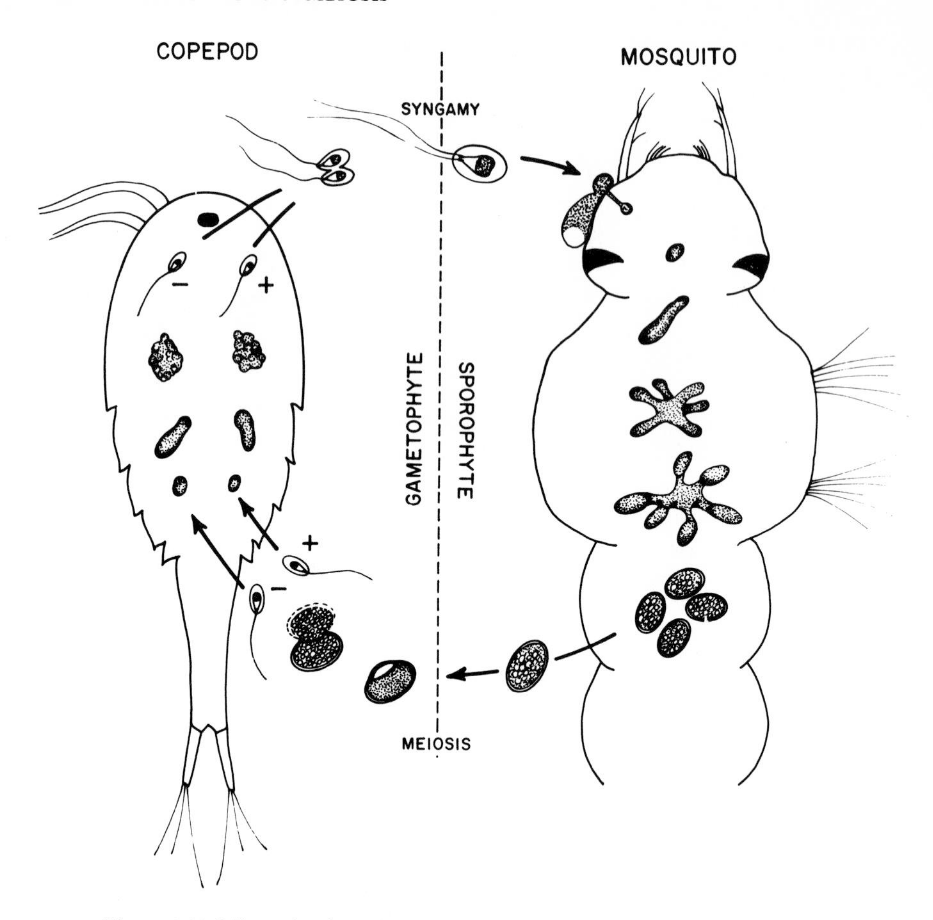

Figure 1.46. Life cycle of *Coelomomyces psorophorae.* SPOROPHYTE (Diplophase): Zygote attaches to host mosquito. Penetration tube from appressorium of cyst enters host cuticle. Fungus develops in hemocoel as a weakly branched mycelium which eventually differentiates into thick-walled resistant sporangia (RS.) GAMETOPHYTE (Haplophase): Under appropriate environmental conditions, the nuclei in RS undergo meiosis and release meiospores of opposite mating type, which infect the alternate host, *Cyclops vernalis.* Each spore develops into a thallus, and eventually, gametangia. Gametes of opposite mating type fuse either in or outside of the copepod to form the mosquito-infecting zygote.

Coelomomyces, which also possesses alternation of generations. Indeed, this cycle worked out by Emerson and Wilson in 1949 (18), has served as the template for predicting the development of *Coelomomyces.* It is worth noting that *Allomyces* has a variety of short cycles that omit the gametophyte entirely. If this were to occur in *Coelomomyces,* one would expect the spores from the resistant sporangium to infect other mosquito larvae rather than a crustacean. Short cycles in *Coelomomyces* have not been proven, but they might well be encountered in the future.

Jaroslav Weiser and Jiri Vavra have alerted insect pathologists to the possibility that other crustacean hosts may be associated with *Coelomomyces*. In 1964 (51) they reported on a species of *Coelomomyces* that parasitizes chironomid larvae and, being keen field observers, noted the presence of large numbers of ostracods associated with the diseased midges. In 1976 Weiser indicated that these crustaceans were serving as the alternate host for *C. chironomi* (50). We have recently found a similar association in a new species of *Coelomomyces* originally discovered in several mosquito species by Romney in desert potholes of Utah (42). In this case the alternate host is the ostracod *Potamocypris smaragdina*. Our attempts to infect other pothole residents, as well as *Cyclops vernalis* with this new species of *Coelomomyces* have failed.

Clearly, future studies of *Coelomomyces* will continue to yield many interesting surprises. The fungus has obviously evolved in concert with the mosquito and is narrowly adapted to the development of the host. Laboratory studies indicate that the environmental factors that control the release of meiospores from resistant sporangia and gametes from copepods may also relate to the field biology of the mosquito. Because of this finely tuned association and the high mortalities it exerts, *Coelomomyces* appears to have real potential as a biological control agent. It is important, therefore, to critically assess the practical value of this fungus as a mosquito killer (either alone or in combinations with insecticides or other pathogens). Such analysis cannot happen until some biological and economic problems are solved. *Coelomomyces* is so closely attuned to its host, that it is difficult to isolate into axenic culture. Reports of the cultures of other "obligate pathogens" as well as some advances with *Coelomomyces* suggest that the situation is not hopeless (43). Until this hurdle is passed, however, proper evaluation of *Coelomomyces* as a control agent will be severely hampered. The economic problem is equally vexing. Although arthropod-borne diseases are a, if not the, major world health problem, support for the study of biological control agents is relatively minuscule. There are a number of different biological control agents that need to be examined in depth, but low budgets slow such research. Furthermore, the health problem is so intense, that there is often a tendency to go to the field before adequate laboratory information is available. When this happens, the results, usually negative, must be questioned. The United States is undoubtedly the major contributor to this area of research, but since the full medical impact of mosquitoes is not felt in temperate regions, the relatively low level of funding is not surprising. Global pesticide contamination, growing resistance of mosquitoes to insecticides, and, finally, a realization that the problems of the tropics are eventually felt in developed countries, all demand a change in the funding picture in the near future.

What role *Coelomomyces* may play remains to be seen. Critics may well point out that if the fungus is successful in fighting mosquitoes, the animals will eventually become resistant to *Coelomomyces* as they have to insecticides. But, *Coelomomyces* has something that chemicals don't: sex. *Coelomomyces* has, through the advantages of genetic recombination, evolved a large number of species and varieties during its long association with the mosquito. We can assume that the various taxa of *Coelomomyces* contain a wide genome that has permitted them to keep pace with their evolving hosts. With this in mind, we have developed techniques of long-term storage of different field collections of *Coelomomyces* in liquid nitrogen with the dream (fantasy?) that, once we have learned how to grow the fungus in culture, we will be able to respond to insecticide resistance problems by hybridization and selection for strains that can meet the mosquitoes' challenge.

Even if *Coelomomyces* fails completely in the practical arena of mosquito abatement, its study will still add some exciting bits of information to the complex spectrum of fungal-arthropod associations. The subsequent chapters in this text give further witness to the interesting problems that are being explored in the rapidly developing field of entomological mycology.

ACKNOWLEDGMENTS

I was introduced to these weird and wonderful molds by a great teacher and mycologist, Professor Ralph Emerson. Support for their study was generously provided by the National Institutes of Health (AI 05882), the University of Washington, and McGill University, but it was through the work and aid of students and associates in the laboratory that the battle between fungi and arthropods was observed.

LITERATURE CITED

1. Ainsworth, G. C. 1956. *Agostino Bassi, 1773–1856.* Nature 177: 255–257.

2. Anderson, J. F., and S. L. Ringo. 1969. *Entomophthora aquatica sp.n. infecting larvae and pupae of floodwater mosquitoes.* J. Invert. Pathol. 13: 386–393.

3. Armstrong, D. A., D. V. Buchanan, and R. S. Caldwell. 1976. *A mycosis caused by Lagenidium sp. in laboratory-reared larvae of the Dungeness crab, Cancer magister, and possible chemical treatments.* J. Invert. Pathol. 28: 329–336.

4. Benjamin, R. K. 1965. *Study in specificity.* Natural History 74: 42–49.

5. Bigelow, H. E., and E. G. Simmons, eds. 1977. Second international mycological congress, abstracts. University of South Florida, Tampa. 786 pp.

6. Bland, C. E., and H. V. Amerson. 1973. *Observations on Lagenidium callinectes: isolation and sporangial development.* Mycologia 65: 310–320.

7. Bland, C. E., and J. N. Couch. 1973. *Scanning electron microscopy of sporangia of Coelomomyces.* Can. J. Bot. 51: 1325–1330.

8. Brock, T. 1961. *Milestones in microbiology.* Prentice-Hall, Englewood Cliffs, N.J. 275 pp.

9. Chapman, H. C. 1974. *Biological control of mosquito larvae.* Ann. Rev. Entomol. 19: 33–59.

10. Chatton, E. 1920. *Les Peridiniens parasites, morphologie, reproduction, ethologie.* Arch. Zool. Exp. Gen. 59: 1–473.

11. Chatton, E., and F. Picard. 1908. *Sur une Laboulbeniacee, Trenomyces histophtorus Chatton et Picard endoparasite des poux de la poule domestique.* Bull. Soc. Mycol. Fr. 25: 117–170.

12. Christensen, C. M. 1965. *The molds and man.* McGraw-Hill, N.Y. 284 pp.
13. Clark, T. B., W. R. Kellen, and J. E. Lindegren. 1963. *Axenic culture of two Trichomycetes from California mosquitoes.* Nature (London) 197: 208-209.
14. Couch, J. N. 1945. *Revision of the genus Coelomomyces parasitic in insect larvae.* J. Elisha Mitchell Sci. Soc. 61: 124-236.
15. ———. 1972. *Mass production of Coelomomyces, a fungus that kills mosquitoes.* Proc. Nat. Acad. Sci. 69: 2043-2047.
16. Couch, J. N., and C. J. Umphlett. 1963. *Coelomomyces infections.* Pages 149-188 *in* E. A. Steinhaus, ed. *Insect pathology,* Vol. 2. Academic Press, New York.
17. Couch, J. N., S. V. Romney and B. Rao. 1974. *A new fungus which attacks mosquitoes and related Diptera.* Mycologia 66: 374-379.
18. Emerson, R., and C. M. Wilson. 1949. *The significance of meiosis in Allomyces.* Science 110: 86-88.
19. Federici, B. A. 1977. *Differential pigmentation in the sexual phase of Coelomomyces dodgei.* Nature (London) 267: 514-515.
20. Federici, B. A., and D. W. Roberts. 1976. *Experimental laboratory infection of mosquito larvae with fungi of the genus Coelomomyces II. Experiments with Coelomomyces punctatus in Anopheles quadrimaculatus.* J. Invert. Pathol. 27: 333-341.
21. Fredeen, F. J. H., and J. A. Shemanchuk. 1960. *Black flies of irrigation systems in Saskatchewan and Alberta.* Can. J. Bot. 38: 723-735.
22. Galt, J. H., and H. C. Whisler. 1970. *Differentiations of flagellated spores in Thalassomyces, Ellobiopsid Parasite of marine Crustaceae.* Arch. Mikrobiol. 71: 295-303.
23. Giebel, P. E., and A. J. Domnas. 1976. *Soluble trehelases from larvae of mosquito, Culex pipiens, and fungal parasite Lagenidium giganteum.* Insect Biocontrol 6: 303-311.
24. Gotelli, G. 1974. *Morphology of Lagenidium callinectes 2. Zoosporogenesis.* Mycologia 66: 846-858.
25. Hibbits, J. 1978. *Marine eccrinales found in Crustaceans of the San Juan Archipelago.* Washington Syesis. In press.
26. Humber, R. A. 1976. *The systematics of the genus Strongwellsea.* Mycologia 68: 1042-1060.
27. Keilin, D. 1921. *On a new type of fungus: Coelomomyces stegomiae, N.G., N.Sp., parasitic in the body cavity of the larva of Stegomia scutellaris Walker (Diptera, Nematocera, Culicidae).* Parasitology 13: 225-234.
28. Laird, M. 1972. *A novel attempt to control biting flies with their own diseases.* Science Forum 5: 12-14.
29. Large, E. C. 1940. *The advance of the fungi.* Jonathan Cape, London. 488 pp.
30. MacLeod, D.M. 1963. *Entomophtorales infections.* Pages 189-231 *in* E.A. Steinhaus, ed. *Insect pathology,* Vol. 2. Academic Press, New York.
31. Manier, J. F. 1950. *Recherches sur les Trichomycetes.* Ann. des Sci. Natl., Bot., Ser. II. 11: 53-162.
32. McEwen, F. L. 1963. *Cordyceps infections.* Pages 273-290 *in* E. A. Steinhaus, ed. *Insect pathology,* Vol. 2. Academic Press, New York.
33. McNitt, R. E., and J. N. Couch. 1977. *Coelomomyces pathogens of Culicidae (mosquitoes).* Pages 123-145 *in* D. W. Roberts and M. A. Strand, eds. *Pathogens of medically important arthropods.* Bull. WHO Suppl. No. 1, Vol. 55.
34. Moss, S. T., and R. W. Lichtwardt. 1976. *Development of trichospores and their appendages in Genistellospora homothallica.* Can. J. Bot. 54: 2346-2364.
35. ———. 1977. *Zygospores of the Harpellales: an ultrastructural study.* Can. J. Bot. 24: 3099-3110.
36. Muspratt, J. 1963. *Destruction of the larvae of Anopheles gambiae Giles by a Coelomomyces fungus.* Bull. WHO. 29: 81-86.
37. Noguer, A., W. Wernsdorfer, R. Kouznetsov, and J. Hempel. 1978. *The malaria situation in 1976.* WHO Chronicle 32: 9-17.
38. Pellai, J. S., T. L. Wong, and J. T. Dodgshun. 1976. *Copepods as essential hosts for the development of a Coelomomyces parasitizing mosquito larvae.* J. Med. Entomol. 13: 49-50.
39. Pipes, W. O., and D. Jenkins. 1965. *Zoophagus in activated sludge—a second observation.* Int. J. Air, Wat. Poll. 9: 495-500.
40. Ramsbottom, J. 1953. *Mushrooms and toadstools.* Collins, London. 306 pp.
41. Roberts, D. W. 1967. *Some effects of Metarrhizium anisopliae and its toxins on mosquito larvae.* Pages 243-246 *in* P. A. van der Laan, ed. *Insect pathology and microbial control.* North-Holland Publ. Co., Amsterdam.

42. Romney, S. V., M. M. Boreham, and L. T. Nielsen. 1971. *Intergeneric transmission of Coelomomyces infections in the laboratory.* Utah Mosq. Abat. Assoc. Proc. 24: 18–19.

43. Shapiro M., and D. W. Roberts. 1976. *Growth of Coelomomyces psorophorae mycelium in vitro.* J. Invert. Pathol. 27: 399–402.

44. Shemanchuk, J. A. 1959. *Note on Coelomomyces psorophorae Couch, a fungus parasitic on mosquito larvae.* Can Entomol. 91: 743–744.

45. ———, and R. A. Humber. 1978. *Entomophthora culicis (Phycomycetes: Entomophthorales) parasitizing black fly adults (Diptera: Simuliidae) in Alberta.* Can. Entomol. 110: 253–256.

46. Soper, R. S., L. F. R. Smith, and A. J. Delyzer. 1976. *Epizootiology of Massospora levispora in an isolated population of Okanagana rimosa.* Ann. Entomol. Soc. Amer. 69: 275–283.

47. Thaxter, R. 1896, 1908, 1924, 1926, 1936. *Contribution towards a monograph of the Laboulbeniaceae.* Parts I, II, III, IV, V. Mem. Amer. Acad. Arts. Sci. Vols. 12, 13, 14, 15, 16A: 188–429, 217–469, 309–426, 427–580, 1-435.

48. Travland, L. B. 1978. *Initiation of infection of mosquito larvae (Culiseta inornata) by Coelomomyces psorophorae.* Submitted to J. Invert. Pathol.

49. Travland, L. B. 1978. *Structures of the motile cells of Coelomomyces psorophorae and function of the zygote in encystment on a host.* Submitted to Canad. J. Bot.

50. Weiser, J. 1976. *The intermediate host for the fungus Coelomomyces chironomi.* J. Invert. Path. 28: 273–274.

51. Weiser, J., and J. Varra. 1964. *Zur Verbreitung der Coelomomyces Pilze in europaischen Insekten.* Z. tropen med. Parasitol. 15: 38–42.

52. Whisler, H. C. 1968. *Developmental control of Amoebidium parasiticum.* Dev. Biol. 17: 562–570.

53. Whisler, H. C. 1968. *Experimental studies with a new species of Stigmatomyces.* Mycologia 60: 56–75.

54. Whisler, H. C., and L. B. Travland. 1974. *The rotifer trap of Zoophagus.* Arch. Microbiol. 101: 95–107.

55. Whisler, H. C., S. L. Zebold, and J. A. Shemanchuk. 1974. *Alternate host for mosquito parasite Coelomomyces.* Nature 251: 715–716.

56. Whisler, H. C., S. L. Zebold, and J. A. Shemanchuk. 1975. *Life History of Coelomomyces psorophorae.* Proc. Nat. Acad. Sci. 72: 693–696.

57. Zebold, S. L., H. C. Whisler, L. B. Travland and J. A. Shemanchuk. 1978. *Host penetration and specificity in the mosquito pathogen, Coelomomyces psorophorae.* Submitted to Nature.

chapter 2

Lipids of Ambrosia Fungi and the Life of Mutualistic Beetles

by L. T. KOK*

ABSTRACT

Many fungi are known to be associated with ambrosia beetles. Included are species in the genera: *Aspergillus, Ambrosiella, Ambrosiomyces, Ascoidea, Botryodiplodia, Cephalosporium, Ceratocystis, Cladosporium, Colletotrichum, Endomyces, Endomycopsis, Fusarium, Graphium, Monacrosporium, Monilia, Mortierella, Penicillium, Pestalozzia, Phialophoropsis, Raffaelea, Sporothrix, Sporotrichum, and Tuberculariella.* Most of them are Fungi Imperfecti. Nutritional dependency of the beetles on their ambrosia fungi is well documented. First recognized by Hubbard in 1897, interactions between ambrosia fungi and their mutualistic beetles have subsequently received much attention. Despite numerous studies on fungi-beetle symbiosis, the specific role of the ambrosia fungi in the nutrition of the mutualistic beetles is not well understood. The lipoidal nature of most growth hormones, pheromones and sex attractants has greatly stimulated the study of lipid biochemistry in plants and insects. Determinations of the lipid content and composition of a wide range of fungi have been reported, but few were on ambrosia fungi. In a study on the nutritional potential of the ambrosia fungi associated with *Xyleborus* beetles, analyses of lipids in the three ambrosia fungi showed that total lipid content was 10.5% in *Cephalosporium* sp., 12.7% in *Graphium* sp. and 15% in *Fusarium solani* on a dry-weight basis. Fatty acids were the major components, followed by polar lipids, while sterols made up less than 0.25% in all three species. The predominant fatty acids were palmitic, oleic and

*Department of Entomology, Virginia Polytechnic Institute and State University, Blacksburg, Virginia 24061.

INSECT-FUNGUS SYMBIOSIS /Batra (ed.) / Allanheld, Osmun, Montclair, NJ

stearic acid. Approximately one-third of the total were polyunsaturated fatty acids. The total fatty acid content in *F. solani* (11%) was about twice that of *Cephalosporium* sp. and *Graphium* sp. The phospholipid pattern in the three ambrosia fungi was similar. Major components were phosphatidylethanolamine (PE), phosphatidylcholine (PC) and phosphatidylinositol (PI). PC and PE totaled more than 60% of the phospholipids in *Cephalosporium* sp. and *F. solani,* and more than 80% in *Graphium* sp. The only sterol detected in and isolated from the three ambrosia fungi was ergosterol. Yields were 0.12% in *Graphium* sp., 0.15% in *Cephalosporium* sp. and 0.24% in *F. solani* for 15 day-old fungi. These results demonstrated similarity in composition and content of lipids in the three ambrosia fungi. Significantly only one sterol, ergosterol, was detected in all three. Such a Δ^7 sterol is required by the continued growth and development of the beetle, *Xyleborus ferrugineus,* in the absence of its ambrosia fungi. This strongly indicates that ergosterol, which is provided by the ambrosia fungi, is a critical nutrient to the beetle. This dependence of *X. ferrugineus* on the ambrosia fungi establishes a major basis for the existing obligatory symbiosis. Similar studies of other ambrosia fungi and their mutualistic beetles may reveal the nutritional basis of their symbiosis.

INTRODUCTION

The symbiotic interrelationship between ambrosial fungi and their mutualistic beetles has been widely investigated over the years. Despite numerous studies on fungi-beetle symbiosis, the specific role of ambrosial fungi in the nutrition of their mutualistic beetles is not well understood. Because of the importance of lipids to insect development, examination of the lipid composition of various ambrosial fungi could provide some insight into the nutrients that are available to the ambrosia beetles. In this presentation, the following are reviewed:

1) Ambrosial fungi and their nutritional importance to the ambrosia beetles
2) Lipid requirements of insects
3) Lipid composition of ambrosial fungi and their role in the life of *Xyleborus ferrugineus*

AMBROSIAL FUNGI AND THEIR NUTRITIONAL IMPORTANCE TO THE AMBROSIA BEETLES

The ambrosial form of many fungi has been well documented. Some of these ambrosial fungi are listed in Table 2.1. In addition, *Botryodiplodia, Colletotrichum, Leptographium, Monacrosporium, Mortierella, Pestalozzia, Raffalea, Sporothrix,* and *Sporotrichum* among others, were included in reviews by Baker (1963), Batra (1967), and Francke-Grosmann (1967).

Ambrosia beetles live in obligatory symbiosis with their mutualistic fungi; the beetles derive nutrients synthesized by the fungi, and the fungi in return are disseminated by the beetles. Beetle dependency on the nutri-

Table 2.1 Examples of Ambrosial Fungi.

Genus	References
Absidia	Nakashima (1971)
Ambrosiella	Brader (1964); French and Roeper (1972)
Aspergillus	Beal and Massey (1945)
Cephalosporium	Baker and Norris (1968)
Ceratocystis	Buchanan (1940); Bakshi (1950, 1951); Mathiesen-Kaarik (1953); Cachan (1957); Wilson (1959); Baker (1963); Norris (1965); Barras and Perry (1972); Yearian et al. (1972)
Chaetomium	Nakashima (1971)
Endomyces	Neger (1909); Verrall (1943)
Endomycopsis	Batra (1963)
Fusarium	Müller (1933); Norris (1965); Norris and Baker (1967)
Graphium	Baker and Norris (1968)
Monilia	Hartig (1872); Müller (1933); Francke-Grosmann (1956, 1958)
Penicillium	Smith (1935); Fischer (1954)
Tuberculariella	Farris and Funk (1965)

tionally rich ambrosial spores, borne terminally on special hyphae, was recognized by Hubbard as early as 1897. Such nutritional associations between ambrosia beetles and their microbial symbionts subsequently received much attention (Francke-Grosmann 1963, 1965, and 1967; Baker 1963; Batra 1963, 1966, and 1972; Graham 1967). Recent studies on the nutritional dependency of ambrosia beetles on their mutualistic fungi include: *Dendroctonus frontalis* (Barras 1973; Barras and Hodges 1974); *Xyleborus dispar* (French and Roeper 1972, 1975) and *Xyleborus ferrugineus* (Norris and Chu 1971; Kok and Norris 1972d and 1973b; Norris 1972 and 1974; Kingsolver and Norris 1977).

A reduction in number of progeny was found in *D. frontalis* in the absence of its mycangial fungi (Barras 1973) and the greatest gain in weight of the beetle coincided with active feeding in the fungal-phloem structure (Barras and Hodges 1974). In *X. dispar*, oocyte development and oviposition occurred only after the postdiapause female had fed on the ambrosial form of the fungus (French and Roeper 1975); and larvae required the ambrosial form of the fungus *Ambrosiella hartigii* to develop and pupate (French and Roeper 1972). In studies on the nutritional interrelationships between *X. ferrugineus* and its ambrosial fungi, Norris and Baker (1967) showed that a fungus was required for reproduction, and Baker and Norris (1968) reported a complex of fungi mutualistically involved in the nutrition of the beetle. In the absence of a mutualistic fungus, second brood pupation of *X. ferrugineus* failed except when ergosterol was substituted for cholesterol in the diet (Norris et al. 1969). Further tests revealed that ergosterol or 7-dehydrocholesterol was adequate as the sole sterol source for the continued growth and reproduction of the fungus-free beetle (Chu et al. 1970; Kok et al. 1970). Norris and Chu (1971) found that maternal aposymbiotic females of *X. ferrugineus* lost their ability to pass required pupation factors to their progeny, and initial reproduction of females maintained on fungus diet was significantly earlier than for those on non-fungus diet (Kingsolver and Norris 1977). Although these studies demonstrate the importance of the ambrosial fungi in the life of their mutualistic beetles, the specific nutritional role of the ambrosial fungi is not well understood.

LIPID REQUIREMENTS OF INSECTS

Lipid components especially important to the development of insects are the sterols, phospholipids and fatty acids. Sterols are crystalline unsaponifiable alcohols with properties resembling those of cholesterol. They serve as precursors for essential steroid metabolites, such as certain hormones, and as structural components of cells. All insects so far examined were unable to synthesize sterols and therefore required a dietary source of sterol for growth and development. Many insects preferentially utilized cholesterol (Table 2.2), but most, if not all, of the functions of sterols in insects can be fulfilled by other sterol types (Fast 1964). Requirements for lipids other than sterol vary with species and no general pattern can be discerned.

Phospholipids constitute varying proportions of body fat and cell membranes. An apparent dietary requirement for phospholipids has been indicated in a beetle, *Leptinotarsa decemlineata* (Grison 1948) and in a moth, *Bombyx mori* (Ito 1960).

Fatty acids are important as energy source and as components of phospholipids and glycerides. In general, insects are able to synthesize long-chain fatty acids from simple precursors (Bloch et al. 1956; Rajalakshmi et al. 1963; Strong 1963; Lambremont 1965; Lamb and Monroe 1968; Stephen and Gilbert 1969), but biosynthesis of polyunsaturated acids has not been demonstrated (Stephen and Gilbert 1969). A dietary requirement for the polyunsaturated acids, linoleic and/or linolenic acid, has been indicated in certain species of Orthoptera and Lepidoptera (Fast 1964).

LIPID COMPOSITION AND ROLE OF AMBROSIAL FUNGI

Studies on the lipid composition of ambrosial fungi involved three species, *Fusarium solani, Cephalosporium* sp. and *Graphium* sp., associated with *Xyleborus ferrugineus* (Kok 1971; Kok and Norris 1972a, 1973a and b). Procedures for isolation of sterols, fatty acids and phospholipids are presented in Fig. 2.1. Lyophilized fungus was extracted in batches of 15 g for total lipids by the methods of Folch et al. (1957). Proportions of the lipid components are shown in Fig. 2.2 and Table 2.3. Fatty acids were the major components, followed by polar lipids, with sterols making up less than 0.25% of the dry weight in all three species. The content of lipids (10.5% to

Table 2.2 Sterol Requirements of Some Insects[a].

Insect	Cholesterol	7-Dehydrocholesterol	Cholestanol	Sitosterol	Stigmasterol	Ergosterol
Lucilia sericata	3	--	--	2	--	1
Dermestes vulpinus	3	0	0	0	0	0
Attagenus piceus	3	2	0	--	--	--
Lasioderma serricorne	3	3	2	3	0	3
Stegobium paniceum	3	3	1	3	--	3
Blattella germanica	3	0	0	3	3	1
Drosophila melanogaster	3	--	2	3	--	3
Bombyx mori	1	--	--	3	2	0

[a]Sterols supporting development: 3 good, 2 moderate, 1 poor, or 0 ineffective and—not studied (Clayton 1964).

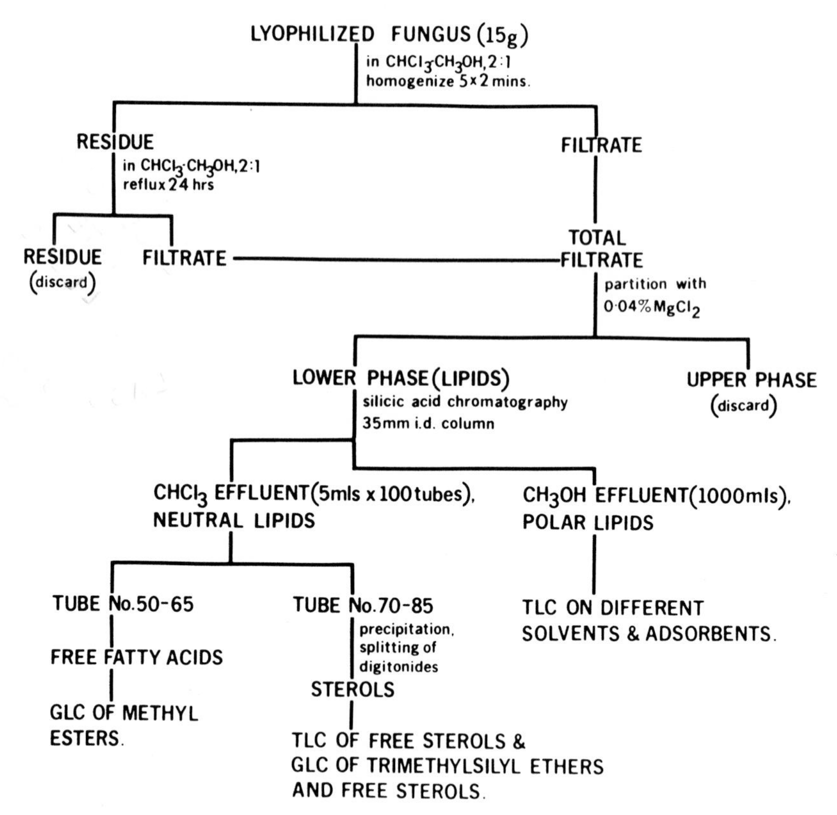

Figure 2.1. A flow diagram showing the procedures for the isolation of lipids of ambrosial fungi. (Kok 1971.)

15.0%) in these three ambrosial fungi was not unusual when compared to those reported for other fungi (Kok and Norris 1973a). It was higher than the 1.5% to 7.5% in several other Moniliales: *Pithomyces chartarum, Stemphylium dendriticum* and *Cylindrocarpon radicicola* (Hartman et al. 1962), and that of *Candida* sp. (Kates and Baxter 1962); comparable to the 11% in *Nadsonia elongata* (Dyke 1964); and lower than those reported in *Rhodotorula* spp. (Enebo et al. 1946; Hartman et al. 1959) and *Lipomyces* spp. (Starkey 1946; McElroy and Stewart 1967). It was well within the range of 1% to 40% cited for filamentous fungi (Pruess et al. 1934; Prescott and Dunn 1940; Bloor 1943). Studies on other *Fusarium* spp. showed 22% to 37% for parent and mutant strains of *F. vasinfectum* (Ahamed et al. 1973) and 8.6% in *F. culmorum* (Nombela-Cano and Peberdy 1971). A much lower lipid content, less than 3%, was found in three pathogenic fungi, *Alternaria dauci, F. solani* f. *phaseoli* and *Sclerotium rolfsii* (Gunasekaran and Weber 1972).

Lipid content of ambrosial fungi cultured in 100-ml Neutral-Dox-Yeast medium per 500-ml flask, incubated for 15 days at 28°C, 70% R.H. in darkness, expressed as per cent dry weight.

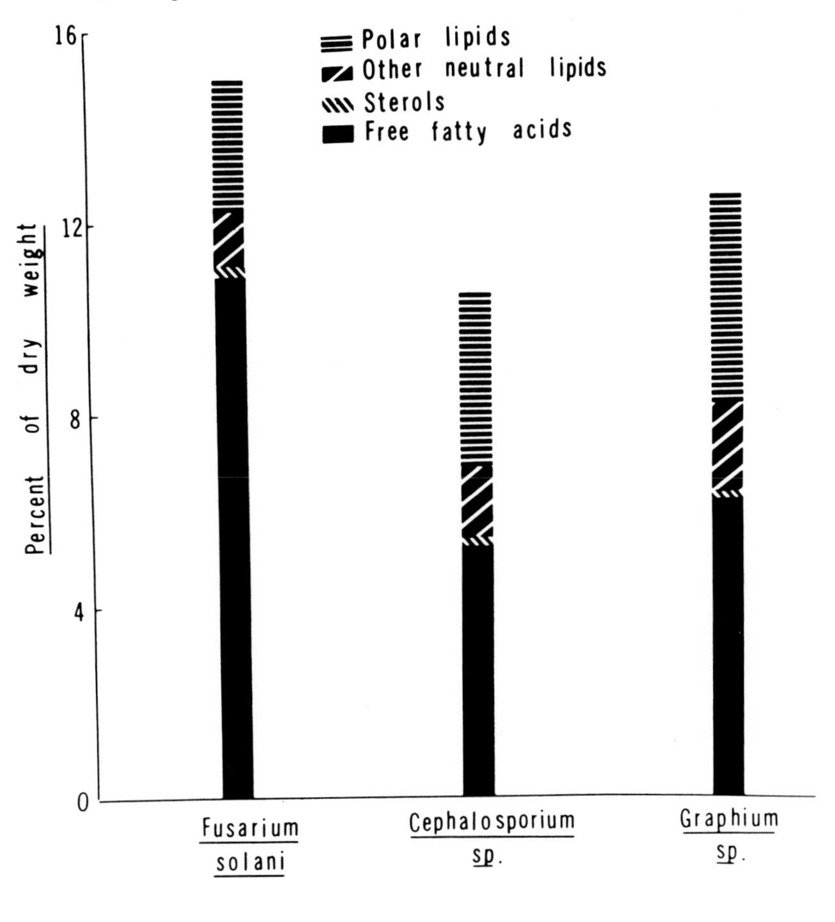

Figure 2.2. Lipid composition of ambrosial fungi. (Kok 1971.)

Table 2.3 Lipid Content (% dry fungal weight) of Ambrosial Fungi[a].

Lipid	Ambrosial Fungi (mycelial age in days)				
	Fusarium solani			*Cephalosporium* sp.	*Graphium* sp.
	(5)	(10)	(15)	(15)	(15)
Fatty acids	11.07	11.07	10.93	5.33	6.33
Sterols	0.17	0.20	0.24	0.15	0.12
Other neutral lipids	2.40	1.93	1.07	1.40	1.75
Polar lipids	3.13	2.87	2.80	3.60	4.53
Total lipids	16.77	16.07	15.04	10.48	12.73

[a]Kok and Norris (1973a).

Table 2.4 Sterols Isolated from Fungi[a].

Sterol	Empirical formula	Fungal source
1. Zymosterol	$C_{27}H_{44}O$	Yeast
2. Desmosterol (24-Dehydrocholesterol)	$C_{27}H_{44}O$	Saprolegniales, Leptomitales
3. Cholesterol	$C_{27}H_{46}O$	Penicillium, Saprolegniales, Leptomitales, Mucorales, and "Fungus-gardens" of termites
4. 14-Dehydroergosterol	$C_{28}H_{42}O$	*Aspergillus*
5. 24-Dehydroergosterol	$C_{28}H_{42}O$	Yeast
6. Fungisterol	$C_{28}H_{42}O$	*Claviceps, Rhizopus*
7. Ergosterol	$C_{28}H_{44}O$	*Mucor, Blakeslea, Phycomyces, Rhizopus; Neurospora, Ceratocystis, Claviceps, Helvella,* Yeasts, Lichens and wood-rotting Basidiomycetes; *Aspergillus, Penicillium, Dematium, Torula, Helminthosporium, Trichophyton, Dermatophytes, Fusarium,* "Fungus-gardens" of ants and termites
8. Ergosta-5,7,24(28)-trien-3β-ol	$C_{28}H_{44}O$	*Phycomyces*
9. Brassicasterol	$C_{28}H_{46}O$	*Trichophyton*
10. 22-Dihydroergosterol	$C_{28}H_{46}O$	*Polyporus,* Mucorales
11. 5-Dihydroergosterol	$C_{28}H_{46}O$	Yeast
12. 24-Methylenecholesterol	$C_{28}H_{46}O$	Saprolegniales, Leptomitales
13. Episterol	$C_{28}H_{46}O$	Yeast, *Phycomyces*
14. Ascosterol	$C_{28}H_{46}O$	Yeast
15. Fecosterol	$C_{28}H_{46}O$	Yeast
16. Cerevisterol	$C_{28}H_{46}O_3$	Yeast, *Amanita, Claviceps*
17. Fucosterol	$C_{28}H_{48}O$	Saprolegniales, Leptomitales
18. Stigmasterol	$C_{29}H_{48}O$	"Fungus-gardens" of termites
19. β-Sitosterol	$C_{29}H_{50}O$	"Fungus-gardens" of termites
20. Lanosterol	$C_{30}H_{50}O$	Yeast, *Aspergillus, Phycomyces*

[a]Kok (1971).

Sterols. Ergosterol, the most common sterol in fungi (Kritchevsky 1963) was the only detectable sterol in the three ambrosial fungi (Kok et al. 1970; Kok and Norris 1973b). The same sole sterol was also found in three species of *Fusarium* (Fiore 1948), 14 species of Basidiomycetes (Milazzo 1965) and *Ceratocystis fagacearum* (Collins and Kalnins 1969). More than one sterol, however, were found in several species of Saprolegniales, Leptomitales and Mucorales (McCorkindale et al. 1969) as well as in the fungus gardens of termites (Cmelik and Douglas 1970). At least 20 mycosterols, isolated from a wide range of fungi, are known (Table 2.4); the majority being C_{28}—C_{29} sterols.

F. solani contained a significantly larger amount of ergosterol than the other two species. Yields on dry-weight basis (Table 2.5) were 0.12% in *Graphium* sp., 0.15% in *Cephalosporium* sp. and 0.24% in *F. solani* (Kok et al. 1970; Kok and Norris 1973b). These values were at the lower end of the 0.1% to 1.0% range of sterol contents reported by Pruess et al. (1931) for a number of yeasts, molds, and mushrooms. They were more comparable to values found in the Saprolegniales (0.01% to 0.25%), but higher than those of the Leptomitales (0.01% to 0.05%) and Mucorales (0.005% to 0.025%) (McCorkindale et al. 1969). The ergosterol content of 14 Basidiomycetes ranged from 0.017% to 0.42% (Milazzo 1965), but dermatophytes had a much wider range, 0.016% to 1.5% (Blank et al. 1962).

Sterol contents determined by the colorimetric method included 0.56% in *F. lycopersici* and up to 1% in *F. lini* (Fiore 1948); 0.18% in *F. vasinfectum*, 0.22% in *Cephalosporium* sp. and 0.07% to 0.84% in a number of yeasts (Appleton et al. 1955). In contrast various strains of *F. vasinfectum* were reported to have extremely high sterol contents ranging from 12% to 31%

Table 2.5 Ergosterol Content of Ambrosial Fungi[a].

Species and Age	Percent Dry-weight
Graphium sp.	
15 days old	0.12a
Cephalosporium sp.	
15 days old	0.16ab
Fusarium solani	
5 days old	0.17ab
10 days old	0.20 bc
15 days old	0.24 c

[a]Mean of 3 replications; means with the same letter are not significantly different at the 5% level, Duncan's multiple range test (Kok and Norris 1973b).

(Ahamed et al. 1973). The unusually high values reported by the latter could be due to their extraction technique and/or purity of the sterol extract.

The mean levels (0.17% to 0.24%) of ergosterol in *F. solani* and the other two mutualistic fungi, 0.12% in *Graphium* sp. and 0.15% in *Cephalosporium* sp., were above those generally required (0.01% to 0.1%) in the diets of insects (Clayton 1964). Thus, in all three ambrosial fungi, ergosterol was present in amounts which might be considered adequate for the growth and development of their mutualistic beetle.

Phospholipids. Major phospholipids of the three ambrosial fungi were phosphatidylethanolamine (PE), phosphatidylocholine (PC) and phosphatidylinositol (PI) (Kok and Norris 1972a and c). PC and PE together totalled more than 60% of the phospholipids in *Cephalosporium* sp. and *F. solani,* and more than 80% in *Graphium* sp. (Table 2.6). The increased percentage in the latter was due to a higher PE content. The PC:PE ratio of 1:1 in the first two species was comparable to that of *Saccharomyces cerevisiae* (Longley et al. 1968) and *Lipomyces lipofer* (McElroy and Stewart 1967). The PE contents were in the upper area of the previously reported 18% to 30% range, but PC contents were much lower than the 41% to 54% of four other Moniliales (Graff et al. 1968). In *Graphium* sp., PE predominated by about a 2:1 ratio over PC and was much higher than the PE content of any fungi studied. The 14% of PI in *F. solani* was about three times that in *Cephalosporium* sp. and *Graphium* sp., but it was not high when compared to the 22.4% in *Lipomyces lipofer* (McElroy and Stewart 1967). Although faint traces were detected in the cardiolipin region in the other two ambrosial fungi, cardiolipin could only be quantitatively determined in *Graphium* sp. This phospholipid had also been demonstrated in the yeasts, *Candida* (Kates and Baxter 1962), *Kloeckera* (Dawson et al. 1962) and *Saccharomyces* (Lester 1963), but not in *Lipomyces*

Table 2.6 Phospholipids of Ambrosial Fungi[a].

Phospholipid	Percent of total phospholipid		
	Fusarium solani	*Cephalosporium* sp.	*Graphium* sp.
PE	28.2 ± 4.2	33.8 ± 2.9	53.3 ± 4.7
PC	34.3 ± 3.8	33.3 ± 1.6	30.4 ± 3.2
PI	14.5 ± 1.3	5.1 ± 1.7	5.7 ± 1.6
Cardiolipin	0	0	2.4 ± 0.7
$Unknown_1$	20.5 ± 2.0	15.3 ± 2.3	3.4 ± 0.7
$Unknown_2$	2.4 ± 0.6	12.5 ± 3.3	4.7 ± 0.3

[a]Kok and Norris (1972a and c).

(McElroy and Stewart 1967) nor the four Endomycetales and the four Moniliales species (Graff et al. 1968). One notable deviation from published results was the absence of any detectable concentration of PS in the three ambrosial fungi. In the study by Gunasekaran and Weber (1972) on the pathogenic *Fusarium solani* f. *phaseoli*, a trace of PS was detected, but PI, PE, and PC were absent.

There was a distinct similarity in phospholipid composition of the three ambrosial fungi and that of *X. ferrugineus* (Kok and Norris 1972c). All the phospholipids present in the fungi were also found in the beetle. In addition, the beetle also had a small amount of PS; probably as a result of the incorporation of serine, present in the nutrient pool supplied by the fungi, into phospholipid. This resemblance is a good indication of the close nutritional interrelationship between the mutualistic fungi and the beetle.

Fatty acids. Gas-liquid chromatography (GLC) of fatty acid methyl esters of ambrosial fungi as shown in Fig. 2.3 demonstrated the presence of myristic, palmitic, stearic, oleic, linoleic, linolenic, and arachidic or behenic acid in all three species (Kok and Norris 1973a). There was a marked similarity in the proportions of individual fatty acids, but differences were noticeable in terms of the total amount of fatty acids. *F. solani*, with fatty acids making up 11% of the fungal dry weight, had about twice the total fatty acid content of *Cephalosporium* sp. and *Graphium* sp. Linoleic acid, one of the two polyunsaturated acids, was the major component in the three species, ranging from 24% to 27% of the total fatty acids; but linolenic acid, the other polyunsaturated acid, constituted only a minor proportion (3% to 4%). Other major components were palmitic, stearic and oleic acid. A small amount of myristic acid was present in all three species, and in addition, a trace of arachidic acid was detected in *F. solani*. The other two species had a small portion of behenic acid. The pattern and quantity of fatty acids were generally unaffected by the period of incubation (5 to 15 days) of the fungi.

The predominant fatty acids, namely, palmitic, oleic, linoleic and stearic acid, in all three ambrosial fungi were either 16 or 18 carbons long (Kok and Norris 1973a). These compositions were similar to those reported in a number of Fungi Imperfecti (Shaw 1966). Quantitatively, the age of mycelia up to 15 days had little effect on the total fatty acid content, which remained around 11% of dry weight, in *F. solani*. These values compared favorably with the 10% reported in *Penicillium griseofulvum* (Light 1965) and 9.8% in *Penicillium pulvillorum* (Nakajima and Tanenbaum 1968). *Cephalosporium* sp. (5.3%) and *Graphium* sp. (6.3%) had about half the amount in *F. solani*, but all the three ambrosial fungi were higher in fatty acid content than that reported in other Fungi Imperfecti: *Pithomyces chartarum*, *Stemphylium dendriticum* and *Cylindrocarpon radicicola* (Hartman et al. 1962) and in a number of yeast-like forms (Shaw 1966).

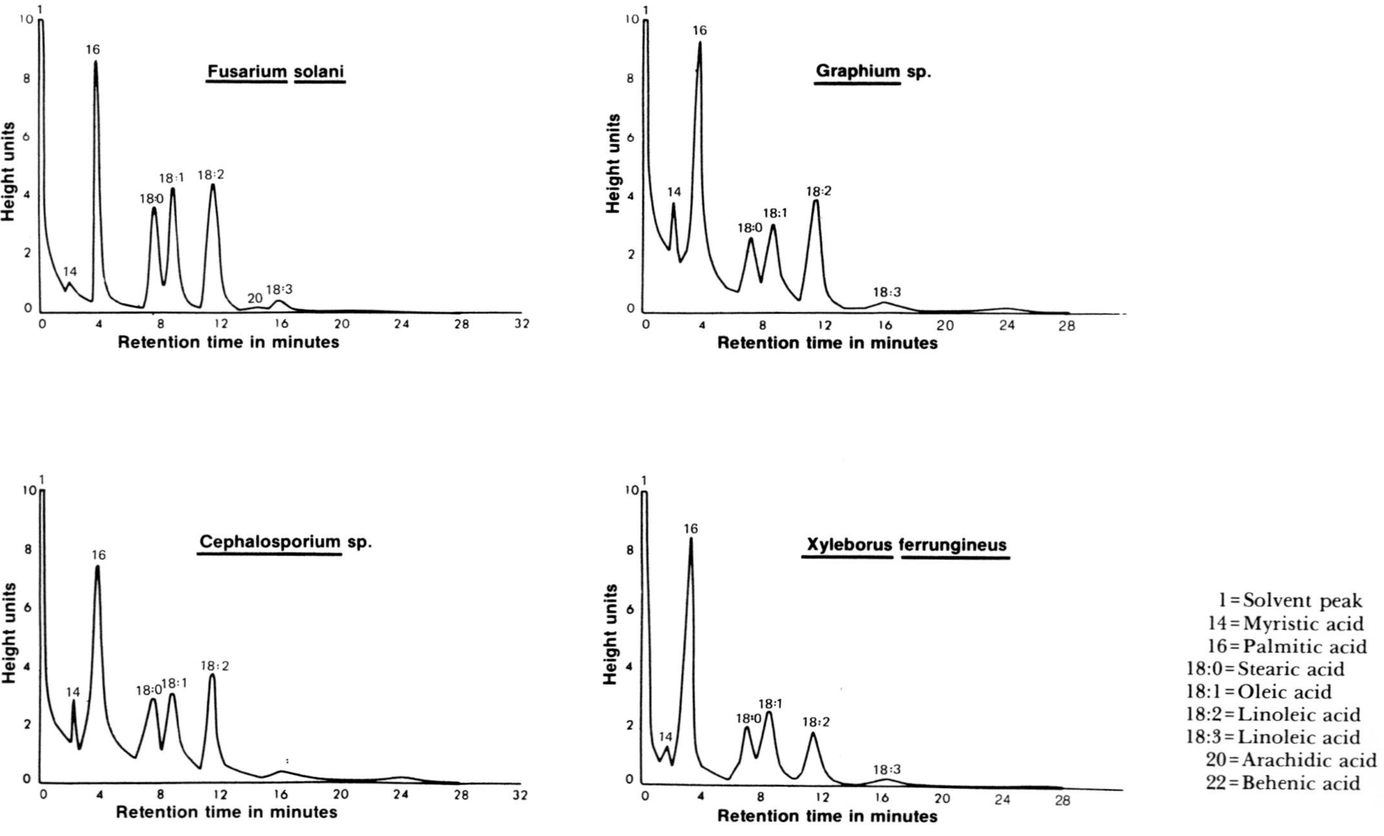

Figure 2.3. Gas-liquid chromatography of ambrosial fungi and *Xyleborus ferrugineus.* (Kok 1971.)

Of the fatty acids, only the polyunsaturated acids, linoleic (18:2) and linolenic (18:3), have been demonstrated as dietary requirements in insects by different workers (Gordon 1959; Dadd 1961 and 1964; Fraenkel and Blewett 1946; Pepper and Hastings 1943; Vanderzant et al. 1957; Vanderzant and Richardson 1964; Earle et al. 1967; Chippendale et al. 1964 and 1965; Nelson and Sukkested 1968). The presence of 20% linoleic and linolenic acid in *X. ferrugineus* (Kok and Norris 1972b) showed that these nutrients are not deficient in this beetle. Their high proportion (30%) of the total fatty acids in the three ambrosial fungi demonstrated that these symbionts can provide the fatty acid requirements of the beetle.

Nutritional studies have shown that fungal-free *Xyleborus* beetles cannot utilize cholesterol as a sole source of sterol. In *X. ferrugineus*, ergosterol or 7-dehydrocholesterol satisfied the deficiencies of cholesterol, i.e. pupation requirements (Chu et al. 1970; Kok et al. 1970); and in *X. posticus*, continued growth, development and reproduction occurred when ergosterol was present in sufficient quantities (Moya 1970). 7-Dehydrocholesterol apparently was not adequate as a sole source of sterol for *X. posticus*. In the presence of mutualistic fungi, these two *Xyleborus* species developed at a rapid rate without any other exogenous sterol source. Thus, it is highly significant that the same sterol, ergosterol, was detected in all three ambrosial fungi. This strongly indicates that ergosterol which is provided by the ambrosial fungi is a critical nutrient to the beetles, and this dependence establishes a major basis for the existing mutualistic symbiosis. The mutualism makes it possible for the beetle, with its unusual sterol requirement, to survive and reproduce.

That 7-dehydrocholesterol is also adequate for *X. ferrugineus* suggests that the insect is either unable to dehydrogenate cholesterol at the C_7 position, or lacks the ability to do this efficiently. The only structural characteristic present in both ergosterol and 7-dehydrocholesterol, but absent in cholesterol, is the Δ^7 bond (Fig. 2.4). Pupation of *X. ferrugineus* must therefore require this double bond (Kok et al. 1970). *X. ferrugineus* is apparently unique among studied insects in its use of cholesterol for egg production and larval growth, but not for pupation. As a sole sterol source, cholesterol has been reported inadequate for both larval growth and pupation in only one insect, *Drosophila pachea* (Heed and Kircher 1965).

Although the structural role of sterols in insects is quantitatively their most important function, the sterols also serve a vital function as precursors for the ecdysones, the steroid hormones that promote molting and adult development in insects. All known insect ecdysones have a Δ^7-6-keto group, and a 14 α-hydroxyl group; the position of other OH groups vary. The Δ^7 bond is an important structural requirement in all known molting hormones. Structures of the known insect ecdysones; the Δ^7 sterols, ergosterol and 7-dehydrocholesterol; β-sitosterol and cholesterol are illustrated in Figure 2.4.

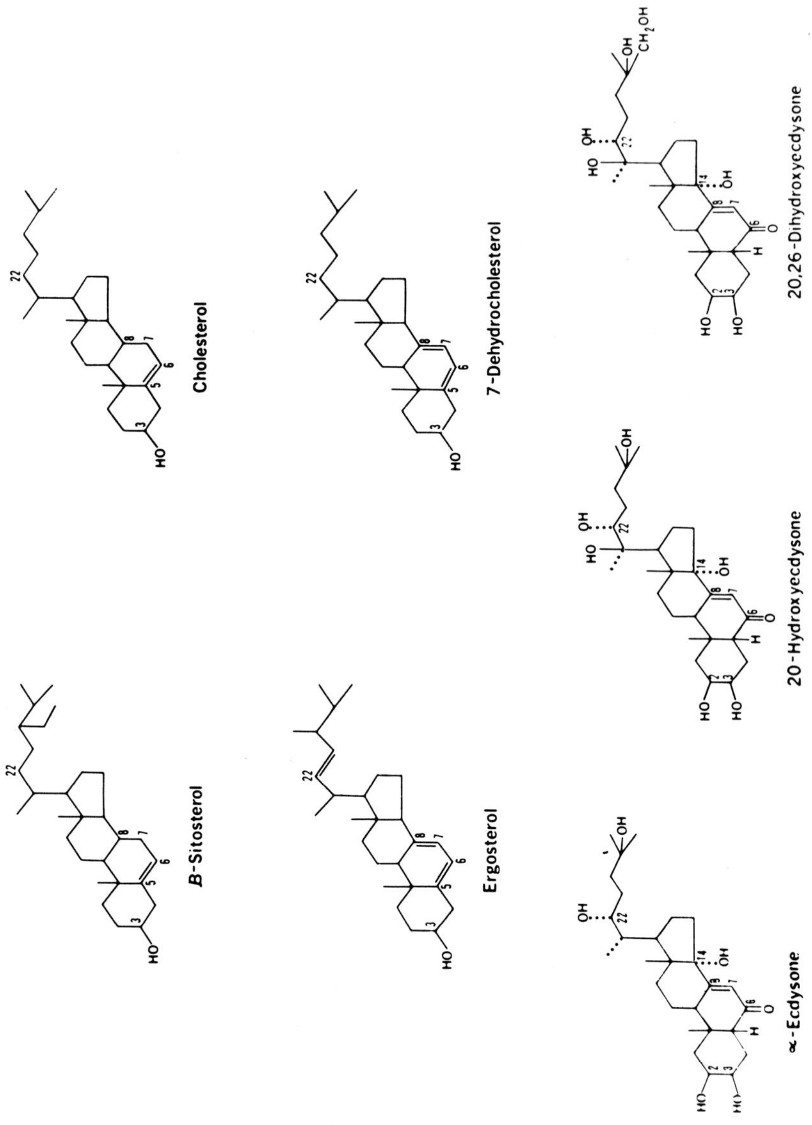

B-Sitosterol

Cholesterol

Ergosterol

7-Dehydrocholesterol

α-Ecdysone

20-Hydroxyecdysone

20,26-Dihydroxyecdysone

Figure 2.4. Structures of the known insect ecdysones and some sterols. (Kok 1971.)

The sterol nutrition situation in *X. ferrugineus* suggests some significant deviation in metabolism from the scheme (Fig. 2.5) proposed for steroids in the tobacco hornworm, *Manduca sexta* (Robbins et al. 1971). A metabolic block in *X. ferrugineus* for conversion of cholesterol to 7-dehydrocholesterol is apparent because cholesterol does not meet pupation requirements. The ability of the insect to use 7-dehydrocholesterol and ergosterol for pupation indicates that these Δ^7 sterols are metabolized into, or allow the biosynthesis of, pupational hormone(s) which may, or may not, be identical to known ecdysone(s) in insects. Two possible pathways of steroid metabolism in *X. ferrugineus* are shown in Figure 2.6 (Kok 1971). Whether the 7-dehydrocholesterol pathway is as important as that of the symbiont-supplied ergosterol is uncertain, but both are utilized to produce the steroidal hormone(s) necessary for pupation.

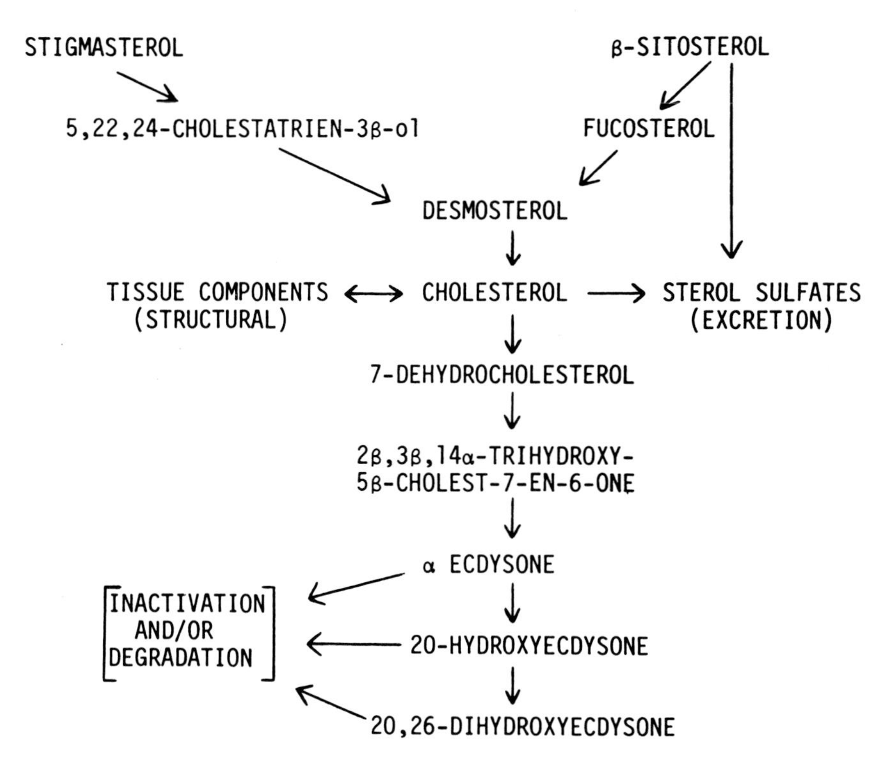

Figure 2.5. Metabolic scheme for steroids in the tobacco hornworm based on known intermediates and metabolic conversions. (Robbins et al. 1971.)

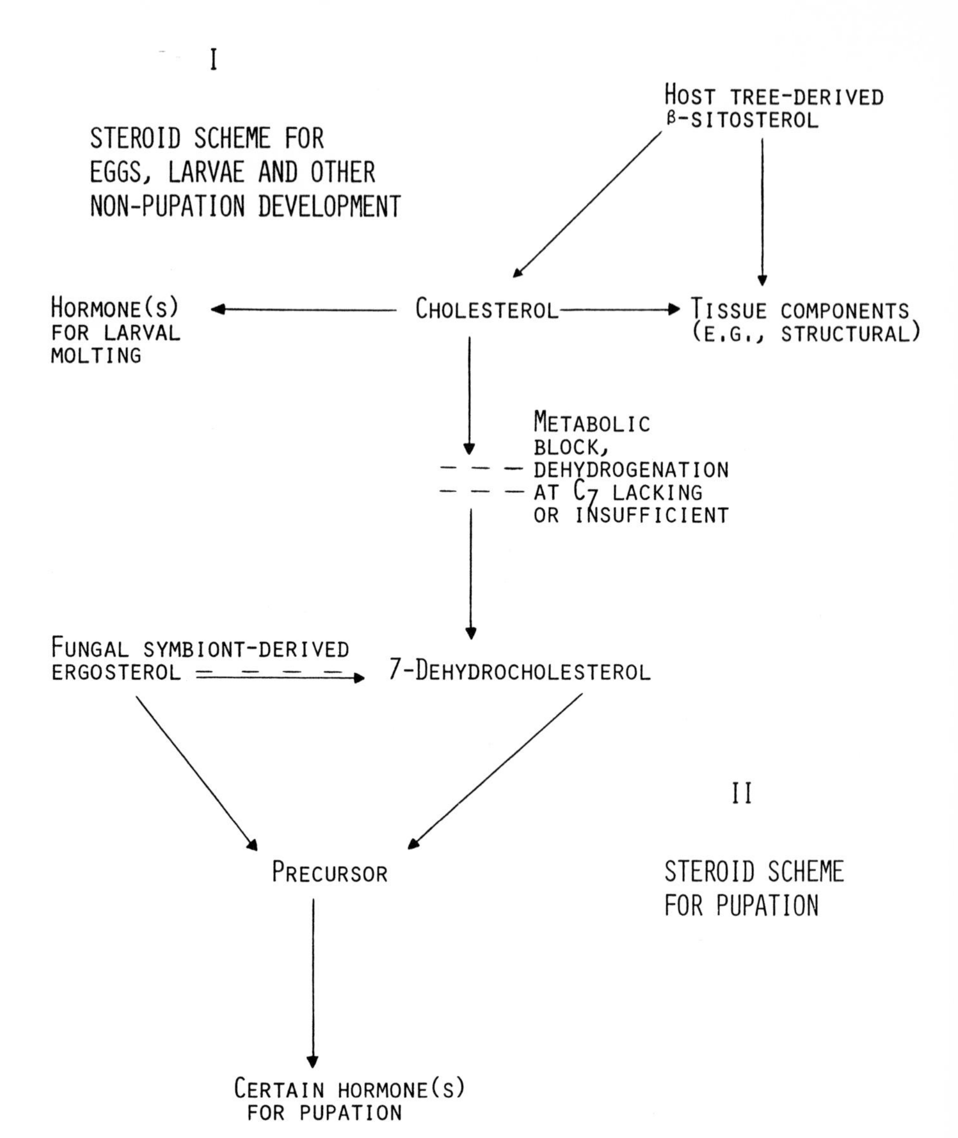

Figure 2.6. Proposed metabolic scheme for steroids in *Xyleborus ferrugineus*. (Kok 1971.)

LITERATURE CITED

Ahamed, N. M. M., S. Meenakshisundaram, and E. R. B. Shanmugasundaram. 1973. Lipids and lipase activity in strains of *Fusarium vasinfectum*. Indian J. Expt. Biol. 11: 37-39.

Appleton, G. S., R. J. Kieber, and W. J. Payne. 1955. The sterol content of fungi. II. Screening of representative yeasts and molds for sterol content. Appl. Microbiol. 3: 249-251.

Baker, J. M. 1963. Ambrosia beetles and their fungi, with particular reference to *Platypus cylindrus*. Fab. Symp. Soc. Gen. Microbiol. 13: 232-265.

Baker, J. M., and D. M. Norris. 1968. A complex of fungi mutualistically involved in the nutrition of the ambrosia beetle *Xyleborus ferrugineus*. J. Invert. Pathol. 11: 246-250.

Bakshi, B. K. 1950. Fungi associated with ambrosia beetles in Great Britain. Brit. Mycol. Soc. Trans. 33: 111-120.

———. 1951. Studies on four species of *Ceratocystis* with a discussion on fungi causing sapstain in Britain. Mycol. Papers 35: 1-16.

Barras, S. J. 1973. Reduction of progeny and development in the southern pine beetle following removal of symbiotic fungi. Canad. Entomol. 105: 1295-1299.

Barras, S. J., and J. D. Hodges. 1974. Weight, moisture, and lipid changes during life cycle of the southern pine beetle. U.S. For. Serv. Res. Note SO-178: 5 pp.

Barras, S. J., and T. Perry. 1972. Fungal symbionts in the prothoracic mycangium of *Dendroctonus frontalis* (Coleoptera: Scolytidae). Z. Angew. Entomol. 71: 95-104.

Batra, L. R. 1963. Ecology of ambrosia fungi and their dissemination by beetles. Trans. Kansas Acad. Sci. 66: 213-236.

———. 1966. Ambrosia fungi: extent of specificity to ambrosia beetles. Science 153: 193-195.

———. 1967. Ambrosia fungi: a taxonomic revision, and nutritional studies of some species. Mycologia 59: 976-1017.

———. 1972. Ectosymbiosis between ambrosia fungi and beetles—I. Indian J. Mycol. Plant Pathol. 2: 165-169.

Beal, J. A., and C. L. Massey. 1945. Bark beetles and ambrosia beetles, with special reference to species occurring in North Carolina. Bull. Duke Univ. School Forestry 10, Ser. 9, 40 pp.

Blank, F., F. E. Shortland, and G. Just. 1962. The free sterols of dermatophytes. J. Invest. Dermatol. 39: 91-94.

Bloch, K., R. C. Langdon, A. J. Clark, and G. Fraenkel. 1956. Impaired steroid biogenesis in insect larvae. Biochim. Biophys. Acta 21: 176.

Bloor, W. R. 1943. Biochemistry of the fatty acids and their compounds, the lipids. Reinhold Publ. Corp., New York. 387 pp.

Brader, L. 1964. Etude de la relation entre le scolyte des rameaux du cafèier, *Xyleborus compactus* Eichh. (*X. morstatti* Hag.) et sa plantehôte. Dissertationes Wageningen. 109 pp.

Buchanan, W. D. 1940. Ambrosia beetle *Xylosandrus germanus* transmits Dutch elm disease under controlled conditions. J. Econ. Entomol. 33: 819-820.

Cachan, P. 1957. Les Scolytoidea mycetophages des forets de Basse Cote d'voire. Rev. Pathol. Vegetale. Entomol. Agr. France 36: 1-126.

Chippendale, G. N., S. D. Beck, and F. M. Strong. 1964. Methyl linolenate as an essential nutrient for the cabbage looper, *Trichoplusia ni* (Hubner). Nature 204: 710-711.

———. 1965. Nutrition of the cabbage looper, *Trichoplusia ni* (Hubner). I. Some requirements for larval growth and wing development. J. Insect Physiol. 11: 211-223.

Chu, H. M., D. M. Norris, and L. T. Kok. 1970. Pupation requirement of the beetle, *Xyleborus ferrugineus*: sterols other than cholesterol. J. Insect Physiol. 16: 1379-1387.

Clayton, R. B. 1964. The utilization of sterols by insects. J. Lipid Res. 5: 3-19.

Cmelik, S. H. W., and C. C. Douglas. 1970. Chemical composition of "fungus gardens" from two species of termites. Comp. Biochem. Physiol. 36: 493-502.

Collins, R. P., and K. Kalnins. 1969. The occurrence of ergosterol in the fungus *Ceratocystis fagacearum*. Mycologia 61: 645-646.

Dadd, R. H. 1961. The nutritional requirements of locusts. V. Observations on essential fatty acids, chlorophyll, nutritional salt mixtures, and protein or amino acid components of synthetic diets. J. Insect Physiol. 6: 126-145.

———. 1964. A study of carbohydrate and lipid nutrition in the wax moth, *Galleria mellonella* (L.) using partially synthetic diets. J. Insect Physiol. 10: 161-178.

Dawson, R. M. C., R. W. White, and N. Freinkel. 1962. The fate of mesoinositol during the growth of an inositol-dependent yeast *Kloeckera brevis*. J. Gen. Microbiol. 27: 331–339.

Dyke, K. G. H. 1964. The chemical composition of the cell wall of the yeast *Nadsonia elongata*. Biochim. Biophys. Acta 82: 374–384.

Earle, N. W., B. H. Slatten, and M. L. Burks, Jr. 1967. Essential fatty acids in the diet of the boll weevil, *Anthonomus grandis* Boheman (Coleoptera: Curculionidae). J. Insect Physiol. 13: 187–200.

Enebo, L., G. L. Anderson, and H. Lundin. 1946. Microbiological fat synthesis by means of *Rhodotorula* yeast. Arch. Biochem. Biophys. 11: 383–395.

Farris, S. H., and A. Funk. 1965. Repositories of symbiotic fungus in the ambrosia beetle *Platypus wilsoni* Swaine (Coleoptera, Platypodidae). Canad. Entomologist 97: 527–536.

Fast, P. G. 1964. Insect lipids: A review. Mem. Entomol. Soc. Canada 37: 1–50.

Fiore, J. V. 1948. On the mechanism of enzyme action XXXII. Fat and sterol in *Fusarium lini* Bolley, *Fusarium lycopersici* and *Fusarium solani* D_2 Purple. Arch. Biochem. 16: 161–168.

Fischer, M. 1954. Untersuchungen uber den kleinen holzbohrer (*Xyleborinus saxeseni*). Pflanzenschutz Ber. 12: 137–48.

Folch, J., M. Lees, and G. H. Sloane-Stanley. 1957. A simple method for the isolation and purification of total lipids from animal tissues. J. Biol. Chem. 226: 497–509.

Fraenkel, G., and M. Blewett. 1946. Linoleic acid, vitamin E, and other fat-soluble substances in the nutrition of certain insects (*Ephestia kuehniella, E. elutella, E. cautella* and *Plodia interpunctella* (Lep.).) J. Exp. Biol. 22: 172–90.

Francke-Grosmann, H. 1956. Hautdrausen als Trager der Pilzsymbiose bei Ambrosiakafern. Z. Morphol. Oekol. Tiere 45: 275–308.

———. 1958. Uber die Ambrosiazucht holzbrutender Ipinden im Hinblick auf das System. Verhandl. Deut. Ges. Angew. Entomol. 14: 139–44.

———. 1963. Some new aspects in forest entomology. A. Rev. Entomol. 8: 415–38.

———. 1965. Uber symbiosen von Xylo-mycetophagen und phloeophagen mit holzhewohnenden Pilzen. G. Becker, ed. Supplement to materials and organisms: holtz und organismen. Duncker and Humblot, Berlin, Heft 1: 503–522.

———. 1967. Ectosymbiosis in wood-inhabiting insects. S.M. Henry, ed. Symbiosis, Academic Press, New York. Vol. 2. 141–205.

French, J. R. J., and R. A. Roeper. 1972. Interactions of the ambrosia beetle, *Xyleborus dispar* (Coleoptera: Scolytidae), with its symbiotic fungus *Ambrosiella hartigii* (Fungi Imperfecti). Canad. Entomol. 104: 1635–1641.

———. 1975. Studies on the biology of the ambrosia beetle *Xyleborus dispar* (F.) (Coleoptera: Scolytidae). Z. Angew Entomol. 78: 241–247.

Gordon, H. T. 1959. Minimal nutritional requirements of the German roach, *Blattella germanica* L. Ann. N.Y. Acad. Sci. 77: 290–351.

Graff, G. L. A., B. Vanderkelen, C. Gueuning, et J. Humpers, Presentee par P. E. Gregoire. 1968. Analyse des phospholipides constitutifs de levures Endomycetacees, Sporobolomycetacees et Cryptococcacees. C. R. Soc. Biol. 162(8–9): 1635–1638.

Graham, K. 1967. Fungal-insect mutualism in trees and timbers. A. Rev. Entomol. 12: 105–26.

Grison, P. 1948. Action des lecithines sur la fecundite du Doryphora. C. R. Acad. Sci., Paris 227: 1172–1174.

Gunasekaran, M., and D. J. Weber. 1972. Polar lipids and fatty acid composition of phytopathogenic fungi. Phytochem. 11: 3367–3369.

Hartig, T. 1872. Der Fichtensplintkafer *Bostrychus (Xyloterus) lineatus*. Allgem. Forst-Jagdztg. 48: 181–183.

Hartman, L., J. C. Hawke, F. B. Shorland and M. E. diMenna. 1959. Arch. Biochem. Biophys. 81: 346–352.

Hartman, L., I. M. Morice and F. B. Shorland. 1962. Further study of the lipids from *Pithomyces chartarum (Sporidesmium bakeri)* and related fungi. Biochem. J. 82: 76–80.

Heed, W. B., and H. W. Kircher. 1965. Unique sterol in the ecology and nutrition of *Drosophila pachea*. Science 149: 758–61.

Hubbard, H. C. 1897. The ambrosia beetles of the United States. U.S. Dept. Agr. Div. Entomol. Bull. 7: 9–35.

Ito, T. 1960. An artificial diet for the silkworm, *Bombyx mori* and the effect of soybean oil on its growth. Proc. XI Int. Congr. Entomol. Vienna 3 (pt. 2): 157–162.

Kates, M., and R. M. Baxter. 1962. Lipid composition of mesophilic and phychrophilic yeasts (*Candida* species) as influenced by environmental temperature. Canad. J. Biochem. Physiol. 40: 1213–1227.

Kingsolver, J. G., and D. M. Norris. 1977. The interaction of *Xyleborus ferrugineus* (Coleoptera: Scolytidae) behavior and initial reproduction in relation to its symbiotic fungi. Ann. Entomol. Soc. Amer. 70: 1-4.

Kok, L. T. 1971. Fungal symbionts of *Xyleborus* spp.: Certain chemical components and their nutritional significance to the ambrosia beetles. Ph.D. dissertation, Univ. of Wisconsin, Madison. 175 pp.

Kok, L. T., and D. M. Norris. 1972a. Phospholipid composition of fungi mutualistic with *Xyleborus ferrugineus*. Phytochem. 11: 1449-1453.

———. 1972b. Lipid composition of adult female *Xyleborus ferrugineus*. J. Insect Physiol. 18: 1137-1152.

———. 1972c. Comparative phospholipid compositions of adult female *Xyleborus ferrugineus* and its mutualistic ectosymbionts. Comp. Biochem. Physiol. 42(B): 245-254.

———. 1972d. Symbiontic interrelationships between microbes and ambrosia beetles. VI. Amino acid composition of ectosymbiotic fungi of *Xyleborus ferrugineus*. Ann. Entomol. Soc. Amer. 65: 598-602.

———. 1973a. Fatty acids of fungi mutualistic with *Xyleborus ferrugineus*. Phytochem. 12: 383 pp.

———. 1973b. Comparative sterol compositions of adult female *Xyleborus ferrugineus* and its mutualistic fungal ectosymbionts. Comp. Biochem. Physiol. 44(B): 499-505.

Kok, L. T., D. M. Norris, and H. M. Chu. 1970. Sterol metabolism as a basis for a mutualistic symbiosis. Nature 225: 661-662.

Kritchevsky, D. 1963. Sterols. M. Florkin and E. H. Stotz, eds., Comprehensive biochemistry. Vol. 10: 1-22.

Lamb, N. J. and R. E. Monroe. 1968. Lipid synthesis from acetate-1-C^{14} by the cereal leaf beetle, *Oulema melanopus*. Ann. Entomol. Soc. Amer. 61: 1164-1166.

Lambremont, E. N. 1965. Biosynthesis of fatty acids in aseptically reared insects. Comp. Biochem. Physiol. 14: 419-24.

Lester, R. L. 1963. The phospholipids of *S. cerevisiae*. Federation Proc. 22: 415.

Light, R. J. 1965. Acetate metabolism in *Penicillium griseofulvum*. Incorporation of 1-C^{14}-2-H^{3}-acetate into 6-methylsalicylic and fatty acids. Arch. Biochem. Biophys. 112: 163-169.

Longley, R. P., A. H. Rose, and B. A. Knights. 1968. Composition of the protoplast membrane from *Saccharomyces cerevisiae*. Biochem. J. 108: 401-412.

Mathiesen-Kaarik, A. 1953. Eine ubersicht uber die gewohnlichen mit Borkenkafern assoiierten Blauepilze in Schweden. Medd. Statens Skogsforskningsinst. 43: 1-74.

McCorkindale, N. J., S. A. Hutchinson, B. A. Pursey, W. T. Scott, and R. Wheeler. 1969. A comparison of the types of sterol found in species of the Saprolegniales and Leptomitales with those found in some other Phycomycetes. Phytochem. 8: 861-867.

McElroy, F. A., and H. B. Stewart. 1967. The lipids of *Lipomyces lipofer*. Canad. J. Biochem. 45: 171-178.

Milazzo, F. H. 1965. Sterol production by some wood-rotting Basidiomycetes. Canad. J. Botany 43: 1347-1353.

Moya, G. E. 1970. Some aspects of the biology and nutrition of four species of *Xyleborus ambrosia* beetles. Ph.D. dissertation. Univ. of Wisconsin, Madison. 129 pp.

Müller, H. R. A. 1933. Ambrosia fungi of tropical Scolytidae in pure culture. Verslag. Afdeel. Ned-Oost-Ind. Ned. Entomol. Ver. 1: 105-125.

Nakajima, S., and S. W. Tanenbaum. 1968. The fatty acids of *Penicillium pulvillorum*. Arch. Biochem. Biophys. 127: 150-156.

Nakashima, T. 1971. Notes on the associated fungi and the mycetangia of the ambrosia beetle, *Crossotarsus niponicus* Blandfor (Coleoptera: Platypodidae). Appl. Ent. Zool. 6: 131-137.

Neger, F. W. 1909. Ambrosiapilze II. Ber. Deut. Botan. Ges. 27: 372-389.

Nelson, D. R., and D. R. Sukkested. 1968. Fatty acid composition of the diet and larvae and biosynthesis of fatty acids from ^{14}C-acetate in the cabbage looper, *Trichoplusia ni*. J. Insect Physiol. 14:293-300.

Nombela-Cano, C., and J. F. Peberdy. 1971. The lipid composition of *Fusarium culmorum* mycelium. Trans. Brit. Mycol. Soc. 57: 342-344.

Norris, D. M. 1965. The complex of fungi essential to growth and development of *Xyleborus sharpi* in wood. Holz und Organismen 1: 523-529.

———. 1972. Dependence of fertility and progeny development of *Xyleborus ferrugineus* upon chemicals from its symbiotes. J. G. Rodriguez, ed. Insect and Mite Nutrition. North-Holland Pub. Co., Amsterdam. 299-310.

———. 1974. Sterols in *Xyleborus* beetle reproduction and metamorphosis. J. Amer. Chemists Soc. 51: 524A.

Norris, D. M., and J. M. Baker. 1967. Symbiosis: effects of a mutualistic fungus upon the growth and reproduction of *Xyleborus ferrugineus*. Science 156: 1120–1122.

Norris, D. M., and H. M. Chu. 1971. Maternal *Xyleborus ferrugineus* transmission of sterol or sterol-dependent metabolites necessary for progeny pupation. J. Insect Physiol. 17: 1741–1745.

Norris, D. M., J. M. Baker, and H. M. Chu. 1969. Symbiontic interrelationships between microbes and ambrosia beetles. IV. Ergosterol as the source of sterol to the insects. Ann. Entomol. Soc. Amer. 62: 413–414.

Pepper, J. H., and E. Hastings. 1943. Biochemical studies on the sugar beet webworm (*Loxostege sticticalis* L.). Montana State Coll. Agr. Exp. Sta. Bull. 413: 1–36.

Prescott, S. C., and C. G. Dunn. 1940. Industrial microbiology. McGraw-Hill, New York. Chaps. 10 and 33.

Pruess, L. M., E. C. Eichinger, and W. H. Peterson. 1934. The chemistry of mold tissue. IV. Composition of certain molds with special reference to lipid content. Zentr. Bakt. Parasitenk, II. 89: 370–377.

Pruess, L. M., W. H. Peterson, H. Steenbock, and E. B. Fred. 1931. Sterol content and antirachitic activatibility of mold mycelia. J. Biol. Chem. 90: 369–384.

Rajalakshmi, S., D. S. R. Sarma, and P. S. Sarma. 1963. Biosynthesis of fatty acid in rice moth, *Corcyra cephalonica* St. Indian J. Exp. Biol. 1: 155.

Robbins, W. E., J. N. Kaplanis, J. A. Svoboda, and M. J. Thompson. 1971. Steroid metabolism in insects. Ann. Rev. Entomol. 16: 53–72.

Shaw, R. 1966. The polyunsaturated fatty acids of microorganisms. Adv. Lipid Res. 4: 107–174.

Smith, J. H. 1935. The pinhole borer of North Queensland cabinet woods. Bull. Div. Entomol. Pl. Pathol. Queensland (N.S.) 12: 1–38.

Starkey, R. L. 1946. Lipid production by a soil yeast. J. Bacteriol. 51: 33–50.

Stephen, W. F., and L. I. Gilbert. 1969. Fatty acid biosynthesis in the silkmoth, *Hyalophora cecropia*. J. Insect Physiol. 15: 1833–1854.

Strong, F. E. 1963. Studies on lipids in some homopterous insects. Hilgardia 34: 43–61.

Vanderzant, E. S., D. Kerur, and R. Reiser. 1957. The role of dietary fatty acids in the development of the pink bollworm. J. Econ. Entomol. 50: 606–608.

Vanderzant, E. S., and C. D. Richardson. 1964. Nutrition of the adult boll weevil: Lipid requirements. J. Insect Physiol. 10: 267–72.

Verrall, A. F. 1943. Fungi associated with certain ambrosia beetles. J. Agric. Res. 66: 135–144.

Wilson, C. L. 1959. The Columbian timber beetle and associated fungi in white oak. Forest Sci. 5: 114.

Yearian, W. C., R. J. Gouger, and R. C. Wilkinson. 1972. Effects of the bluestain fungus, *Ceratocystis ips*, on development of *Ips* bark beetles in pine bolts. Ann. Entomol. Soc. Amer. 65: 481–487.

chapter 3

The Mutualistic Fungi Of Xyleborini Beetles*

by DALE M. NORRIS†

ABSTRACT

Information on fungal-insect mutualism began in 1836 with Schmidberger's description of an "ambrosia" lining the tunnels of the beetle *Apate dispar* (= *Xyleborus dispar*). Hartig in 1844 confirmed that this ambrosia was a fungus which he named *Monilia candida*. Thus, it seems especially appropriate to update this subject as pertinent to *Xyleborus* and the closely related *Xylosandrus* ambrosia beetles. Current knowledge of fungal symbiotes includes only a few of the more than 1,500 described species (mostly in *Xyleborus*) in these two genera. Though fungi in several genera (*Fusarium, Ambrosiella, Cephalosporium, Graphium, Ceratocystis, Monilia, Monacrosporium, Botryodiplodia, Ambrosiamyces, Cladosporium, Paecilomyces, Pestalozzia*) have been reported as symbiotes of beetles in one or both genera, one fungus commonly is dominant. Recent studies in our laboratories have shown that *Fusarium solani* is the dominant filamentous fungal symbiote of five species of *Xyleborus* collected in Central America. Similar findings are available for the much studied *Xylosandrus compactus* (= *Xyleborus compactus*) and *Xylosandrus germanus* (= *Xyleborus germanus*). Earlier studies on *X. compactus* led Brader to create a new genus: *Ambrosiella* for this beetle's major symbiote. Batra then also classified the ambrosia of *Xyleborus dispar* as *Ambrosiella hartigii*. Recent work of French and Roeper on *X. dispar* followed Batra's classification of the symbiotic fungus. Thus, though other genera such as *Monacrosporium*, the current genus of the ambrosial fungus of *Xyleborus fornicatus*, have received very recent emphasis, most major ambrosial

*Contribution from the College of Agriculture and Life Sciences; and funded partially by the Wisconsin Department of Natural Resources, Schoenleber Foundation and Grants No. AI 06195 and RR-00779 from the U.S. National Institutes of Health.

†Department of Entomology, University of Wisconsin, Madison, Wisconsin 53706.

fungi of these two genera of beetles seem to be settling taxonomically into *Fusarium* or *Ambrosiella*. The observed series of taxonomic events regarding the fungi of *Xylosandrus compactus* could possibly suggest that *Ambrosiella* may prove to designate forms of pleomorphic *Fusarium* spp. Our continuing physiological and morphological investigations with fungi of *Xyleborus* spp. supports *Fusarium* as the major genus of ambrosial fungi among these beetles.

INTRODUCTION

Thousands of insect species exploit the more abundant native energy substrates (e.g., cellulose and lignin) on earth through reciprocally beneficial associations with microorganisms termed symbiotes. Thus at the very fingertips of starving humans around the world are insects gardening their symbiotes in wood (Fig. 3.1) and harvesting food fit for the gods, "ambrosia." Our symposium thus addresses a subject of far greater importance than scientific curiosity about some wonders of nature. Investigations of such mutualistic symbioses should further awaken us to potential technologies for efficiently converting wood to food for domesticated animals, if not for humans.

Our research on this subject focuses on ambrosia beetles in the tribe Xyleborini, and especially those in the genus *Xyleborus,* of the family Scolytidae. In a real sense we have domesticated some *Xyleborus* spp. and their microbial symbiotes to laboratory culture on artificial substrates including chemically defined diets. This domestication may enable us to decipher their evolved strategies for encouraging a complex of filamentous fungi, yeasts and bacteria to convert wood into "ambrosia" (Norris 1972; 1975).

It seems especially appropriate to update the subject of mutualism between ambrosia beetles and microbes using Xyleborini and especially

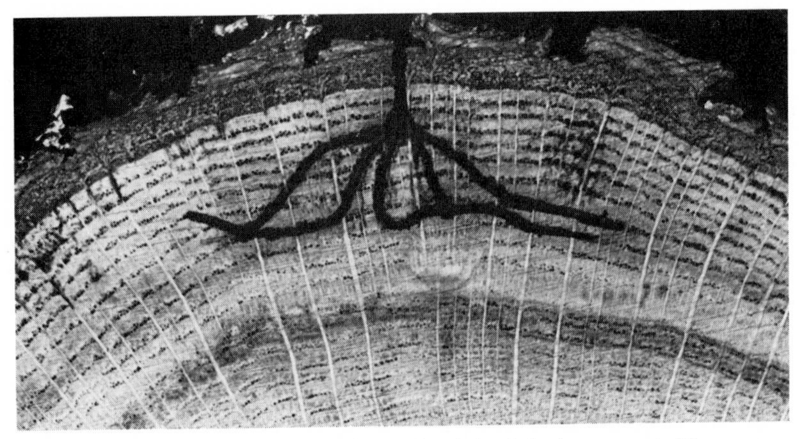

Figure 3.1. Sinuous tunnel system of an ambrosia beetle in the sapwood of *Quercus.*

Xyleborus beetles for emphasis, because the original non-microbial designation of "ambrosia" by Schmidberger (1836) and the recognition of their fungal nature by Hartig (1844) both involved the beetle *Apate dispar* (= *Xyleborus dispar*). Hubbard (1897) also relied heavily on fungi associated with several *Xyleborus* spp. in presenting the first discussion of principal growth forms of ambrosia.

XYLEBORINI AMBROSIA BEETLES

The more than 1500 described species in Xyleborini, mostly in *Xyleborus*, are proof that this type of mutualism is ecologically and evolutionarily competitive with other forms of insect life. This is especially true in the tropics but Xyleborini are found more or less wherever trees grow. Browne (1961) claimed that nearly 50% of known scolytid beetles in Malaya are members of Xyleborini which consists of the major genus *Xyleborus* and the minor genus, *Xylosandrus*.

Focusing on *Xyleborus*, it is best regarded taxonomically as a lumping of numerous species-groups and subgenera (Browne 1961). The entire genus shows a social organization of extreme polygamy. Sexual dimorphism is strongly developed especially in the prothorax. The ratio of females (diploids) to males (haploids) is high (e.g., 20-30:1) in natural broods. Males are produced asexually and are most frequently, but not always, smaller than females. Males are flightless, short-lived inseminators of progeny sisters or virgin mothers. The males normally never leave the vicinity of the common brood chamber, and thus extreme inbreeding usually occurs, but excellent vigor is maintained. Each virgin female thus represents a potential inbred line which may lead to a newly recognized species. This situation seems to offer a word of caution to systematists who are "splitters." On the other hand, they are likely to be busy if they choose *Xyleborus* as their group.

The maternal female is a very dedicated, efficient mother. She must live and tend the ambrosia garden and her progeny until they reach adulthood. The maternal female either eats dead and weakened progeny or entombs them in a short branch tunnel. Thus, she routinely takes rather extreme preventive and curative measures directed at maintaining a vigorous brood. If the mother dies, the brood dies. Death of unmothered progeny is from "suffocation" by microbes, or other secondary causes. Progeny alone cannot keep the ambrosia under control even though they feed mostly, if not exclusively, on these microbes. We still do not adequately understand the mechanisms used by the mother beetle to control the growth of ambrosia.

In a tropical environment a maternal female raises one brood of progeny in about 30 days, and may die in her nest (brood gallery) after her offspring have left. However in laboratory conditions females have been kept alive for more than one year. When given special care and a new chemically defined nutrient substrate every three weeks, the mean longevity of ectosymbiote-free females exceeded 30 weeks. Each of these females produced numerous cycles of progeny.

Most species usually attack unhealthy or newly felled trees and thus are scavengers, but some attack such exotic substrates as wine casks. Some species are highly host selective and stay within a single family of plants. They also may only attack a particular part (e.g., terminal branch) of a plant.

THE AMBROSIA

One of the more interesting facets of the mutualism between ambrosia beetles and fungi in wood is the insect-induced "ambrosia" growth forms of the symbiotic fungi. Hubbard (1897) stated that this growth on the insect's gallery walls "glistens upon the walls like hoarfrost." He also said that "the entire surfaces of the walls of the brood chamber are plastered over with ambrosia fungus—it consists of short erect stems, terminating in spherical conidia." Since Hubbard's descriptions, disagreement about the identifications and "true natures" of ambrosial fungi has prevailed. Some history of these points as related to some Xyleborini is presented in Table 3.1. Early studies largely described the fungal symbiotes as being in the form genus *Monilia* Persoon ex Fries (Hartig 1844; Muller 1933). Some more recent classifications have placed particular fungi in the new genus *Ambrosiella* (Brader 1964). However, numerous ambrosial fungi have been placed in *Fusarium* (Muller 1933; Ngoan et al. 1976).

Our main interest in this symposium is to describe the pleomorphic nature of these symbiotic fungi, and to discuss the roles of the insects in initiating and maintaining the ambrosia "gardens" in their gallery system in the wood. Continuing our focus on *Xyleborus,* a major symbiotic fungus of *X. ferrugineus, Fusarium solani,* grows in classical monilioid form as a pure culture on the wall of an occupied tunnel in artificial diet (Baker and Norris, 1968). Abrahamson (1969) in our laboratory showed that urates (major nitrogenous excretory products of *Xyleborus*), urea or aspartic acid especially caused monilioid growth by *F. solani* in several agar media at initial pHs between 6 and 7. This fungus produced ambrosia consisting mostly of single budded terminal cells on hyphal sprouts when grown on many nitrogen sources. This form was common in "well cropped" ambrosia gardens inhabited by active maternal beetles. The multicelled monilioid chains occurred mostly in gardens in older or newly abandoned galleries which were not being cropped by beetles.

Cephalosporium sp. is a secondary symbiote of every *Xyleborus* species we have studied and especially of *X. ferrugineus.* When grown in pure culture in beetle tunnels in artificial diets it yielded an ambrosia consisting of long, upright, largely unbranched conidiophores bearing many small spherical conidia. Under otherwise comparable dietary conditions it

Table 3.1 Ambrosia Fungi of Xyleborini Beetles.

Insect species	Fungal species	References
Xyleborus dispar (=*Abate dispar*)	*Monilia candida*[a]	Hartig (1844), Neger (1909), Francke-Grosmann (1958)
	Ambrosiella hartigii[a]	Batra (1967), French and Roeper (1972)
Xyleborus morigerus	*Monilia*	Neger (1911)
Xyleborus habercorni	*Monilia* spp., *Fusarium* sp.	Müller (1933)
Xyleborus bicornis	*Monilia* spp., *Fusarium* sp.	Müller (1933)
	Penicillium sp.	Fischer (1954)
Xyleborus saxeseni	Yellow monilioid fungus	Francke-Grosmann (1958)
Xyleborus monographus	Yellow monilioid fungus	Francke-Grosmann (1958)
Xyleborus semipacus	*Ambrosiella xylebori*	Brader (1964)
Xyleborus affinis	*Cephalosporium pallidum*	Verrall (1943)
Xyleborus pecanis	*C. pallidum*	Verrall (1943)
Xyleborus fornicatus	*Monacrosporium ambrosium*	Gadd and Loos (1947)
Xyleborus velatus	*Ascoidea asiatica*	Batra and Francke-Grosmann (1964)
Xyleborus ferrugineus	*Fusarium solani*, *Cephalosporium* sp., *Graphium* sp.	Baker and Norris (1968)
Xyleborus destruens	*Fusarium* sp.	Müller (1933)
Xylosandrus germanus (=*Xyleborus germanus*)	*Ceratocystis ulmi*	Buchanan (1940)
	Monilioid fungus, similar to that of *X. dispar*	Francke-Grosmann (1958)
	Fusarium lateritium, *F. oxysporum*	Kessler (1974)
Xylosandrus compactus (=*Xyleborus compactus*, =*X. morstatti*)	*Monilia*, *Fusarium* sp.	Müller (1933)
	Colletotrichum coffearum, *Pestalozzia coffeicola*	Meiffren (1957)
	Ambrosiella xylebori, *Cephalosporium rubescens*, *Fusarium lateritium*	Brader (1964)
	Botryodiplodia theobromae	Gregory (1954)
	Fusarium solani	Ngoan *et al.* (1976)

[a]Both names refer to the same fungus. (Ed.)

allowed *X. ferrugineus* to produce about one-half as many progeny as did *F. solani* (Table 3.2); however, it alone provided the nutrient requirements for reproduction and complete progeny development which the beetles normally derive from their complex of symbiotes. *Graphium* sp. also met these nutrient requirements of the beetle, but only allowed about 30% of the progeny production that *F. solani* did (Table 3.2). In the artificial diet without one of these fungi, females failed to produce progeny. Thus,

Table 3.2 Progenies of *Xyleborus ferrugineus* Produced in 35 Days on a Meridic Diet Inoculated with One of Three Symbiotic Fungi, or Left Fungus-free.

Fungus	Number of test females	Number with progeny	Stages of progeny: Eggs	Larvae	Pupae	Adults	Total progeny
Fusarium solani	15	15	15	36	6	40	97
Cephalosporium sp.	15	14	2	12	2	30	46
Graphium sp.	14	9	9	6	4	9	28
No fungus	16	0	0	0	0	0	0

reproduction and development of progeny are dependent especially upon the fungal symbiotes.

Xyleborus beetles have co-evolved with their complex of symbiotic microbes to the point that initiation of the female's reproductive processes are dependent upon a particularly rich meal of certain essential amino acids (e.g., lysine, methionine, arginine and histidine), provided by symbiotes, which other studied insects apparently do not require for reproduction (Bridges and Norris, 1977). Without this proper amino acid or protein intake, the *Xyleborus* do not reproduce even if inseminated. Ovarian and oocyte maturations thus are symbiote-dependent, and not sperm-dependent. Insemination usually results in a larger number of matured oocytes, and most are fertilized to yield diploid (female) progeny. Thus, the arrhenotokous parthenogenetic reproduction is symbiote-regulated. For further information on other specific nutritional dependencies which tie *Xyleborus* to their symbiotes, see Kok, Chapter 2, this symposium.

MYCANGIA OR MYCETANGIA

Discussion of the means of ambrosia beetle transmission of its ectosymbiotic microbes from old to new woody substrates began with Neger (1911) who thought it occurred in the gut. Nunberg (1951) suggested that ectodermal prothoracic tube-like glands in some *Trypodendron (= Xyloterus)* spp. might serve for the storage and transmission of spores of ambrosia. Nunberg's suggestion later was confirmed by Francke-Grosmann (1956). Batra (1963) proposed the term "mycangium" for this organ. Giese (1965) proposed the term "mycetangium." Mycangia are commonly described or classified on the basis of their location and/or structural characteristics (e.g., oral pouches).

Our major emphasis in this section is that very significant co-evolution between the beetle and its ectosymbiotes has occurred in terms of insect organs for the perpetuation of these symbiotes. The microbes are not just carried as resting spores, but they multiply while in mycangia. Such growth of symbiotes must be at the expense of the beetle. However, the symbiotes seemingly do not invade portions of the insect's body other than mycangia.

Focusing now on Xyleborini beetles, Francke-Grosmann (1956) first described mycangia of such insects. She found (Table 3.3) sclerotized pouches in the base of each elytron of adult female *Xyleborus saxeseni;* and an intersegmental pouch between the pro- and mesonota in adult female *Xyleborus dispar, Xylosandrus germanus* and *X. compactus.* The membranous oral pouch at the base of each mandible (Table 3.3) in adults was first described by Fernando (1959) in *Xyleborus fornicatus.* Thus, three distinct types of mycangia have been found in Xyleborini.

The ambrosial fungi of *X. ferrugineus* are presented as evidence that each symbiote grows with a characteristic form in the mycangium (Fig. 3.2, 3.3). Baker and Norris (1968) showed this by raising *X. ferrugineus* on pure cultures of *F. solani, Cephalosporium* sp. and *Graphium* sp. The growth of each in the mycangium is perhaps best referred to as propagules; however, some mycelioid forms were present with *F. solani* and especially *Graphium* sp. Thus, the form of an ambrosial fungus is controlled in the mycangium and does not significantly resemble either its growth on the beetle's tunnel wall or on routine laboratory media (e.g., PDA). These fungi are extremely pleomorphic, and the environmental conditions obviously determine the dominant form at any given time.

Table 3.3 Types of Mycangia in Xyleborini Beetles.

Species	Mycangium	References
Xyleborus fornicatus	Membranous pouch at base of each mandible	Fernando (1959)
Xyleborus mascarensis	Membranous pouch at base of each mandible	Francke-Grosmann and Schedl (1960)
Xyleborus ferrugineus	Membranous pouch at base of each mandible	Baker and Norris (1968)
Xyleborus saxeseni	Sclerotized pouches in base of elytra in female	Francke-Grosmann (1956)
Xyleborus gracilis, X. schreineri and *X. sentosus*	Sclerotized pouches in base of elytra in female	Schedl (1962)
Xyleborus dispar and *Xylosandrus germanus*	Intersegmental pouch between pro- and mesonotum in female	Francke-Grosmann (1956)
Xylosandrus compactus (=*Xyleborus compactus* =*X. morstatti*)	Intersegmental pouch between pro- and mesonotum in female	Lhoste and Roche (1959)
Xylosandrus discolor	Intersegmental pouch between pro- and mesonotum in female	Schedl (1962)
Xylosandrus semiopacus	Intersegmental pouch between pro- and mesonotum in female	Francke-Grosmann (1958)

Abrahamson (1969) in our laboratory examined the contents of mycangia and found fatty acids, phospholipids, free sterols, sterol esters and triglycerides; and large amounts of the free amino acids, proline, alanine and valine. Proline was especially abundant. Moderate amounts of free arginine, histidine and aspartic acid also were present (Abrahamson and Norris 1970).

Studies of symbiotic fungal growth on media with lipids in the same ratio of concentrations as estimated from mycangia yielded no significant growth (Abrahamson 1969). It was concluded that lipids are not major nutrients for the growth of symbiotic fungi in mycangia.

However, an apparent major nutrient cause of the propagule growth

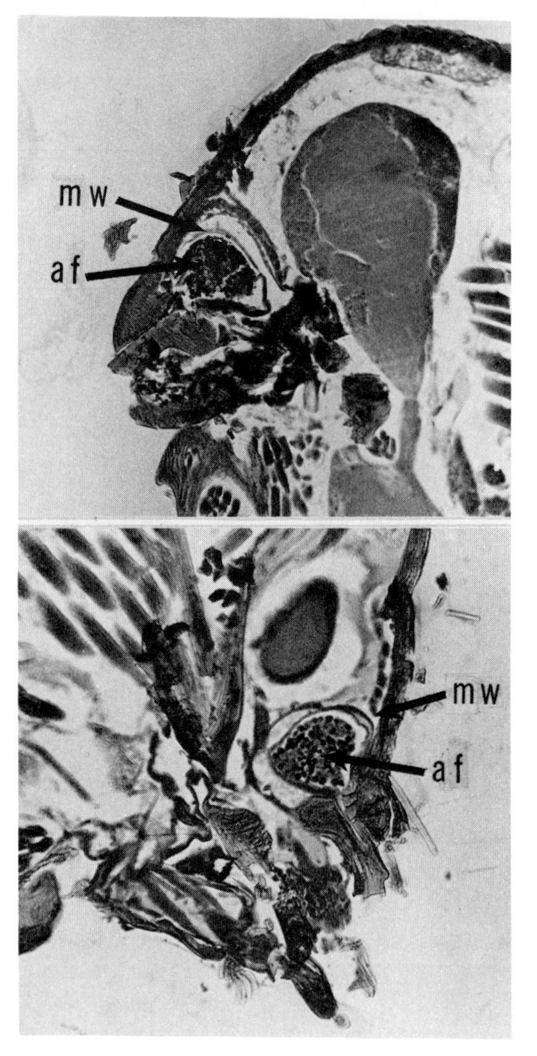

Figures 3.2–3.3. Longitudinal sections through the head of adult *Xyleborus ferrugineus* showing an oral mycangium (mw). *3.2 (top).* Mycangium with a mass of fine hyphae with spores, 1.5×3 μm, (af) of *Graphium* sp. *3.3 (bottom).* Mycangium with spherical spores, 4.5 to 5 μ, (af) of *Cephalosporium* sp. (mw, mycangial wall; af, ambrosial fungus).

form of symbiotic fungi in mycangia is the extreme abundance of the free amino acid, proline, in *X. ferrugineus* adult females (Abrahamson and Norris 1970). It was by far the dominant free amino acid in the insect. Proline and glutamic acid as major nitrogen sources each caused symbiotic fungi to grow especially as propagules in artificial media at about neutral pH (Abrahamson 1969). It is of further significance that the haemolymph of *X. ferrugineus* also is near neutrality (pH 6.8, unpublished data), and that when symbiotic fungi were grown at more acidic pHs (e.g., 5.8, 4.8 and 4.4), even with proline or glutamic acid abundant, the mycelial form predominated (Abrahamson 1969). The propagule forms of symbiotic fungi in mycangia of *X. ferrugineus* thus seem especially attributable to an extreme abundance of the free amino acid proline in the insect's haemolymph and body secretions; and to a pH near neutrality in the mycangium. This molecular and ionic situation provides the beetle with the ability to keep its mycangial-borne symbiotes in relatively slow growing and non-parasitic states.

During this dispersal phase of the mutualistic complex of microbes and insect, the beetle apparently is in the command position. This submission of the microbes to the beetles seems to be an evolutionary price worth paying for a relatively secure trip to a new suitable woody substrate, and tender devoted care, once there.

CONCLUSIONS

Thousands of species of insects have evolved into extremely obligatory states with complexes of microbes which can degrade cellulose and lignin. This trade of certain attributes found in other insects for indirect-nutritive access through microbes to two of the larger readily renewable energy sources on earth seems to have rewarded the *Xyleborus* very well. *Xyleborus* constitutes one of the largest genera of insects, and successfully attacks hundreds of species of woody plants throughout much of the world, and especially in the subtropical and tropical areas.

The genetic and ecological security provided by the described mutualism apparently has allowed *Xyleborus* to: (1) place facultatively sexual reproductive processes in the primary control of their symbiotic microbes; (2) make the progeny maturation process dependent on the survival and constant brood-tending behavior of the maternal female; (3) condition progeny metamorphosis upon symbiotes; (4) degenerate the male to a short-lived, non-flying, non-dispersing, inseminator of sisters or mother; and (5) ignore outbreeding of the genome.

The symbiotic fungi also have relinquished the following as the possible price of this mutualism: (1) determination of vegetative growth form; (2) other means of dispersal; (3) choice of woody substrate; (4) other means of inoculation; and (5) sexual reproduction.

If this discussion addresses no other point it seemingly makes a sizeable case for the merits of communal existence among consenting, extremely compromising, organisms; in this specific case between Xyleborini beetles and a very imperfect group of fungi.

LITERATURE CITED

Abrahamson, L. P. 1969. Physiological interrelationships between ambrosia beetles and their symbiotic fungi. Ph.D. Thesis, Univ. of Wisconsin, Madison. 122 pp.

Abrahamson, L. P., and D. M. Norris. 1970. Symbiotic interrelationships between microbes and ambrosia beetles (*Coleoptera: Scolytidae*). V. Amino acids as a source of nitrogen to the fungi in the beetle. Ann. Entomol. Soc. Amer. 63: 177–180.

Baker, J. M., and D. M. Norris. 1968. A complex of fungi mutualistically involved in the nutrition of the ambrosia beetle *Xyleborus ferrugineus*. J. Invert. Pathol. II: 246–250.

Batra, L. R. 1963. Ecology of ambrosia fungi and their dissemination by beetles. Trans. Kansas Acad. Sci. 66: 213–236.

———. 1967. Ambrosia fungi: A taxonomic revision, and nutritional studies of some species. Mycologia 59: 976–1017.

Batra, L. R., and H. Francke-Grosmann. 1964. Two new ambrosia fungi *Ascoides asiatica* and *A. africana*. Mycologia 56: 632–636.

Brader, L. 1964. Etude de la relation entre le scolyte des rameaux du cafeier, *Xyleborus compactus* Eichh. (*X. morstatti* Hag.) et sa plantehote. Dissertationes Wageningen. 109 pp.

Bridges, J. R., and D. M. Norris. 1977. Inhibition of reproduction of *Xyleborus ferrugineus* by ascorbic acid and related chemicals. J. Insect Physiol. 23: 497–501.

Browne, F. G. 1961. The biology of Malayan Scolytidae and Platypodidae. Malay. Forest Rec. 22: 1–255.

Buchanan, W. D. 1940. Ambrosia beetle *Xylosandrus germanus* transmits Dutch elm disease under controlled condition. J. Econ. Entomol. 33: 819–820.

Fernando, E. F. W. 1959. Storage and transmission of ambrosia fungus in the adult *Xyleborus fornicatus* Eichh. (Coleoptera: Scolytidae). Ann. Mag. Nat. Hist. [13] 2: 478.

Fischer, M. 1954. Untersuchungen uber den kleinen Holzbohrer (*Xyleborinus saxeseni*). Pflanzenschutz Ber. 12: 137–148.

Francke-Grosmann, H. 1956. Hautdrusen als Trager der Pilzsymbiose bei Ambrosiakafern. Z. Morphol. Oekol. Tiere 45: 275–308.

———. 1958. Uber die Ambrosiazucht holzbrutender Ipiden im Hinblick auf das System. Verhandel. Deut. Ges. Angew. Entomol. 14: 139–144.

Francke-Grosmann, H., and W. Schedl. 1960. Ein orales Ubertragungsorgan der Nahrpilze bei *Xyleborus mascarensis* Eichh. (Scolytidae). Naturwissenschaften. 47: 405.

French, J. R. J., and R. A. Roeper. 1972. In vitro culture of the ambrosia beetle *Xyleborus dispar* (Coleoptera: Scolytidae) with its symbiotic fungus, *Ambrosiella hartigii*. Ann. Entomol. Soc. Amer. 65: 719–721.

Gadd, C. H., and C. A. Loos. 1947. The ambrosia fungus of *Xyleborus fornicatus* Eichh. Brit. Mycol. Sci. Trans. 30: 13–18.

Giese, R. L. 1965. The bioecology of *Corthylus columbianus* Hopkins. Holz und Organismen 1: 361–370.

Gregory, J. L. 1954. Shot hole borers of cacao. Rept. 6th Commonwealth Entomol. Conf. 1954. 293 pp.

Hartig, T. 1844. Ambrosia des *Bostrychus dispar*. Allg. Forst-u. Jagdztg. 13: 73.

Hubbard, H. G. 1897. The ambrosia beetles of the United States. U.S.D.A. Bull. 7: 9–30.

Kessler, K. J. Jr., 1974. An apparent symbiosis between *Fusarium* fungi and ambrosia beetles causes canker on black walnut stems. Plant Disease Reporter 58: 1044–1047.

Lhoste, J., and A. Roche. 1959. Contribution a la connaissance de l'anatomie interne de *Xyleborus morstatti*. Cafe, Cacao, The 3: 76–86.

Meiffren, M. 1957. Les maladies du cafeier en Cote d'Ivoire. Centre Rech. Agron. Bingerville. 103 pp.

Muller, H. R. A. 1933. Ambrosia fungi of tropical Scolytidae in pure culture. Verslag. Afdeel. Ned-Oost-Ind. Ned. Entomol. Ver. 1: 105–125.

Neger, F. W. 1909. Ambrosiapilze II. Ber. Deut. Botan. Ges. 27: 372-389.

———. 1911. Zur Ubertragung des Ambrosiapilzes von *Xyleborus dispar*. Naturw. Z. Land-Forstw. 9: 223-225.

Ngoan, N. D., R. C. Wilkinson, D. E. Short, C. S. Moses, and J. R. Mangold. 1976. Biology of an introduced ambrosia beetle, *Xylosandrus compactus,* in Florida. Ann. Entomol. Soc. Am. 69: 872-876.

Norris, D. M. 1972. Dependence of fertility and progeny development of *Xyleborus ferrugineus* upon chemicals from its symbiotes. Pages 299-310 *in* J. G. Rodrigeus, ed. Insect and Mite Nutrition. North-Holland, Amsterdam.

———. 1975. Chemical interdependencies among *Xyleborus* spp. ambrosia beetles and their symbiotic microbes. Organismen und Holz 3: 479-488.

Nunberg, M. 1951. Nieco o gruczolach znajdujacych sie w przedtulowiu kornikow (Scolytidae) i wyrynnikow (Platypodidae) (Coleoptera). (Contribution to the knowledge of prothoracic glands of Scolytidae and Platypodidae.) Ann. Musei. Zool. Polon. 14: 261-265.

Schedl, K. E. 1962. Scolytidae and Platypodidae Afrikas. Rev. Entomol. Mocambique 5: 1-1352.

Schmidberger, J. 1836. Naturgeschichte des Apfelborkenkafers *Apate dispar*. Beitr. Obsbaumzucht Naturgesch. Obstbaumen schadlichen Insekten 4: 213-230.

Verrall, A. F. 1943. Fungi associated with certain ambrosia beetles. J. Agr. Res. 66: 135-144.

chapter 4

The Fungi Symbiotic with Anobiid Beetles

by GERHARD JURZITZA*

ABSTRACT

Most Anobiidae (Coleoptera) live in an intracellular symbiosis with yeastlike fungi. Larvae and adults harbor the symbiotes in pouches at the beginning of the midgut. The mycetocytes are large, their plasma is filled with symbiote cells. The central nucleus is indented, following the contours of the symbiote cells. The brush border is greatly reduced. In the basal part of the mycetocyte the fungi multiply by budding, in the apical their cell wall is lysed. Their protoplasts are partly resorbed, partly expelled into the lumen together with few intact cells. The symbiotes are transmitted to the offspring by smearing the egg shell with fungus cells. The hatching larva eats part of it and, thus, becomes infected. Several authors tried to cultivate the symbiotes, but they were successful only in approximately 25 percent of the tested species. Recently, five symbiotes were brought in culture. Three are *Torulopsis* species; two show morphological and cytological similarity to certain *Taphrina* species and were, therefore, named *"Symbiotaphrina"* by the author. The role of the symbiotes in the host metabolism was studied by comparing growth of normal and aposymbiotic larvae in artificial diets deficient in essential ingredients. They are able to supply vitamins, choline, a sterol and essential amino acids. In protein deficient diets, normal larvae can use uric acid as additional nitrogen source. Under normal ecological conditions, vitamin supply is of minor importance. The author postulates that the main function of the symbiotes is the recycling of excreted nitrogen and the adaptation of plant proteins to the demands of the host. This enables the wood-destroying Anobiidae to develop in such an extremely nitrogen deficient substratum as wood.

INTRODUCTION

The symbiosis. Endosymbioses are widespread in insects and they are restricted to species with highly specialized nutrition, such as blood and

*Botanisches Institut I der Universität, Kaiserstrasse, Karlsruhe, Germany.

INSECT-FUNGUS SYMBIOSIS /Batra (ed.) / Allanheld, Osmun, Montclair, NJ

plant sap suckers and wood eaters. Nearly all Anobiid species live in an endosymbiosis with yeastlike fungi. Most species live in wood, either undecayed or in all stages of decomposition. Some species are specialized in fruiting bodies of fungi, two species have become pests of stored products, and a few live in different parts of weeds. As far as we know the last group is devoid of symbiosis. Karawaiew (1899) was the first who described the evaginations at the beginning of the larval midgut of the drugstore beetle, *Stegobium paniceum* (L.), and the tearshaped microorganisms which inhabit them. Escherich (1900) identified them as yeasts. From studies published by Buchner (1921), Heitz (1927), Breitsprecher (1928), Müller (1934), Graebner (1954), Cymorek (1957, 1960, 1964), Foeckler (1961), Grinbergs (1962), Milne (1963), Francke-Grosmann (1966) and Jurzitza (1970, 1972, 1976, 1977) we are well informed about the Anobiid symbiosis.

Morphology of cecae or blind sacs. At the point where fore- and midgut meet, there is a group of blind sacs which are well provided with tracheae. In the larvae, their structure is fundamentally the same as that of the normal midgut epithelium (Figs. 4.1, 4.2). A cell monolayer consists of many mycetocytes and few sterile columnar cells; both are fixed on a basal membrane (Fig. 4.3). The mycetocytes are large and broad, their microvilli border is rudimentary. Their protoplasm is reduced to a netlike system in

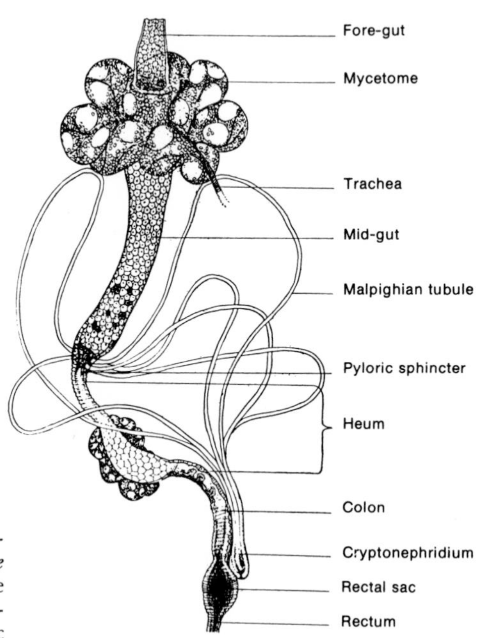

Figure 4.1. Semidiagrammatic representation of gut of *Lasioderma serricorne* larva. Note the blind sacs or pouches (here labeled mycetome) and trachea. Malpighian tubules are attached to the symbiotic organs. (Courtesy of D. L. Milne, J. Entomol. Soc. South Afr. 26:49–51. 1963.)

which a great number of symbiote cells are embedded. In the electron microscope, the protoplasm proves to be less dense than that of the sterile columnar cells. It contains few short strands of the rough endoplasmic reticulum; mitochondria are small, round or oval, with only few cristae. Golgi complexes are scarce. All this indicates that the metabolic activity of the mycetocytes is low. The centrally located nucleus has an irregular shape due to numerous indentations which follow the contours of the symbiotes.

In the basal part of the mycetocytes, the symbiotes appear intact, and budding cells are not uncommon. In the apical part, however, the symbiotes are lysed. They are surrounded by many electron transparent vesicles which may contain the enzymes that attack the yeast cell walls. In some symbiotes, the wall shows perforations through which the protoplast may protrude. In others, the walls become thinner and may even disappear. A large vacuole appears in the formerly dense fungus protoplasm. Some of the protoplasts may be resorbed by the mycetocyte; others are expelled into the blind sac lumen together with few intact symbiotes.

The sterile columnar cells are located singly or in small groups between the mycetocytes. They are shorter (Breitsprecher 1928, Foeckler 1961) or longer (Jurzitza 1977) than the mycetocytes. Their structure is the same as that of normal midgut epithelial cells. The protoplasm is dense, organelles

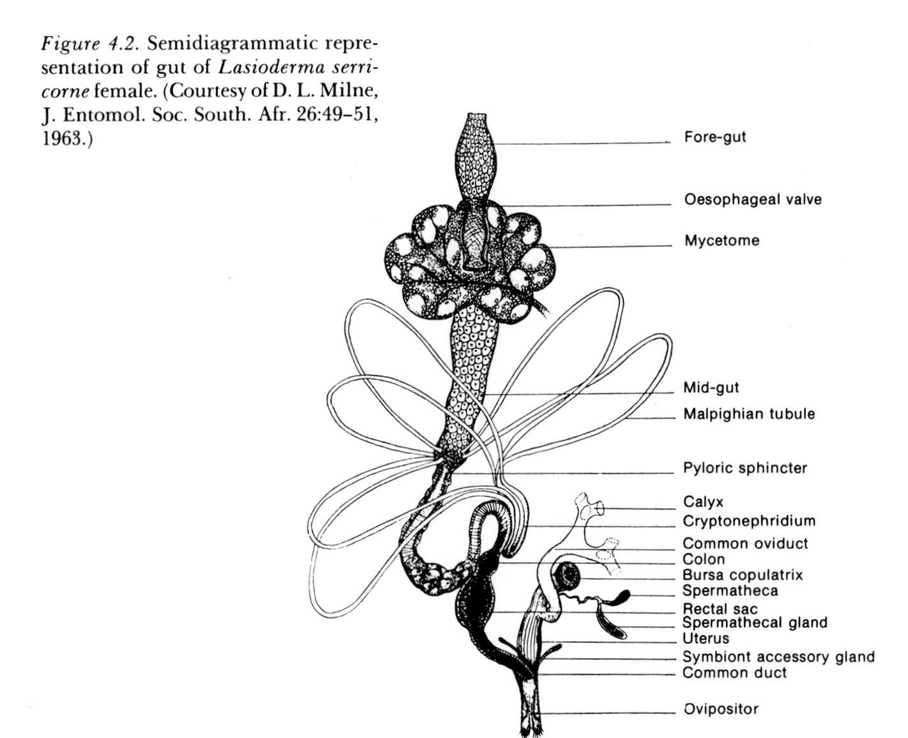

Figure 4.2. Semidiagrammatic representation of gut of *Lasioderma serricorne* female. (Courtesy of D. L. Milne, J. Entomol. Soc. South. Afr. 26:49–51, 1963.)

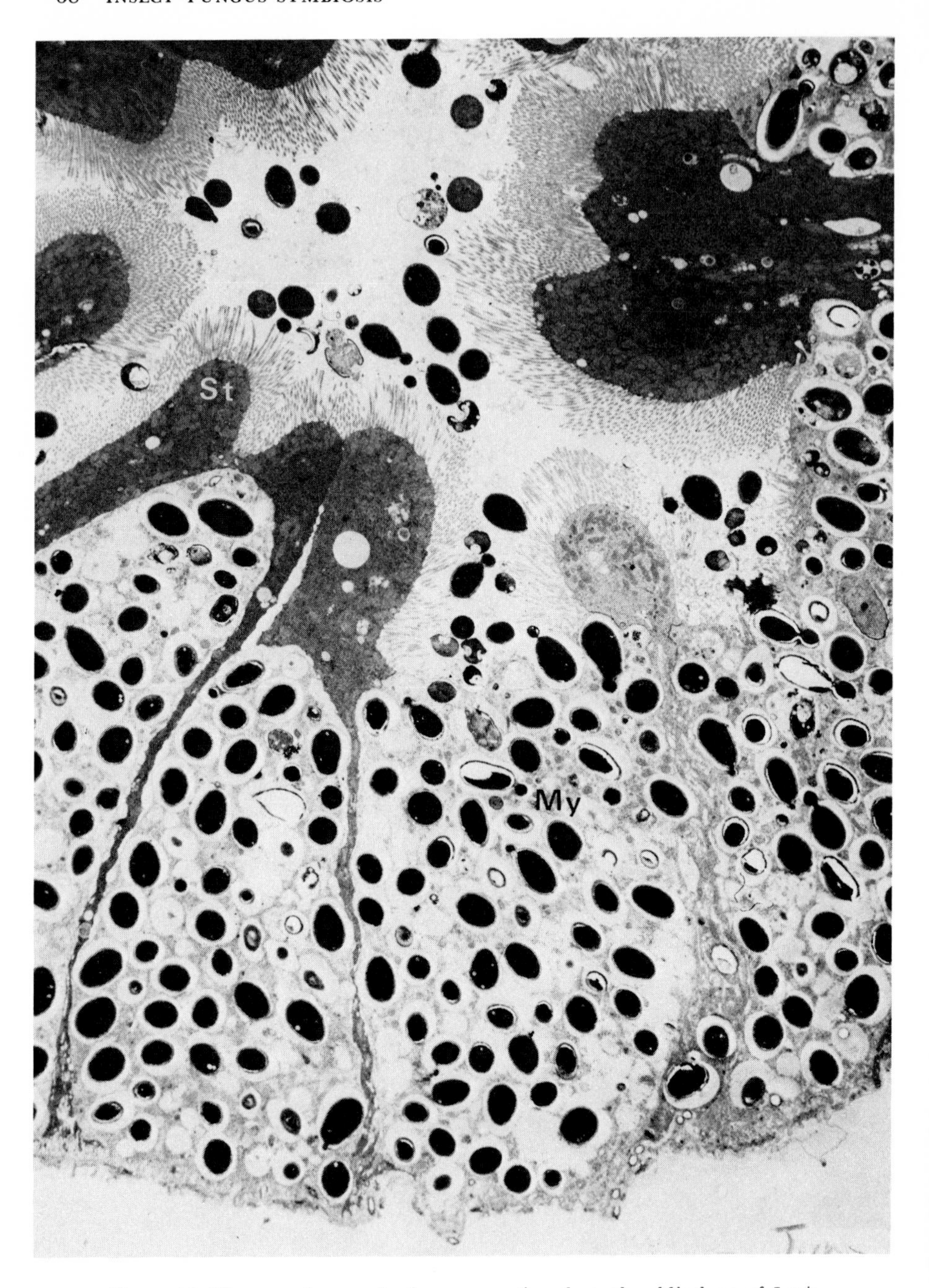

Figure 4.3. Electronmicrograph of a cross section through a blind sac of *Lasioderma* larva. Note broad mycetocytes (My) and narrower sterile (St) columnar cells, the latter with denser protoplasm. In the apical part of mycetocytes, symbiote cell wall is lysed and its cells are expelled into the lumen. (Courtesy Springer-Verlag, New York; Jurzitza, 1977.) (×2500.)

and membrane systems are well developed and give the impression of an intense secretory metabolism. Among them, close to the basal membrane, small embryonic cells with large nuclei and few organelles are located.

Blind sacs in the adults are smaller and more delicate. Information about their ultrastructure is not yet available.

The symbiotes are transmitted to the offspring by contamination of the egg shell (Fig. 4.6). In the adult female, an intersegmental membrane of the ovipositor forms two "intersegmental tubules" which are located on the sides of this organ. Their lumina are stuffed with symbiote cells. Two additional yeast-filled pouches are found at the end of the ovipositor. When an egg passes through the ovipositor, its surface is smeared with a symbiote containing a secretion(s) (Buchner 1921; Francke-Grosmann 1966; Jurzitza 1972). The hatching larva eats part of the egg shell and, thus, becomes infected with the symbiotes. One week later, the still small midgut diverticula of the larva contain the symbiotes.

The endosymbiotes. Karawaiew (1899) who first saw the tearshaped symbiotes of *Stegobium paniceum,* thought they were flagellates. He misinterpreted budding cells, when mother and daughter cell were hanging together at the pointed ends, as a copulation. One year later, Escherich (1900) tried to cultivate the organisms in question. He obtained well-growing cultures of a yeast with oval cells and, therefore, stated the symbiotes to be yeasts. Heitz (1927) made attempts to cultivate the symbiotes of several Anobiidae, among them *Stegobium.* He found that the symbiotes of this species grow very slowly on artificial media and retain their tear shape also in culture. Therefore he doubted that Escherich's cultures have been identical with the symbiotes. Symbiote culture was also reported by Müller (1934), Pant and Fraenkel (1950, 1954), Graebner (1954), Van Der Walt (1961), Kühlwein and Jurzitza (1961), Grinbergs (1962), Jurzitza (1964, 1970), and Bismanis (1976). The authors were successful only in five out of twenty species involved. Most of the isolated strains grow very slowly on artificial media.

Only few authors discuss the systematic position of the Anobiid symbiotes. A strain isolated by Müller (1934) from *Ernobius mollis* L. was described as *Torulopsis ernobii* by Lodder and Kreger-Van Rij (1952). Graebner (1954) isolated and studied a strain from *Stegobium paniceum* and named it *Torulopsis buchnerii.* Kühlwein and Jurzitza (1961), who had isolated the same symbiote, believed it to be different from *Torulopsis* in cytological and morphological characters. They redescribed it as *Symbiotaphrina buchnerii* and suggested placing this new genus provisionally in the Taphrinaceae because of its similarity to some *Taphrina* species. Independently, Van Der Walt (1961) isolated the symbiote of *Lasioderma serricorne.* Also basing upon cytology, morphology and behavior in culture, he supposed that it might be a *Taphrina.* In 1964, Jurzitza isolated another symbiote strain out of the same species. Several characteristics indicated that a relationship exists between the symbiote of *Stegobium* and *Lasioderma*. Both of them produce a reddish pigment which is light induced in the *Lasioderma* symbiote but also occurs in the darkness in the *Stegobium* fungus. The pigments could be isolated in both strains and were identified as carotenoids by thin layer chromatography (Jurzitza, unpublished). Bismanis (1976) describes pigment production also in his strain of the *Stegobium* symbiote, but, contrary to Jurzitza, he identifies it as a non-carotenoid pigment. Some immunological tests confirmed the relationship of both strains. For these reasons Jurzitza (1964) described the symbiote of *Lasioderma serricorne* as *Symbiotaphrina kochii*. Recently, Bismanis (1976) renewed the discussion on the systematics of the *Stegobium* symbiote. Without mentioning the papers of Van Der Walt (1961) and Jurzitza (1964), he criticized the redescription as *Symbiotaphrina* and, using the arguments which may be the matter of further discussion, he reestablished Graebner's name *"Torulopsis buchnerii."* Two further Anobiid symbiotes were named *Torulopsis karawaiewi* and *T. xestobii* by Jurzitza (1970).

Information about the ultrastructure of Anobiid symbiotes is, until now, available only for the *Lasioderma* symbiote (Jurzitza 1977). There seem to be no important differences from other yeasts except that no vacuole exists in the symbiotic state (Fig. 4.4) but appears when the cells are lysed (see some cells in Fig. 4.3, 4.5). The round or oval nucleus is surrounded by some mitochondria. The endoplasmic reticulum is scarce, the protoplasm is dense and stuffed with ribosomes.

The role of the endosymbiotes in the host metabolism. Endosymbioses are widespread in insects and restricted to species with highly specialized nutrition, such as blood and plant sap suckers and wood eaters. Anobiidae belong to the latter group. Most species live in sound or more or less decayed timbers, and some develop in fruiting bodies of fungi. Another ecological group feeds from fresh weeds. *Lasioderma redtenbacheri,* a representative of this group (Cymorek 1964), has no symbiotic organs and is devoid of symbiotes. Two additional species, both living in symbiosis, are pests of stored products. The drugstore beetle, *Stegobium paniceum,* may originate from wood-destroying ancestors; the cigarette beetle, *Lasioderma serricorne,* from fresh weed feeders. This conclusion is drawn from larval morphology of both species.

Koch (1933a, b; 1934) was the first who succeeded in disrupting the symbiosis of the drugstore beetle. He washed the eggs with disinfecting solutions, thus killing the symbiotes which adhere to the eggshell (Fig. 4.6). So the hatching larvae remained uninfected. Koch tried to rear them. In a diet very suitable for normal larvae, the aposymbiotic ones were unable to grow. But when Koch added dried yeast, yeast extracts or wheat germs to the diets, the sterile larvae grew nearly as well as the normal ones, and aposymbiotic adults resulted. From these results, Koch concluded that the

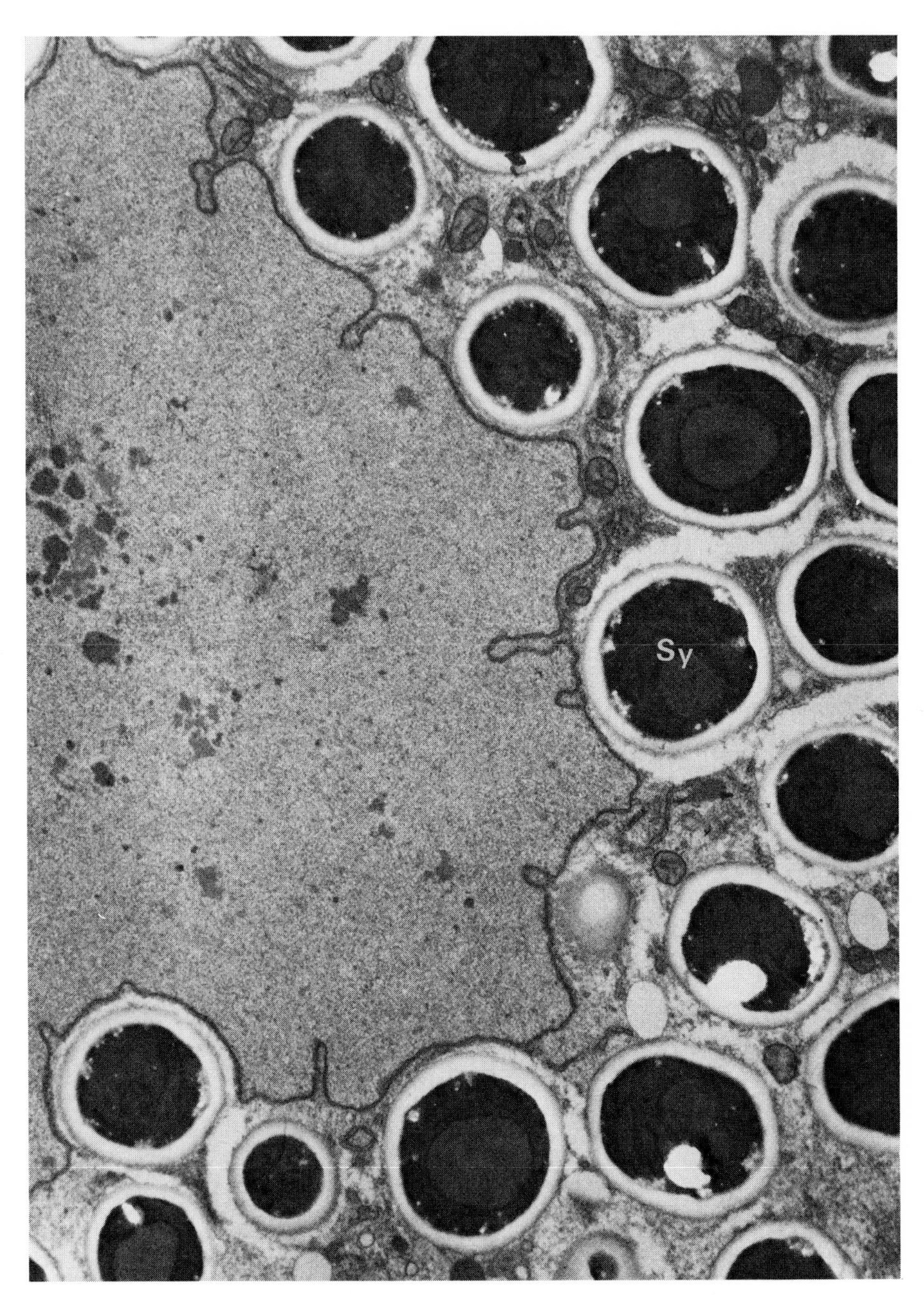

Figure 4.4. Part of a mycetocyte of *Lasioderma*, near the nucleus which is indented and follows contours of the symbiote yeast (Sy). The transparent areas around yeast cells are due to cell shrinking. In the animal plasma mitochondria and endoplasmic reticulum are visible. (Jurzitza, 1977.) (×15000.)

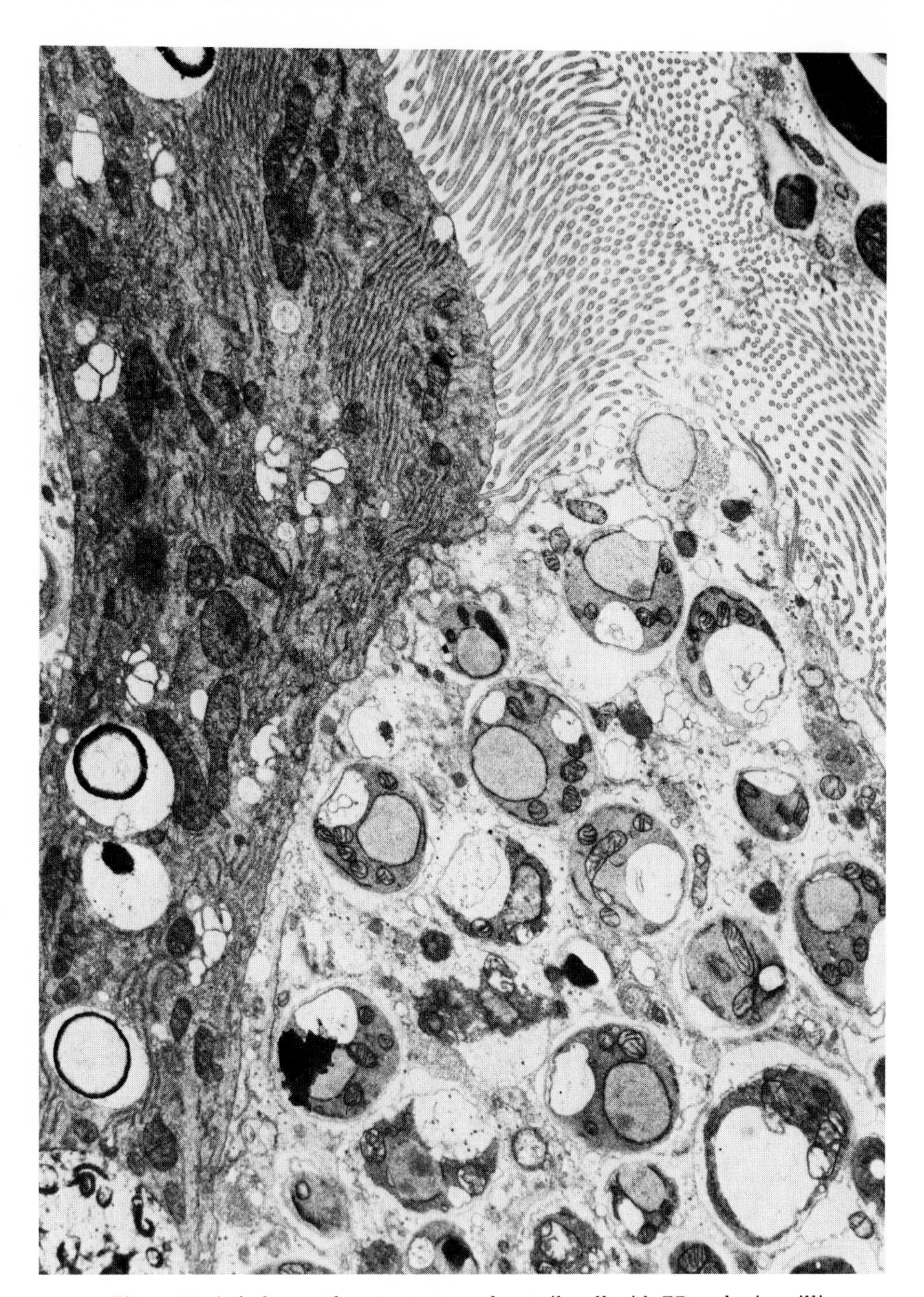

Figure 4.5. Apical part of a mycetocyte and a sterile cell with ER and microvilli border. Cell walls of most of the symbiote cells are partly or entirely lysed. At this stage, the symbiote cells have vacuoles which are lacking in intact cells. (Jurzitza, 1977.) (×10,000.)

symbiotes would supply their hosts with vitamins in which the diet might have been deficient. Fraenkel and Blewett (1942, 1943a, b, c) were able to confirm this idea under laboratory conditions. Using Koch's method, they obtained aposymbiotic larvae which were reared in defined diets deficient in one of the following substances: thiamine, riboflavin, pyridoxine, biotin, nicotinic acid, pantothenate, folic acid, choline, or cholesterol. In most of these diets, only the normal larvae were able to continue their development while aposymbiotic ones grew only slowly or not at all. Pant and Fraenkel (1950, 1954) isolated the symbiotes, both of *Stegobium* and *Lasioderma*. They were able to reinfect sterile larvae by smearing the egg shells with a symbiote suspension. In the same way, they were able to interchange the symbiotes of these two species. Even in the wrong host the symbiotes were able to supply vitamins. From these experiments we know that differences exist between symbiote species in the amount of supply of single substances. Jurzitza (1969d, 1974) was able to confirm these results. He was able to demonstrate that *Lasioderma serricorne* is able to develop even in a completely vitamin-free diet, although slower than in the presence of these substances, due to symbiotic action.

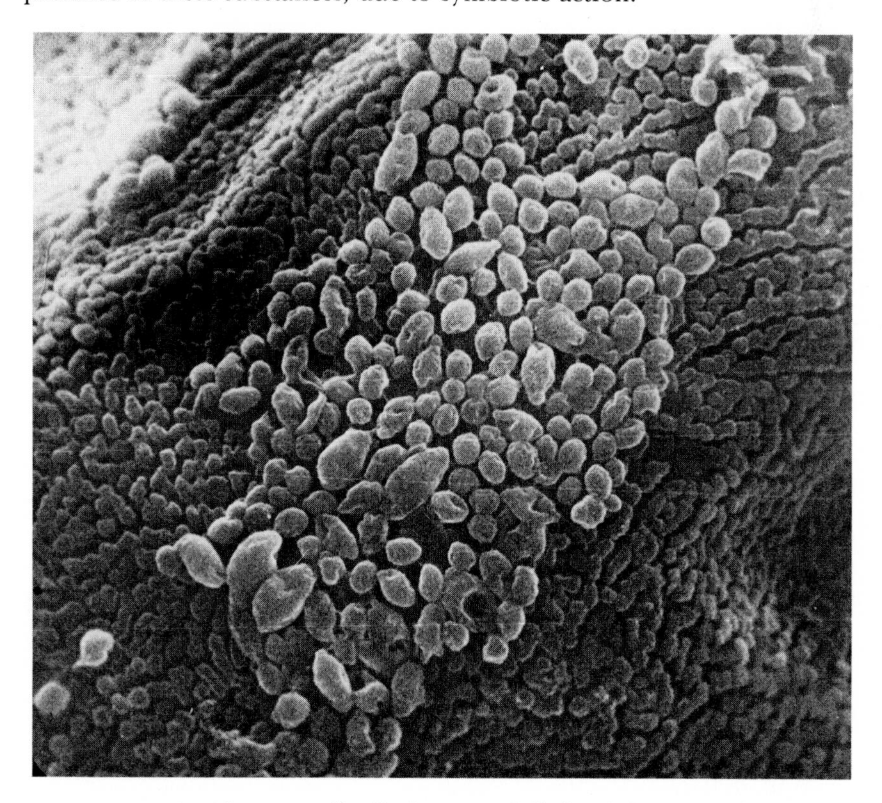

Figure 4.6. Symbiote yeast cells adhering to egg shell of *Lasioderma*. Size of symbiote cells 2–3 μm. (× 2200.)

Contrary to this idea about the function of the symbiotic fungi, Becker (1942a, b, c) discussed the possibility that the endosymbiotes might play a role in the nitrogen metabolism of wood-destroying insects. He thought that they might compensate the protein deficiency in the food by fixing atmospheric nitrogen and supplying their hosts with proteins. Following this suggestion, Pant, Gupta and Nayar (1960) compared the growth of normal and aposymbiotic *Stegobium* larvae in diets deficient in single amino acids. They found that sterile larvae were able to develop in diets deficient in one out of nine amino acids not essential for them, but growth ceased when essential ones were lacking. On the contrary, normal larvae developed even in the latter diets, although in some of them growth was markedly retarded. These results were confirmed by Jurzitza (1969b, c) for *Lasioderma.* The symbiotes of this species are able to furnish all of the 10 amino acids essential for their host, but only lysine and phenylalanine in amounts sufficient for optimal growth.

Jurzitza (1972) compared growth of normal and aposymbiotic larvae of *Lasioderma* in diets with suboptimal concentrations of proteins. He learned that normal ones were less affected by the protein deficiency than aposymbiotic ones. Furthermore, he was able to show that the effect of protein deficiency could be partly compensated in normal but not in aposymbiotic larvae when uric acid was added to the diets. It is also easy to demonstrate that the symbiote of this species is able to grow on media with uric acid as the only nitrogen source. So Jurzitza suggested a recycling of the excretory products of the larvae by the symbiotes. This may diminish the loss of nitrogen and thus help the larvae to survive in a substrate with such an extremely low protein content as wood. This idea is confirmed by an observation made by Milne (1963). According to him, in the larva of *Lasioderma serricorne* two of the malpighian tubules are fixed to the symbiotic blind sacs.

CONCLUSIONS

From the reported results, the conclusion must be drawn that the endosymbiotes are able to supply their hosts with vitamins and a sterol and with essential amino acids. They can use its excretory products to synthetize these substances. But all the experiments reported were performed under artificial conditions and give no idea about the real importance of these abilities under ecological conditions.

Jurzitza (1969a, d) reared the larvae of *Lasioderma* in powders of dried plant material and wood; to this diet vitamin free casein and/or complete vitamin mixtures were added. In powders of dried leaves, growth both of normal and aposymbiotic larvae was the same. In other substrates, for example in fermented tobacco, growth of aposymbiotic larvae was retarded. But this inhibitory effect of symbiote loss could be compensated by the addition of vitamin-free casein but not by the complete vitamin mixture.

Only in wood powders both additions had growth-promoting effects. From these results it is evident that the main importance of the symbiotes lies in the protein metabolism of the host. Vitamin supply, insignificant in many natural substrates of *Lasioderma,* may become important in species which inhabit extreme substrates such as old, dry timbers and man-influenced stored products such as flour and flour products.

ACKNOWLEDGMENT

The author's thanks are due to Dr. L. R. Batra, Research Mycologist, Beltsville Agricultural Research Center, for his careful revision of the manuscript.

LITERATURE CITED

Becker, G. 1942a. Ökologische und physiologische Untersuchungen über die holzzerstörenden Larven von *Anobium punctatum* De Geer. Z. Morph. Ökol. Tiere 39: 98-152.

Becker, G. 1942b. Beobachtungen und experimentelle Untersuchungen zur Kenntnis des Mulmbockkäfers (*Ergates faber* L.). Z. angew. Entomol. 29: 1-30.

Becker, G. 1942c. Untersuchungen über die Ernährungsphysiologie der Hausbockkäferlarven. Z. vergl. Physiol. 29: 315-388.

Bismanis, J. E. 1976. Endosymbionts of *Sitodrepa panicea.* Canad. J. Microbiol. 10: 1415-1424.

Breitsprecher, E. 1928. Beiträge zur Kenntnis der Anobiidensymbiose. Z. Morph. Ökol. Tiere 11: 495-538.

Buchner. P. 1921. Studien an intrazellulären Symbionten. III. Die Symbiose der Anobiiden mit Hefepilzen. Arch. Protistenk 42: 320-331.

Cymorek, S. 1957. Beitrag zur Kenntnis der Pochkäferarten *Anobium punctatum* Deg., *Anobium hederae* Ihss., *Anobium inexspectatum* Lohse (Col., Anobiidae). Entomol. Blätter 53: 87-94.

———. 1960. *Anobium fulvicorne* Sturm var. *rufipenne* Duft. - eine eigene Art, var. *demelti* nov. var. eine neue Farbform. Entomol. Blätter 55: 264-275.

———. 1964. Notizen über das Vorkommen, den Wirt und die Lebensweise von *Lasioderma redtenbacheri* Bach (Col. Anobiidae). Entomol. Blätter 60: 154-161.

Escherich, K. 1900. Über das regelmässige Vorkommen von Sprosspilzen in dem Darmepithel eines Käfers. Botan. Zbl. 20: 350-358.

Foeckler, F. 1961. Reinfektionsversuche steriler Larven von *Stegobium paniceum* L. mit Fremdhefen. Z. Morph. Ökol. Tiere 50: 119-162.

Fraenkel, G., and M. Blewett. 1942. Biotin, Riboflavin, Nicotinic Acid, B_6 and Pantothenic Acid as Growth Factors for Insects. Nature 150: 177.

———. 1943a. Intracellular symbionts of insects as sources of vitamins. Nature 152: 506.

———. 1943b. The vitamin B-complex requirements of several insects. Biochem. J. 37: 686-692.

———. 1943c. The sterol requirements of several insects. Biochem. J. 37: 692-695.

Francke-Grosmann, H. 1966. Zur Übertragung der Symbionten bei *Anobium rufipes* Fabr. Zool. Beitr., N. F., 12: 17-25.

Graebner, K. E. 1954. Vergleichend morphologische und physiologische Studien an Anobiiden—und Cerambycidensymbionten. Z. Morph. Ökol. Tiere 42: 471-528.

Grinbergs, J. 1962. Untersuchungen über Vorkommen und Funktion symbiontischer Mikroorganismen bei holzfressenden Insekten Chiles. Arch. Mikrobiol. 41: 51-78.

Heitz, E. 1927. Über intrazelluläre Symbiose bei holzfressenden Käferlarven. Z. Morph. Ökol. Tiere 7: 279-305.

Jurzitza, G. 1964. Studien an der Symbiose der Anobiiden. II. Physiologische Studien am Symbionten von *Lasioderma serricorne* F. Arch. Mikrobiol. 49: 331-340.

———. 1969a. Untersuchungen über die Wirkung sekundärer Pflanzeninhaltsstoffe auf die Pilzsymbiose des Tabakkäfers *Lasioderma serricorne* F. 2. Die entwicklung normaler und aposymbiontischer Larven in Tabak mit verschiedenem Nikotingehalt. Z. angew. Entomol. 63: 233-236.

———. 1969b. Die Rolle der hefeartigen Symbionten von *Lasioderma serricorne* F. (Coleoptera, Anobiidae) im Proteinmetabolismus ihrer Wirte. 1. Das Wachstum normaler und aposymbiontischer Larven in Diäten mit Proteinen, Proteinderivaten und Aminosäuregemischen als N-Quellen. Z. vergl. Physiol. 63: 165-181.

———. 1969c. Der Vitaminbedarf normaler und aposymbiontischer *Lasioderma serricorne* F. (Coleoptera, Anobiidae) und die Bedeutung der symbiontischen Pilze als Vitaminquelle für ihre Wirte. Oecologia (Berl.) 3: 70-83.

———. 1969d. Aufzuchtversuche an *Lasioderma serricorne* F. in Drogen—und Holzpulvern im Hinblick auf die Rolle der hefeartigen Symbionten. Z. Naturforsch. 24b: 760-763.

———. 1970. Über Isolierung, Kultur und Taxonomie einiger Anobiidensymbionten (Insecta, Coleoptera). Arch. Mikrobiol. 72: 203-222.

———. 1972. Rasterelektronenmikroskopische Untersuchungen über die Strukturen der Oberfläche von Anobiideneiern (Coleoptera) und über die Verteilung der Endosymbionten auf den Eischalen. Forma et Functio 5: 75-88.

———. 1976. Die Aufzucht von *Lasioderma serricorne* F. in holzhaltigen Vitaminmangeldiäten. Ein Beitrag zur Bedeutung der Endosymbiosen holzzerstörender Insekten als als Vitaminquellen für ihre Wirte. Material und Organismen 1976 (Beiheft 3): 499-505.

———. 1977. Elektronenmikroskopische Untersuchungen an den symbiontenführenden Mitteldarm-Blindsäcken der Larve von *Lasioderma serricorne* F. (Coleoptera, Anobiidae). Z. Parasitenk. 54: 193-207.

Karawaiew, W. 1899. Über Anatomie und Metamorphose des Darmkanals der Larve von *Anobium paniceum*. Biol. Zbl. 19: 122-30; 161-71; 196-220.

Koch, A. 1933a. Über künstlich symbiontenfrei gemachte Insekten. Verh. dtsch. zool. Ges. 35: 143-150.

———. 1933b. Über das Verhalten symbiontenfreier *Sitodrepa*larven. Biol. Zbl. 53: 199-203.

———. 1934. Neue Ergebnisse der Symbioseforschung. 1. Die Symbiose des Brotkäfers *Sitodrepa panicea*. Prakt. Mikroskopie 13: 3-12.

Kühlwein, H., and G. Jurzitza. 1961. Studien an der Symbiose der Anobiiden. 1. Die Kultur des Symbionten von *Sitodrepa panicea* L. Arch. Mikrobiol. 40: 247-260.

Lodder, J., and N. J. W. Kreger-Van Rij. 1952. The yeasts. North-Holland Publishing Co., Amsterdam. 713 pp.

Milne, D. L. 1963. A study of the nutrition of the cigarette beetle, *Lasioderma serricorne* F., (Coleoptera: Anobiidae) and a suggested new method for its control. J. Entomol. Soc. South Afr. 26: 43-63.

Müller, W. 1934. Untersuchungen über die Sumbiose von Tieren mit Pilzen und Bakterien. Über die Pilzsymbiose holzfressender Insektenlarven. Arch. Mikrobiol. 5: 84-147.

Pant, N. C., and G. Fraenkel. 1950. The function of the symbiontic yeasts of two insect species, *Lasioderma serricorne* F. and *Stegobium (Sitodrepa) paniceum* L. Science 112: 498-500.

———. 1954. Studies on the symbiotic yeasts of two insect species, *Lasioderma serricorne* F. and *Stegobium paniceum* L. Biol. Bull. 107: 420-432.

Pant, N. C., P. Gupta, and J. K. Nayar. 1960. Physiology of intracellular symbiotes of *Stegobium paniceum* L., with special reference to amino acid requirements of the host. Experientia (Basel) 16: 311-312.

Van der Walt, J. P. 1961. The mycetome symbiont of *Lasioderma serricorne*. Antonie van Leeuwenhoek 27: 362-366.

chapter 5

Fungus-Culturing by Ants

by NEAL A. WEBER*

ABSTRACT

Fungus-culturing by ants involves a complex behavioral pattern. Only the New World ants of the tribe Attini have developed a mutualistic relationship in which the vegetative mycelium of a particular fungus is passed from one generation of ants to another solely through the recently fecundated female. She carries fragments of the parental garden containing pieces of mycelium in an infrabuccal pocket. A short tunnel is constructed in which this pellet is ejected in damp soil. She then manures it with her liquid fecal droplets and the mycelium grows quickly. Eggs are soon laid and within the first few days a garden about 1 mm in diameter arises with eggs embedded in it. Resulting larvae are fed on the fungus or with broken eggs. Within two months at 25°C the first brood has matured and these workers take over the care of the garden and brood for the remainder of the colony life.

The major attines belong to the genera *Acromyrmex* and *Atta* and are polymorphic leaf-cutters. The larger workers forage for green leaves and flowers, forming well-defined paths to bushes and trees. Sections of foliage frequently 10 mm in diameter are cut and carried to the nest. Here medium to small-sized workers cut the sections to 1-2 mm pieces. In the process, salivation and biting reduces them to pulpy masses. These are embedded in the upper rims of the garden cells. Then an ant picks up strands of hyphae from the adjacent mycelium and dots the fragments with these, forming islands of growth. Liquid anal droplets are added throughout the entire garden life history and contribute proteinases that were derived from the fungus.

The mycelium proliferates as a monoculture, nonsporulating and with masses (staphylae) of inflated hyphae, common to the higher attines. These are compact, nutritive and used as the sole food of larvae and adults. The mycelium commonly

*Department of Biological Science, The Florida State University, Tallahassee, Florida 32306. (Professor of Zoology, Emeritus, Swarthmore College, Swarthmore, Pennsylvania 19081.)

INSECT-FUNGUS SYMBIOSIS /Batra (ed.) / Allanheld, Osmun, Montclair, NJ

grows over the brood as it does over carcasses of insects in the gardens of some primitive attines. The latter have gardens on insect feces and vegetal detritus.

The fungus of several attines is a basidiomycete and is wood rotting. The first ant fungus identification was from *Acromyrmex* and described as *Rozites gongylophora* Möller in 1893. The first sporophores to be produced in culture were identified as *Lepiota* by Locquin. Later this was referred to as *Leucocoprinus*. The fungus grown by *Cyphomyrmex rimosus* Spinola is a yeast as cultured by ants, growing also as a mycelium over the brood or in artificial culture. Other species of fungi may be grown by other ant species. Several proposed names may be those of contaminants. It is necessary to culture artificially alleged ant fungi, then return them to the ants for verification. Sporulating fungi are always rejected.

The success of this mutualistic relation is shown by isolated laboratory ant colonies subsisting solely for more than ten years on the fungus, derived from the original inoculum brought by the female. Artificial cultures from 23 species of attines have been maintained by the New York Botanical Garden as the Weber Ant Fungi Collection.

INTRODUCTION

The fungus-culturing behavior of ants has often been referred to as an example of mutualism or symbiosis. Mutualism (Random House Dictionary of the English Language, New York 1971)—"A relationship between two species of organisms in which both benefit from the association." Symbiosis—"Biol. the living together of two dissimilar organisms, especially when this association is mutually beneficial."

In this account, the preferred term for the relationship will be mutualistic since it is slightly more precise, as defined above, although it is also recognized as symbiotic.

Ants make use of plants world-wide. Some ants nest in plant stems or within leaves, others take nectar or gather seeds. The so-called ant gardens, early described by Ule in the Amazon (see Weber 1943; Wilson 1971, 1975), are creations of humus and fibrous root masses of epiphytes up in trees that involve nest making by *Camponotus* and *Crematogaster* ants. These gardens are their creations, with a special flora growing from them, but the plants themselves are not confined to these situations. Unique, however, is the association of the American tribe of attine ants, the Attini, and particular fungi that have not been identified elsewhere. These fungus-culturing or fungus-growing ants (*Saúva* in Portuguese, see Preface), the world's first gardeners, have evolved a behavioral pattern far more intricate than any of those ant and plant relationships above noted.

The tribe belongs to the largest subfamily of ants, the Myrmicinae, and is confined to the Americas. It appears to have evolved in northern South America and spread south to latitude 44° in Argentina and north to the United States (40° in the East, about 39° in the West; Weber 1972a).

The superficial resemblances between some attines and other myrmicine ants may be marked. The ants of the primitive attine genera and of the smaller workers of the leaf-cutters are like some other tropical genera. The equally widespread *Pheidole*, that includes some seed-gathering species, have a large and widespread species, *P. cephalica* F. Smith (formerly called *opaca*). A new subspecies of it, *P.C. apterostigmoides* Weber, was collected close to a nest of the attine, *Apterostigma urichi* Forel. The two were similar in size, appearance and slow-moving habit. Investigators have speculated that there may be an evolutionary relationship between the attines and such genera as *Pheidole* and another seed-gatherer, *Pogonomyrmex* and its allies (Weber 1958).

References in the literature to fungus-growing insects frequently apply to other regular associations between insects and fungi. These include especially certain Old World termites (e.g. *Macrotermes* or *Bellicositermes)* and certain Holarctic wood-inhabiting beetles or wasps. It is true the fungi grow regularly in the nests or burrows of these insects, and they themselves may have anatomical structures, such as mycangia, that render the transfer possible from generation to generation. While attine ants are also fungus-growers, they are fungus-cultivators as well.

The attine colony originates from and remains centered about the queen. She becomes a functional queen when she is fertilized and starts her garden and colony. The workers of attines are largely alike in appearance and function in the colonies of the primitive attine genera, thus are called monomorphic. A genus such as *Trachymyrmex*, one of the commonest and most widespread, shows distinct variation in worker size and is transitional to the clearly polymorphic leaf-cutters of the genera *Acromyrmex* and *Atta*. The basic behavioral pattern in fungus-culturing was summarized as follows (Weber 1972c):

> A colony of attine ants begins with a recently fecundated female carrying hyphae from the parental garden in a pellet in an infrabuccal pocket. All future food of the colony will be derived from this nucleus. She digs a cavity in the ground, ejects this pellet and manures it with her liquid excrement. As the hyphae proliferate, eggs are laid on them and the colony is launched. She continually licks both the hyphae and the brood. Thus, both salivary and anal excretions play a vital role in the beginning of a colony and this pattern is repeated by the resulting workers.

THE ANTS

Distribution of Attines: best known species. As noted, the attines extend through the entire American tropical region and into the Temperate Zones of both hemispheres. Maps showing this distribution of the genera (Weber 1972a) show that species of *Cyphomyrmex, Trachymyrmex, Acromyrmex* and *Atta* have the greatest ranges and *Acromyrmex* exceeds all in ecological and altitudinal range. The genera *Sericomyrmex, Myrmicocrypta* and *Apterostigma* are on tropical mainlands; *Mycetosorites* and *Mycetarotes* are known only from small and scattered areas. A genus of the South American grasslands, *Mycetophylax*, has one species that is found on several seashores of the southern Caribbean islands. Of special interest to mycologists is the genus *Myrmicocrypta* of the tropical mainland, whose species *M. buenzlii* Borgmeier of Trinidad (an island zoogeographically mainland) cultured a fungus forming sporophores reared by Robbins et al. (Weber, loc. cit.). These were identical to those from *Cyphomyrmex costatus* Mann of Panama (see below), a very different genus. The Panamanian *M. ednaella* fungus developed a nearly mature sporophore that was unfortunately unidentifiable.

The genus *Cyphomyrmex,* occurring from California and the Gulf Coast of the United States to Argentina, has the single yeast culturer *C. rimosus.* Other species grow the usual attine mycelium but without well-defined gongylidia.

The first species to be investigated by mycologists was *Acromyrmex discigera* Mayr of Brazil (Möller, 1893). This was at first identified as *Atta* but agreement has been general in the 20th century that it is an *Acromyrmex.* The fungi of other species of both genera were examined by various South Americans, especially of *Atta sexdens* L. *rubropilosa* Forel and *Atta cephalotes* L. Weber (1938) included Wheeler's 1907 description of the fungus of *Cyphomyrmex rimosus.*

At the present time the genera noted above and under Distribution of Attines are those of particular promise for fungus studies. However, the widespread *Trachymyrmex* (Argentina to United States) may be easy to keep and its fungus would be of particular interest in studying fungal evolution within a genus with this widespread distribution.

Through the collaboration of W. J. Robbins of Rockefeller University and the New York Botanical Garden, a collection of cultures was maintained at the latter institution. This was known as the Weber Ant Fungi Collection (Robbins 1969, Table 5.1). As field work on the attines and their fungi was extended to various U. S. localities and foreign countries, new cultures were added. Table 5.1 represents the fungus cultures from attine ant species in the 1963–1973 period. The ants range the entire spectrum from most primitive to most specialized and from the American North Temperate Zone through the tropics to the South Temperate Zone. So far as known, it is a unique collection and was made available to various investigators.

Morphology and roles of queens, males and worker castes. Females upon fecundation may be known as queens to distinguish them from unfertilized females. Ordinarily the females are fertilized on a nuptial flight and independently start their colonies. They are by far the largest members of the colony. They lose their large, membranous wings at this time, leaving the thorax as a prominent hump that held the wing muscles. These are soon metabolized for the production of eggs, leaving a large air space. Queens are larger than workers in all three segments (head, thorax and gaster). Only the *Atta* soldier head is proportionately larger and it has a

Table 5.1 Ant Fungi in Pure Culture (maintained by the New York Botanical Garden as Weber (W) Collection, see text for additional data).

No.	Field Notes	Ant species	Collection date
W1	4260	*Trachymyrmex septentrionalis*	1963
W2	4312	*Atta cephalotes*	1964
W3	4350	*Acromyrmex (A.) octospinosus*	1964
W4	4314	*Trachymyrmex urichi*	1964
W5	4331	*Sericomyrmex urichi*	1964
W6	4330	*Myrmicocrypta buenzlii*	1964
W7	4325	*Mycetophylax conformis*	1964
W8	4404	*Atta cephalotes*	1965
W9	4441	*Acromyrmex (A.) octospinosus*	1965
W10	4454	*Acromyrmex (A.) lobicornis*	1965
W11	4455	*Acromyrmex (Moellerius) striatus*	1965
W12	4460	*Trachymyrmex septentrionalis*	1966
W13	4461	*Trachymyrmex septentrionalis*	1966
W14	4468	*Myrmicocrypta ednaella Mann*	1966
W15	4469	*Apterostigma mayri*	1966
W16	4470	*Atta colombica tonsipes*	1966
W17	4471	*Cyphomyrmex costatus*	1966
W18	4472	*Cyphomyrmex costatus*	1966
W19	4475	*Cyphomyrmex costatus*	1966
W20	4477	*Atta cephalotes isthmicola*	1966
W21	4525	*Trachymyrmex cornetzi*	1967
W22	4529	*Apterostigma auriculatum*	1967
W23	4532	*Cyphomyrmex rimosus*	1967
W24	4528	*Azteca alien fungus* (discarded)	1967
W25	4546	*Trachymyrmex relictus Borgmeier*	1967
W26	4685	*Atta laevigata F. Smith*	1970
W27	4701	*Atta laevigata F. Smith*	1970
W28	4683	*Acromyrmex (Moellerius) landolti*	1970
W29	4702	*Acromyrmex (Moellerius) landolti*	1970
W30	4668	*Atta sexdens*	1970
W31	4710	*Atta sexdens*	1970
W32	4761	*Cyphomyrmex rimosus*	1972
W33	4806	*Acromyrmex aspersus (F. Smith)*	1972
W34	4875	*Trachymyrmex* sp.	1973
W35	4912	*Acromyrmex heyeri*	1973
W36	4915	*Acromyrmex heyeri*	1973
W37	4921	*Atta colombica tonsipes*	1973

much smaller gaster. The gasters of all attine queens are larger than those of workers and do not expand markedly during egg-laying. There is ordinarily one queen to a colony.

Males are winged throughout their life and are large, but smaller than queens. Their heads are disproportionately smaller than those of queens or workers from lack of large mandibular muscles. They are the darkest members of the colony and, as maturing pupae, may readily be distinguished by this characteristic alone. Male and female pupae are also easily distinguished by their large size and prominent wing pads. Males play no active role within the nest, and as they mature as adults they tend to move to peripheral cells and tunnels, as do the females. Males and virgin females are present in the nest only for a short time in the year, usually leaving at the onset of the rainy season in the tropics. The days and hours of flight of representative species in both the tropics and the temperate zones are noted in Weber (1972a).

The workers make up the entire adult population of the nest, except for the queen, during most of the year. They are all of similar size in most primitive species or vary slightly. In *Acromyrmex* and *Atta* they are highly polymorphic, in contrast to the monomorphic genera (Weber 1972a).

The minima in *Acromyrmex* and *Atta* are 1.5–3 mm long, measured as the ant walks normally with head slightly extended forward. They are less spiny than the larger castes and more smooth and shiny. They make up 60–67% of the ants in the normal garden. The media are 4–6 mm long, are generally darker with larger spines than the minima, and may make up about 30–33% of ants in the garden (Weber 1972c). The maxima are 7–9 mm long and have the longest spines; they make up 1 or 2% of a garden but appear to be more numerous because of their size. A larger percent is outside the nest engaged in the activities described below. The soldier caste occurs only in *Atta* and makes up less than 1% of the colony (Weber 1972c). Few or none are in or on the average garden. The head is disproportionately large, and the thoracic spines are greatly reduced and extend from massive humps.

The general functions of the worker castes are as follows, but there may often be exceptions:

· *Minima*: Nursemaids for the brood, caretakers of the fungus garden, initial swabbers of the leaf sections as they are brought to the nest.
· *Media:* Gatherers of substrate, cutters of the leaf sections to a millimeter or two in size, preparers of these for the garden. They take general care of the garden and brood, and are excavators of new chambers.
· *Maxima:* Protectors of the colony, gatherers of substrate, maintainers of the trail system, excavators.
· *Soldier (Atta):* Protectors of the colony, patrollers of the trails, and occasional removers of trail debris.

To carry out these functions the castes are usually located as follows: minima in the cells of the garden when not riding on leaf sections; media everywhere, since they are the generalized workers; maxima on the trails and in the external parts of the nest; soldiers at the nest entrances and in the entrances to the single gardens.

The queen is located at the physiological center of the nest. She is the originator of the pheromones that integrate the colony and she lays eggs.

She can move from one garden to another as growth of the colony takes place, but the garden she is in is always a brood garden. She does not take up a position in peripheral gardens that lack brood or in gardens that are disintegrating.

The males are present as adults for one or a few months before and during the nuptial flights. They and the maturing female brood collect in peripheral chambers and tunnels before the flights. The males are essentially parasites in the nest and only function on the day they leave the nest. The females may feed themselves to a limited extent, but both sexes are clumsy with their large wings and get in the way of the workers.

The behavior of the minima can best be examined under the microscope in small sections of the garden. They work over all parts of the garden minutely, exploring all parts with their antennal apices, licking the hyphae and feeding by abrading the filaments or gongylidia with their mouthparts. They constantly groom the larger members of the colony as described below. When on leaf sections on the trails, they have their heads appressed to the surface of the leaf, no matter how rough the ride, and swab it intently. They groom and feed the larvae. Their small size makes them efficient in licking and caring for the eggs and small larvae. When the garden is disturbed during examination, a minima will grasp a particle of substrate embedded in the garden and curl about it, not relinquishing its hold.

The media perform the general functions described below. The maxima function effectively when the nest is disturbed, rushing out and attacking. The mandibles of the maxima in both *Acromyrmex* and *Atta* can cut human skin, and the ants cling tenaciously. They may discharge the contents of the rectal glands, dotting a person's clothing with brown, pungent droplets. These and other glands discharge pheromones that doubtless arouse the colony to launch a general attack, which involves the media (but not the minima). The maxima assist in dragging pieces of twigs and flower parts on the trails to the nest.

The *Atta* soldier mandibles may make a 4–5 mm cut in the human skin that bleeds freely. They may engage the clothing or leather shoes at one site and continue cutting a half-moon arc. The soldiers are fed by trophallaxis and may stand still for minutes at a time. They are regularly groomed by minima and media since their massive heads make them clumsy. If the colony is attacked by phorid and other parasite flies, they may rear up and threaten, but are ineffective on steep slopes, which may cause them to tumble down.

Grooming behavior and senses of worker castes. By grooming is meant the care of the integument. Grooming plays a vital role in the attine colony because of the ideal conditions within the attine nest for the proliferation of alien and hostile organisms. Temperatures and humidity are ideal for many organisms, and there is a constant source of organic food. The large numbers of ants in close quarters in the nest could mean a constant inflow of alien organisms such as bacteria, protozoa, viruses, and spores of other fungi. Any member of the colony picked up at random from inside or outside the nest, however, is immaculate when viewed under the microscope.

This immaculate condition is caused by constant grooming on the part of the individual and its nest mates. Grooming may occupy a major share of their existence. A special structure, the pecten, is a row of short bristles along the inside of the fore tibia, opposing at the base a curved spine that bears other fine bristles. Usually the coarse, curved spine projecting from the base of the tibia and visible to the naked eye is called the pecten; rather it functions as a comb together with the tibial bristles.

The most frequent use of the pecten is to clean the antennal terminal segments as they are drawn through it. This procedure effectively removes dirt and other debris. These terminal antennal segments must be kept very clean to be effective receptors of chemical (pheromonal) and other stimuli. They are constantly played over other ants, the garden, the brood and new substrate.

An ant can clean its own antennae and also its own legs, drawing them through the opposing fore leg pecten. The legs and antennae are also drawn between the partly opened mandibles and the other mouthparts,

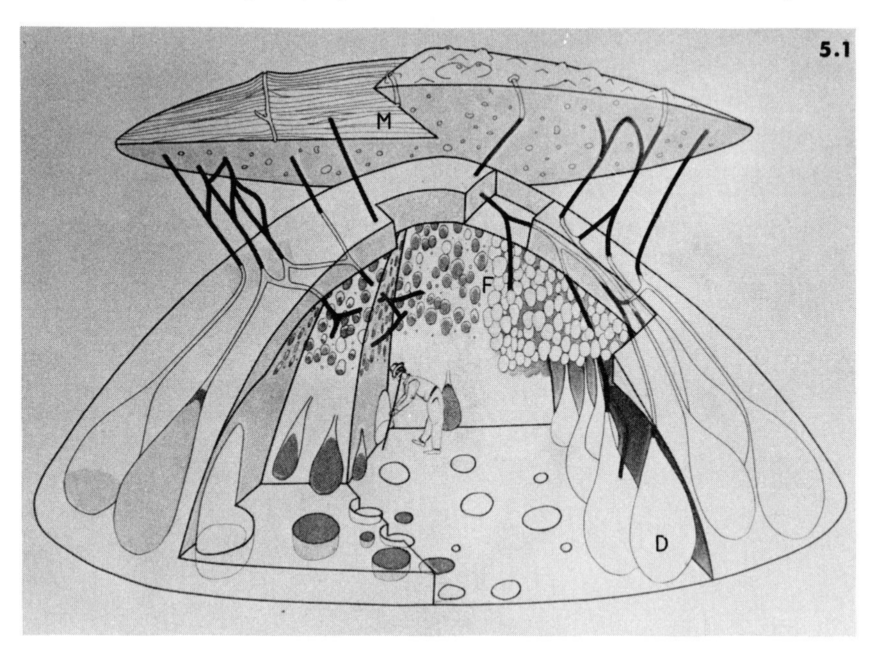

Figure 5.1. A stylized mature nest of *Atts vollenweideri,* based on actual excavations. M=mound or ant hill; F=fungus garden chambers; D=dump chambers for exhausted substratum. (Courtesy of Dr. J. C. Jonkman, Royal Netherlands Embassy, Washington, D.C.)

thus cleaning them more thoroughly and moistening them with saliva. Detritus from this cleaning tends to collect in the infrabuccal pocket (Fig. 5.14) under the pharynx and it is ejected periodically. The body in general is scraped with the short, dense bristles of the ventral surface of all tarsal segments, and the total effect is to keep the ant externally clean.

In cleaning one another, the ant being cleaned remains passive while one or more members of the colony go over it thoroughly with their submandibular mouthparts. The mandibles may be opened or closed and are not used. The glossa of the groomer is appressed to the integument and is applied thoroughly to irregular crevices as well as to flat surfaces. It is in the form of a minutely ridged swab that is moist with salivary secretions.

The sexual forms, and especially the queen, are thoroughly cleaned by the workers, and this care may be extended to the wings. The queen remains in one place for long periods. In *Atta* she is so bulky and large that she is incapable of cleaning all parts of her own integument. The smallest workers go down into her narrowest crevices, such as between the head and the thorax, and lick them carefully.

Communications. Ants of a colony communicate with one another by chemical (pheromonal), tactile, auditory and visual means. The above account of grooming illustrates a combination of pheromonal and tactile means of keeping in touch with one another and would have an integrative function. The antennae are probably the chief receptors.

As the ants walk about, the antennae are held outspread and are directed from side to side, occasionally independently. When the ants are comparatively inactive, the antennae tend to be used in unison. In all cases it is the apices that are gently and carefully brought close to the object being perceived.

A stridulatory organ consists of a filelike structure on the dorsal surface of the first gastric segment and a scraper on the underside of the posterior margin of the postpetiole. When the gaster is moved up and down, a faint stridulation or squeaking noise is produced. The primary known function of stridulation is as an alarm to attack other workers. Stridulation can be easily heard by picking up a worker by the head and holding it close to one's ear. Each of the six species of *Atta* or *Acromyrmex* tested had a characteristic pattern distinguishable on oscillograms. The loudest was *Acromyrmex octospinosus* Reich, and *Atta cephalotes* produced the highest frequencies.

Compound eyes are adequately developed in most attines, but their precise use in the nest is unknown. Ants work equally well in complete darkness and in light, and worker visual communication appears limited. On the nuptial flight, however, the combination of stridulation and sight may be important in bringing the sexes together. The three ocelli in the sexual forms must be coordinated with the compound eyes at least in distinguishing light from dark.

The eyes of attines are precocious in development and are the first part of the body to become pigmented in the pupa as separate ommatidia. This feature can be used to determine the age of pupae. Later these expand to form one apparent continuous dark mass. There is a convex outer integumentary lens for each facet.

Vestiges of ocelli are found quite commonly in the *Atta* soldier and also in a few large *Acromyrmex* workers. When they occur, it is most often the anterior. In almost all specimens the anterior ocellus is double, indicating that ancestral insects had two pairs of ocelli of which the anterior later fused.

Gathering substrate; trails. Substrate, the material on which the fungus is grown, is taken from the immediate vicinity of the young nest, in the form of vegetal detritus. In the more primitive species it also included insect droppings. This fecal material itself is apt to be particles of leaves or woody material that have passed through the digestive tract of Orthoptera, Lepidoptera and Coleoptera. Some attines such as *Myrmicocrypta, Sericomyrmex* and *Trachymyrmex* may also take woody fruit. Faint trails of the latter two genera may extend a few centimeters from the nest opening.

As the colony grows in *Acromyrmex* and *Atta,* spidery trails may be distinguished that are followed by the minima and first media. As these colonies grow the trails become fewer and more distinct (Fig. 5.2), leading to nearby areas of particularly suitable leaves. These may be the flush or new growth.

Mature trails of the species of the two genera are easily distinguished from one another. Those of *Acromyrmex* are narrow, only one or two centimeters wide; since they tend to be overhung by vegetation they are easily overlooked when they pass through grassy areas. They radiate out from the nest in all directions, as many as eight in the pampas mound nest species. Trails of the mature nests of *Atta cephalotes* may be as much as 30 cm wide and free of vegetation. Ordinarily the *Atta* species have several trails of 5–10 cm wide and may be 200 meters long. Those of *A. vollenweideri* Forel tend to be narrower and more numerous, and those of *A. sexdens piriventris* Santschi may be similar. Substrate used by the 10-year colonies is described later. The substrate used by the more primitive genera of small worker and colony size and that used by the higher attines in general are described here. An evolutionary pattern is evident: from vegetal debris and insect droppings, to fruit and flower parts, to green leaves.

Insect fecal pellets and particles of vegetal debris are used by the ants of smallest colony size and of most generalized habit, for example by *Cyphomyrmex rimosus* and by small species of *Trachymyrmex*. When the pellets are of caterpillars (Lepidoptera), the leaf fragments are dissected out. Carcasses of insects are regularly used on the gardens of many species. The mycelium grows normally over them. Particles of rotted wood fragments may be used by all species and especially by *Apterostigma*.

Particles of fallen tropical fruits, often quite woody, are used by all and especially by *Sericomyrmex*. Flower parts and especially stamens are used by all, and even the smallest species may cut off these and incorporate them in the garden. The stage next to the leaf-cutters is clearly shown by species of *Sericomyrmex* and *Trachymyrmex* that cut one or two millimeter sections of green leaves but mostly use other substrate.

The highest genera, the leaf-cutters, collect the green leaves of a wide variety of plants growing in the vicinity of the nest (Fig. 5.3). They often cut crops such as coffee, cacao and vegetables. Ornamentals such as roses are generally taken, and the importance of this substrate in starting new laboratory colonies is illustrated by the following record:

Grams of fresh rose flowers (including sepals and receptacles) used by an *Atta cephalotes* colony in its first year:

Aug.	60	Feb.	51
Sept.	70	Mar.	77
Oct.	85	Apr.	93
Nov.	125	May	288
Dec.	67	June	495
Jan.	85	July	564

Total: 2,060 grams

When colonies are maintained in laboratories away from tropical plants, the ants take plants of many families and genera. The fungus forms the same inflations and clusters (staphylae) as in nature. Freshly transported gardens from the tropics have particles of their original substrate growing the same mycelium, indistinguishable from laboratory particles.

The exceptions to the above, however, are the tropical grass-cutting species of *Atta* and *Acromyrmex* whose gardens have not been maintained satisfactorily on northern substrates, including local grasses. The mycelium on their native grasses grows the densest of all in the species of ants investigated.

When afforded a choice of substrate, the ants show clear preferences, and these are the ones noted as usual with regularity by the 10-year colonies (see p. 108).

Species of plants collected in nature in recent studies (Cherrett 1968; Lugo et al. 1973; Rockwood 1975) show that no plants are consistently defoliated. Instead the ants move from one plant to another periodically, especially taking the "flush" or new growth. During the rainy season fewer plant species are attacked than in the dry season (Rockwood 1975). Fallen flowers and flower parts are regularly taken in the dry season, and *Acromyrmex octospinosus* may regularly take flowers of cacao growing on trunks and branches (Weber 1972a).

Figures 5.2–5.3. Atta cephalotes. 5.2 (top). A trail in Trinidad. *5.3 (bottom).* Workers cutting pieces of leaves.

The effect of rain was early recognized (Weber 1941, 1966b) and more recently described in detail (Lewis et al. 1974; Rockwood 1975). The effect of the dry season was described as follows by Weber (1957b):

> The ants [*Atta sexdens*] foraged at a reduced rate for leafy substrate, did no excavating, and abandoned the upper tier of cells that during the rest of the year would be occupied by fungus gardens. The fungus itself consisted of a sparser than usual filamentous growth and a particularly dense form of bromatia or staphylae, clusters derived from the enlarged ends of some hyphae. These, like minute white golf balls, were carried about by the ants and used as food. They would be less subject to desiccation than loose filaments. Most of the activity of the colony was sub-surface and devoted to the rearing of the annual sexual brood, forms enormous when compared with the workers.

In an extensive study of seasonality (Rockwood 1975) single colonies of two *Atta* species gathered almost nothing but flowers for a month during the dry season.

The leaf-cutters as well as the more primitive attines, however, may take dry leaves both in the laboratory and in nature. For example, 40 grams of dry leaves of *Fraxinus americanus* were cut and taken into an *Atta cephalotes* 10-year colony on one day. Unpublished observations and photographs from this laboratory show how the ants take dead twigs into the garden chamber and how a dense coating of mycelium with staphylae grow on them. This is consistent with the known cellulose-digesting capacity of the ant fungus.

Nests and gardens. The simplest nest of all is the irregular, pre-formed cavity occupied by the yeast-culturer *Cyphomyrmex rimosus*. The nests may be in the soil, under bark of rotted wood, in humus about roots of epiphytes (although observed as high as 30 meters in a 60-meter tree in rain forest, Weber 1941), or in a dead snail shell on the ground. The brood is segregated from the garden.

Other species of the genus also nest in the soil or in humus among roots of epiphytes. The type nest of *C. longiscarpus* Weber in Colombia was unusual in being in the form of a sack of agglutinated humus suspended from rootlets in a densely forested steep ravine. The sack was 2–7 mm thick and surrounded a single garden 4 cm high and 3 cm wide.

The ants of such genera of small ants as *Mycocepurus, Mycetophylax* and *Myrmicocrypta* form one or several chambers in the soil. The tunnel entrance of one *Mycocepurus tardus* Weber nest in clay was less than 1 mm in diameter and led irregularly to a 2 cm cell at a depth of 20.5 cm. That of *Mycetophylax conformis* in sand was similarly tenuous and led by a tortuous route to a total depth of 65 cm, having several small cells branching off enroute. The nests of *Myrmicocrypta buenzlii* in clay consisted of one or several chambers of comparatively large size, some 10 cm in diameter, and at depths of 2.5 to 9.0 cm. A nest of the common *Apterostigma mayri* Forel had the single cell about 7 cm in diameter at a depth of 4.5 cm. Other *Apterostigma* nests were barely beneath the hard shell of rotted wood. Common sites include the humus at the base of epiphytic plants and in humus under debris on the ground.

The nests of *Sericomyrmex* generally resemble those of *Trachymyrmex,* sometimes having a single cell, sometimes two or three. A nest of *S. harekulli* Weber had the first cell 8.5 cm high by 9.5 cm in diameter and the third was 50 cm underground. A nest of *S. urichi* had seven gardens at depths from 10 to 22 cm. The cells containing the garden varied from 3 cm to 12 cm in height and from 5 to 9 cm in diameter.

The nests of the small species of *Trachymyrmex* such as *T. bugnioni* Forel in clay have correspondingly small cells, one colony having a 2 cm single chamber at a depth of 20 cm. Slightly larger species, such as the widespread *T. cornetzi* Forel, may have a turret entrance and a main cell some 4.5 cm in diameter at a depth of 7 cm. Always more shallow is that made initially by the new queen. The larger species, such as the widespread *I. urichi* Forel, will have multiple chambers in a vertical series. A nest of *I. urichi* had five chambers, one above the other to a depth of 30 cm; another had six, at depths from 2.5 to 27.0 cm. They were from 3 to 12 cm in diameter. One of *I. zeteki* Weber had a single chamber at a depth of 8 cm that was 8 cm broad and 6 cm high. Such rather shallow nests of a single large cell may occur in other large species, but more commonly as the colony grows more cells will be added at greater depths. In some cases of the northern *T. septentrionalis* McCook (Fig. 5.4) they are in a vertical series, while in others they branch off at several levels. The garden in sand may be supported by rootlets or rest in a sand cell, and the fungus grows over the brood (Fig. 5.6).

The largest nests are those of *Atta.* These are formed in soil, and the different species have different ecological requirements, e.g. *A. cephalotes* nesting in clay under tropical shade, *A. sexdens* at the forest-grasslands ecotone and *A. vollenweideri* in the open.

The fungus garden. The fungus garden of all species is maintained free of the surroundings except when it has to rest on the bottom of the cell or is suspended from rootlets. This air space creates room for the ants to move about on the garden surface and supplies oxygen and moist air necessary for the metabolism of the fungus itself. When the garden is in a cavity in the soil and resting on stones, the soil is excavated between the stones.

The gardens of *Acromyrmex* and *Atta* look alike, but those of the latter are more symmetrical and are formed in earth cells. The *Acromyrmex* gardens are much more irregular (Fig. 5.5). Some *Acromyrmex* gardens are by far the largest of all attines, being 30 cm or more in diameter. These are found in the thatch mound nests of several species in Argentina and adjacent countries. The nest of one *A. hystrix* Emery colony in Venezuela was estimated to contain an estimated total bulk of 30 liters in its gardens and was immediately above high tide level in the Orinoco delta (Weber

Figures 5.4–5.5. 5.4 (top) Crescentic mound of *Trachymyrmex septentrionalis,* a common species in the southeastern United States. *5.5 (bottom)* Fungus garden of *Acromyrmex octospinosus* from Trinidad.

Figures 5.6–5.7. 5.6 (top) Nest of *Trachymyrmex phaleratus* in Guyana, showing the aboveground clay turret and the subterranean fungus garden. *5.7 (bottom)* Young colony of *Atta sexdens* from Panama, with a large queen, minima, and maxima workers, and fungus garden.

1947, 1958). Another nest in this area was at a height of two meters between adjacent trunks of palm trees. The one garden of 16.5 × 10.2 cm was carefully placed on a large leaf on the ground and dissected in search of the queen. Between 19:30 hours and the following 06:00 all parts were carried back by the ants to a site above the former one and made into a new garden. As noted by Weber in 1947, "The experiment demonstrated that the ants do not necessarily have to weed out alien fungi as was manifestly impossible under these conditions."

Dimensions of numerous *Atta* gardens in nature are summarized from the literature in Weber (1972a). The largest garden of a mature nest of *Atta colombica tonsipes* Santschi was estimated to be some 3,300 ml (as 200 cu in) in Martin et al. (1967). A standard size garden of various species in the laboratory when confined to plastic chambers of 2,250 ml was 1,700-1,800 ml.

A representative living *Atta cephalotes* garden of 1,800 ml weighed 203 g, including about 12–15 g of live ants. This ratio of 8 or 9 ml to 1 g is representative of other attine gardens (Weber loc. cit.). A dried *A. cephalotes* garden of an estimated 120 ml (no ants) weighed 6 g. On this basis the 1,800 ml live garden when dried would be about 90 g. These ratios correspond with those Jonkman (1977) found in *Atta vollenweideri* (Figs. 5.1, 5.8 and 5.9).

Green leaves were weighed fresh in the 10-year colonies described below. To convert the fresh weight to dried weight a conversion factor of 4 to 1 and up to as much as 14 to 1 has been found in *Sambucus* leaves (unpubl.). Heavier leaves would be closer to 3:1.

Response of the fungus to the ants. The response of the fungus to the activities of the ants is to proliferate rapidly as mycelium. The manner of growth is the same as in artificial culture in the absence of the ants. In both cases the hyphae in the higher attines tend to form clusters of inflations or staphylae.

As the substrate of the garden reaches an age of two or three months, it turns yellow. The mycelium is gradually less prominent except for the staphylae. These become more compact, like miniature golf balls, and this is as true of the garden in the late stages of the dry season, when leaf-cutting is at a minimum, as in the laboratory under constant conditions.

Exhausted substrate. Theoretically the quantity of exhausted substrate cast out by the ants should equal the input of fresh substrate, less the loss due to ant and fungal metabolic activities and dehydration. This material in the leaf-cutters is similar in texture both in nature and in the laboratory, normally being golden brown, granular and light in weight, floating readily on water. It quickly darkens as it is modified by bacterial and fungal action and absorbs moisture. Sample weights in grams of 10 ml when dried were 1.96 (*Atta colombica tonsipes*); 1.95 [*Acromyrmex (A.) lobicornis,*

Figures 5.8–5.9. Nest of *Atta vollenweideri* in Argentina before and after excavation. Fungus gardens occupy the numerous chambers.

Emery]; 2.0 (*Sericomyrmex urichi,* including dead ants) and 2.1 (the fine grass sections of *Acromyrmex (Moellerius heyeri).* In nature a colony tends to use the same site repeatedly to cast out refuse in quantity (Weber 1956a). For example, the same colony of *Atta colombica tonsipes* carried more than two liters in June 1954 and June 1955 over the same smooth rock, where the ants allowed it to tumble down the slope. Commonly the ants may carry the particles up a nearby tree for a few decimeters, then allow it to drop down to form a cone about the trunk. They may carry it over lianas, then drop it largely at one site.

This material has been found (Haines 1973) to promote the root system of many surface plants by its useful mineral content. About half the original cellulose content of the original leaves was determined (Martin and Weber 1969) to have been metabolized by the fungus.

Abundance and success of the colonies. A census of the colonies of all attine species in a limited area of 18 × 77 meters was carried out in Trinidad (Weber 1972a). Eight species of seven ant genera, including the two leaf-cutters of the island, had recognizable nests that were counted during four years in the period 1965 to 1970. Five colonies of *Atta cephalotes* were initiated here in 1965, of which one was the 10-year colony described below. Representative nests of the three most common species of the 18 × 77 m area had been excavated here in 1964 and had the following populations:

· *Myrmicocrypta buenzlii*—1,700 workers estimated (1,558 counted), two females, distributed in three gardens with a total volume of 800 ml.
· *Sericomyrmex urichi* Forel—2,000 workers (1,691 counted), one queen, seven gardens with a total volume estimated to be 2,000 ml. An unrelated Trinidad colony was one of the 10-year colonies described below.
· *Trachymyrmex urichi*—1,424 adults including 763 workers, 562 males, 71 females of an estimated 2,160 population. Four gardens totalled 2,400 ml.

Three other small genera had adult populations and gardens whose volume, based on other studies, could be estimated conservatively as follows:

· *Cyphomyrmex rimosus*—200 ants, 10 ml garden.
· *Mycocepurus smithi* Forel—500 ants, 50 ml garden.
· *Trachymyrmex cornetzi*—1,000 ants, 50 ml garden.

The total number of colonies of the eight species was 159 in 1965, 67 in 1967, 162 in 1968 and 34 in 1970, varying markedly according to the visibility at the time. The 1970 period was during a long, dry season when most colonies were inactive. Several of them have workers only two or three millimeters long and the crater entrances are easily destroyed. These censuses therefore represent a sampling of what in many tropical localities is a great population of ants of only one of many ant tribes present in such areas. The attines clearly are one of the most numerous of all neotropical ants in terms both of individuals and colonies. Censuses in smaller areas (15.24 meters on a side) of *Acromyrmex (Moellerius) landolti* Forel (Weber loc. cit.) showed 18 to 23 separate colonies.

The mature colony of an *Acromyrmex* species probably has 50,000 to 100,000 workers. That of mature colonies of different *Atta* species has one to three million workers.

Incipient colonies of *Atta sexdens rubropilosa* in Brazil and *Atta colombica tonsipes* in Panama (Autuori 1950; Weber 1977a) occur in great numbers annually, but the survival rate is low. This is true of the *Atta* species in general (Fig. 5.7). A two-year-old colony of *A. cephalotes* from the above census area reached a worker population of 250,000 to 333,000.

In 81 square meters of the area there were 51 nests of five species, an average of one nest per 1.6 m^2. In another crowded area, there were 36 craters in 20.25 square meters or one nest per 0.56 m^2.

Bonetto (1959) found 45 nests of four species of *Acromyrmex (A. lobicornis, A. heyeri* Forel, *A. striatus* Roger, and *A. hispidus* Satschi) in one hectare in Argentina between a railroad and a highway. He considered them to be unusually numerous there. The nests of these four species are much larger and more populous than those of *A. landolti.*

Culturing techniques of the ants. As noted briefly above, culturing by the ants of a colony is initiated by the minute swabbing of the surface of the leaf section by the minima while it is being brought into the nest by larger castes. The significance of this behavior is described below. This swabbing is continued after the leaf sections are brought on to the surface of the garden. In times of rapid leaf-cutting, the sections may be piled at the bases of the gardens, then carried more "leisurely" to the garden surface. Here the section is cut into smaller pieces, a millimeter or two in diameter. These particles are then placed over the garden surface, especially on the rims of the outer and upper coarse cells. As described in Weber (1956b),

> The piece is rotated between the mandibles and fore tarsi and licked repeatedly. Frequently it receives a fecal droplet, the ant curving its abdomen forward to the particle as it is held. The ant then forces it into the garden with its feet and mandibles so that it comes to rest embedded in the mycelium. The garden receives continual attention from the ants. Mouthparts are appressed to the mycelium and the hyphae grow in a loosely intertwining felt on the surface or permeate the substrate less densely. Hitherto unrecorded observations show that the ants actually plant the new substrate with tufts of mycelium, with the result that within 24 hours at $24 \pm 1°C$ the implanted tufts have started to cover the surface, growing in one species at the rate of 13μ per hour. A leaf section 0.8×1.5 mm may have 10 mycelial tufts planted on it by an ant within five minutes. Not only does the planting accelerate spread of the mycelium but the ant saliva may possibly be growth-promoting. This would account for the covering of mycelium that a well-licked larva may develop.

The mycelium proliferates rapidly in all directions except upward. It is prevented from growing into the air by the continual licking that the surface of the garden receives. The continual application of fecal or rectal and salivary excretions causes growth of the mycelium, so that each particle of substrate is indistinguishable from its neighbor. Frequently the particles are picked up and moved about, being fused into the new site within a matter of hours.

The ants work all night on their gardens. A common routine is for them to move the cut sections to the base of the gardens during the day, then spend the night cutting and placing the particles. By the morning the new material has created dark edges to the septa of the upper and outer garden cells, as noted, and this is darker than the older parts of the same septa. It is darker because the mycelium has not had time to proliferate densely. Older parts of the garden are therefore paler than the new and, as described below, the basal and oldest part becomes yellowed. This base has a scanty mycelium, the staphylae becoming prominent as white pinpoints to the naked eye.

When the ants are expanding new gardens, they may take masses of many fused particles of substrate, held together by the mycelium, and transfer them to the new site. In nature, *Atta* excavates a new chamber in the soil, leaving rootlets in place, then workers bring in masses of garden particles and place them on the rootlets. The rootlets or exposed stones have first been cleaned meticulously. New gardens can therefore arise quickly in new chambers. The latter are excavated further, as the garden grows, until the size of the garden reaches that characteristic of the *Atta* species, commonly 10-15 cm in diameter.

If normal particles of an alien attine garden are placed experimentally near that of a colony, the ants immediately perceive their alien nature through the antennal apices that are wafted in that direction. With mouthparts below the mandible tightly contracted, the ant then removes the alien part with its mandibles and casts it out. The same treatment is afforded an alien species of fungus, either as a vegetative mycelium or as a sporulating part.

In a normal garden, the ants never culture an alien fungus or an alien part of the normal fungus garden of a different genus of attine. Occasionally the ants will accept a part of the garden of another ant species in the same genus but will eventually discard it if its own mycelium is in ample supply. However, colonies in Trinidad in 1934-1935 showed variable results (Weber 1945):

Cyphomyrmex rimosus workers failed to eat *Mycetophylax littoralis* and *Myrmicocrypta buenzlii* fungus. A *Cyphomyrmex bigibbosus* female ate bromatia of *Cyphomyrmex rimosus* though its own garden would have been very different.

Mycocepurus trinidadensis workers fed on *Sericomyrmex urichi* and *Atta cephalotes* fungi but refused *Apterostigma urichi* fungi.

Myrmicocrypta urichi workers ate *Mycetophylax littoralis* fungi. *Myrmicocrypta buenzlii* workers tended *M. urichi* fungus.

Mycetophylax littoralis ants fed on the fungus of *Mycocepurus trinidadensis*.

Apterostigma wasmanni ants fed on the fungus of *Sericomyrmex urichi*.

Trachymyrmex urichi ants cultivated *T. bivittatus, T. ruthae* and *Acromyrmex octospinosus* fungi. *T. ruthae* workers ate *Sericomyrmex urichi* and *Atta cephalotes* fungi.

Acromyrmex octospinosus workers ate *Trachymyrmex urichi* and *Atta cephalotes* fungi.

Atta cephalotes workers ate *Sericomyrmex urichi* and *Acromyrmex octospinosus* fungi.

Response of the ants to the fungus. The ants are in contact with the mycelium from the moment eggs are laid. Eggs are placed on the fungus and the larva hatches in a mycelial mesh. The larva is placed by the workers with its head exposed, and the nurses place the mycelial strands on the larval mouthparts. Pupation takes place in the same situation, and all stages of the brood are normally coated by the mycelium at all times. The masses of yeast cells forming the fungus garden of *Cyphomyrmex rimosus* are replaced by a brood cover, a mycelial mesh as in other attines but of a modified form that is a direct outgrowth from the yeastlike cell.

When a worker emerges as a young adult (callow) it is fed by regurgitation from older workers or feeds directly on the fungus. Throughout the life of the ant, most of its active hours (in most members of the colony) are spent in contact with the mycelium, exploring it with the apices of the antennae, licking it, or rearranging the fungus garden.

Those workers who are leaf-cutters leave the nest to go out on trails. When they return to the garden they feed on the fungus directly or by regurgitation (trophallaxis) from other workers. Those small workers who accompany the larger ants on the trails often ride back on leaf sections and start the swabbing process that is a conspicuous part of preparing the substrate for the fungus. Other ants in the nest are similarly engaged in all stages of cutting the leaf to the millimeter or so size particles that are then ready for planting on the mycelium. This activity takes place on the garden so that all members of the colony for much or all of their life (in the case of the smallest) are in intimate contact with the mycelium.

Biochemical role of the ant excretions and the ant fungi. Investigations from the biochemical point of view have gone far in showing exactly what are the contributions of the ants and the ant fungi to this mutualistic relationship. It was long known that the primary requirement for continuation of this mutualism was an adequate input of substrate, characteristically living or dead plant parts, and that the ant excretions were significantly involved. It was also known that the fungus could grow on wood and digest it in artificial culture (Weber 1957a). The chronological stages are instructive.

Through the initiative of Martin the biochemistry of the mutualism was intensively explored. Martin et al. (1969) concluded that the fungus cultured by *Atta colombica tonsipes* provided the ants with a complete diet.

Carbohydrates made up 27% of the fungus dry weight, free amino acids 4.7%, protein-bound amino acid 13%, and lipids 0.2%. The carbohydrates were trehalose, mannitol, arabinitol and glucose. The lipids contained ergosterol as the major sterol. Weber (1947 et seq.) had noted that the ants provided a chemical milieu for maintaining their gardens, and the present investigations concluded that there was an absence of antibiotics involved.

Martin and Weber (1969) showed that the fungus of the above ant can utilize cellulose as a carbon source in synthetic culture media and did not utilize lignin as a nutrient significantly. At least 45% of the cellulose present in the garden was consumed. Martin (1970) at first concluded that the contribution of the fungus to the ant was the enzymatic apparatus for degrading cellulose, a view later modified. Martin and Martin (1970a) found that the ant fungus relationship was a biochemical alliance. The fungus was deficient in the full complement of proteolytic enzymes necessary to grow well on substrates, while the ant feces contributed proteases. Thus the fungus contributes a cellulose-degrading ability, the ants allantoic acid, allantoin, ammonia and 21 amino acids. The ant contribution enables the fungus to compete successfully with other fungi and bacteria. These conclusions, however, were importantly modified in later work (q.v.).

The explanation for the known presence of insect carcasses in the gardens of primitive ants came from further analyses of the ant fecal droplets (Martin et al. 1973). Studies showed the presence of a-amylase and chitinase, the latter contributing to the degradation of insect cuticle by lysing potentially competing chitinous fungi.

Later studies (Martin 1974) emphasized the biochemical role of the ant feces, thus explaining the behavior repeatedly described by Weber (1947, et seq.). The treatment given by the ants to the surfaces of leaf sections, including the deposition of fecal droplets, was explained by Martin: ". . . (this) maceration process is critical in permitting the initial invasion of the tissue by the hyphae and in facilitating the subsequent ramification of the fungus within the tissue." The liquid feces were determined to be proteolytic. Quantities of fecal droplets were obtained by immersing the ants in ether, whereupon the contents of the recta were discharged. These were then analyzed by a variety of biochemical methods, and tested for proteolytic activity. Three enzymes were characterized in *Atta colombica tonsipes* and *A. texana* (Buckley), a serine proteinase and two metalloendopeptidases. However, the diet of the ant does not include pectin, xylan or cellulose. The explanation for the ultimate source of the enzymes was finally summarized as follows (Boyd and Martin 1975 p. 1815):

> The properties of these enzymes had little in common with the properties reported for other insect proteinases, but rather had a striking resemblance to those of proteinases isolated from microbial sources. This finding was rather unexpected,

since it had been our bias that the faecal enzymes were probably digestive enzymes secreted in the midgut and concentrated in the rectum.

They then cultured the fungus of *Atta sexdens* on a synthetic medium and analyzed it. Three fungal proteolytic enzymes were recovered that were similar to those recovered in the ants' fecal droplets. They conclude, therefore, that it is the ant fungus rather than the ants that originated these substances, vital to the mutualism.

THE FUNGI

Definitions of terms. The definitions of the principal terms used here, taken from Weber (1972a, p. 87) are as follows:

· *Hypha,-ae*—any strand or filament of fungus.
· *Mycelium,-a*—a mass of interconnected hyphae.
· *Gongylidium,-a*—a swelling of the middle or at the end of a hypha; derived from the Greek *gongylis,* turnip (Wheeler 1907); also called a kohlrabi head (*Kohlrabiknops*) (Möller 1893) because of its similarity to the shape of that vegetable.
· *Staphyla,-ae*—a cluster of gongylidia (Fig. 5.13) from the Greek for a cluster of grapes (Weber 1957b). Equivalent to ambrosia bodies (Wheeler 1907 and others) and kohlrabi bodies (Möller 1893, and subsequent writers).
· *Bromatium,-a*—cheese-like mass of yeast cells cultured by *Cyphomyrmex rimosus.* Wheeler (1907) used this term also for staphylae of other ants.
· *Cells of fungus garden*—air spaces between septa or partitions of fungus gardens.

Morphology. The finest hyphae occur in the primitive genera and are simple threads some two to six microns in diameter. These become irregularly clavate in *Apterostigma,* some *Cyphomyrmex* (not *rimosus*), *Mycocepurus* and other genera. They form loose intertwining tangles and foreshadow the staphylae or compact masses of gongylidia in nests of the four highest genera (*Trachymyrmex, Sericomyrmex, Acromyrmex, Atta*). The latter have coarser hyphae, often 6 to 10 or more microns in diameter. Individual gongylidia may be 25 to 50 microns. They are sometimes terminal and almost spherical or subterminal and more pear-shaped. One hypha may have several inflations along its length. The hyphae in *Atta* show stages in loose clusters of more primitive type of inflations to the full development of gongylidia. Weber (1957b) found the most compact staphylae were those in the lower part of the mature-to-senile fungus garden and predominated in nature in the dry season. They resemble white golf balls and were carried about by the ants.

A special development of simple hyphae is found in some *Apterostigma* as a veil as noted below; this occurs in artificial culture free of ants and in the ant nest in nature. As described for several ant species in Panama (Weber 1941), the veil is the thickness of the hyphal diameter and does not appear to be interwoven. Between it and the garden proper is a clear space with few connections of hyphae between those covering the granules of the garden and the veil. The ants move about in this space and tend the gardens freely, moving delicately not to disturb the veil. One garden of 45×32×17 mm suspended from the underside of a fallen log had a complete veil with three circular perforations 3 to 4 mm in diameter for the passage of the ants. The veil had many fine and loose staphylae-type hyphal aggregates on the outside. Veils in other species were free of these.

The yeast cells of *Cyphomyrmex rimosus* and its taxa were described by Wheeler (1907), who made an error of ten-fold in the dimensions given for the cells. He did not describe the hyphae growing from the cells. The yeast cells are commonly about 12 μ in diameter. One figured in Weber (1972a) was 15 μm with a bud of 11 μm. A pyriform cell 9 $\mu \times 15$ μm had a hypha 23 μm long and 203 μm wide with one septum near the apex. The hyphae rarely may branch.

Artificial cultures. Artificial cultures of ant fungi were initiated by Möller in south Brazil in 1893 and by Weber in Trinidad and Guyana (British Guiana) in 1934–1935. Möller's work resulted in the production of sporophores described below, and that of Weber of Panama 1955 material had a similar result, also noted below. However, the vast majority of cultures (Table 5.1) from many ant species by the above and other investigators produced no such results. These cultures were valuable for other purposes. Möller figured and described the fungus of *Cyphomyrmex strigatus* Mayr and of *Apterostigma* species but concentrated on that of *Acromyrmex*.

Standard mycological or bacteriological agar culture media were found to be satisfactory for culturing ant fungi in Trinidad and Guyana (Weber 1945). Erlenmeyer flasks were sterilized with a layer of potato dextrose agar (cf. Fig. 2, Weber 1945). A tuft of the garden fungus would be lifted from a vigorous, freshly removed part of the garden with a sterile needle and quickly placed on the sterile agar surface. In Guyana the best time was in the stillness of the humid dawn in the rain forest, no laboratory being available. If only a staphyla was removed, the chances of contamination appeared to be minimal and pure cultures resulted. These growths then developed new areas of filamentous mycelium with the same masses of staphylae soon appearing on them. This technique was extensively used in following years (Weber 1972a) and always verified by returning the cultured fungus to the same ant species to determine acceptance. Only cultures were maintained that were always immediately acceptable to the ants. Contaminants were easily determined; they were invariably faster growing, forming a bacterial slime, or were alien fungi with conidiophores and were not accepted by the ants. The technique was introduced to Martin and his associates for *Atta colombica tonsipes* fungus and was later modified by them to create faster growing liquid cultures on synthetic media for biochemical analyses of the fungus.

For culturing the yeast grown by *Cyphomyrmex rimosus*, it was sufficient to pick up on a needle a bromatium, or cheese-like mass of cells, grown in an undisturbed garden. Sometimes contaminations were introduced which overwhelmed the ant fungus. If the culture was pure, a morel-like structure resulted (Fig. 2, Weber 1945) that bore short hyphal extensions. Both were freely eaten by the ants.

The cultures were repeatedly offered to laboratory colonies of a dozen attine species and the results summarized (Weber 1945). They were acceptable to the original species, thus verifying the fact that the culture was clearly from the presumed host. In some cases it was accepted by other ant species.

Growth rates at different temperatures. Cultures from 1955 Panamanian species were tested in the University of Wisconsin Bacteriology Laboratory with the cooperation of Professor K. B. Raper. The results (Table 5.2) clearly show that a temperature close to 25° C is the best for these two agars, corresponding to the subsurface shaded tropical soil temperature (see Weber 1959). A temperature of 20° C is too low for these tropical fungi and 30° C is too high.

Sporophores. The first ant fungus sporophores to be produced in culture came in 1955 from a small and hitherto biologically unknown ant from Panama (Weber 1957a). The ant, *Cyphomyrmex costatus* Mann, was known previously only as a museum specimen and from nests briefly described later (Weber 1941). These were under small stones on the forest floor, and each consisted of a single cavity containing a fungus garden of four to six milliliters. The ants themselves were 2 mm long. The garden consisted of golden brown vegetal particles 2 to 3 mm long and less than 1 mm thick, apparently from droppings of fruit being eaten by monkeys in high trees overhead, and from other vegetal debris picked up by the ants on the ground. The fungus cultures were started the day after the colonies were collected. Potato dextrose agar was first used, then after removal to Wisconsin, a variety of other nutrient agars was employed (see Weber, 1972b: 108). For about a month the fungus developed its usual fluffy, gray mycelium in an Erlenmeyer 125 ml flask of sterilized oats incubated in the dark at 21 to 25.6° C. Four mushroom buttons with constricted apices developed on one side but they aborted after eight days. Two months after initial inoculation, a new cluster of 15 incipient sporophores developed in a rosette on the opposite side of the flask. These sporophores were about 2 mm high, elongating and attaining a reclinate to upright position. Ten days later two mature basidiocarps were obtained from the second crop of buttons. These were characterized by Alexander Smith as follows and were identified as an undescribed *Lepiota* by Marcel Loquin:

Table 5.2 Relative Growth of Ant Fungi on Saboraud's Dextrose (SAB) and Potato Dextrose (PDA) Agars at Selected Temperatures.

Source Ant	20 C SAB	20 C PDA	25 C SAB	25 C PDA	30 C SAB	30 C PDA
Atta sexdens	+++	++	+++++	+++++	++++	++++
A. cephalotes isthmicola	+++	++	+++	+++	+++	+++
A. colombica tonsipes	++	++	+++++	++++	+++	++
Acromyrmex octospinosus	++	++	+++++	++++	++	++
Trachymyrmex septentrionalis	++	++	+++++	±	++++	+++
T. bugnioni	++	++	+++	+++	±	+
T. cornetzi	++	++	+++	+++	--	--
Sericomyrmex amabilis	+++	++	++++	+++	±	±
Mycocepurus tardus	+++	+++	++++	++++	+++	+++
Apterostigma mayri	+++	+++	++	++++	+++	+++
A. dentigerum	+++	++++	+++	++++	++	++

Pileus 25 mm broad, with a dark dull brown disc, scaly toward the margin but not conspicuously striate, marginal area white except for scales; lamellae free, white, close, drying white; stipe 6.8 cm x 3½ mm smooth, annulus movable and dark-colored, stipe drying dark brown; spores 6.5 μm$\times$4.5–5 μm, elliptic to slightly obovate in face view, subelliptic in side view (with a slight suprahilar-depression), thick-walled, smooth, with an apical germ pore and a small lens-shaped cap over it, hyaline in KOH, dark reddish brown in Melzer's; basidia four-spored, broadly clavate, 20–23 μm$\times$9$\times$12 μm sterigmata 304 μm long, 0.75 μm at base; pleurocystidia, none found; cheilocystidia abundant, 50–80 μm$\times$10–15 μm, elliptic-pedicellate, with a flexuous apical obtuse outgrowth or outgrowth seen as only a short protrusion (all stages present), hyaline, thin-walled, smooth; gill trama hyaline, interwoven; pileus cuticle a loose palisade of clavate-pedicellate cells with fuscous walls in KOH and also the content dark wood brown, many colored filaments also present in tangles, content wood brown and wall incrusted with a somewhat similarly colored pigment; clamp connections absent. Microscopic sections of the gills showed clavate basidia 10$\times$30 μm, bearing four sterigmata 3–4 μm long. These were stout, curved and pointed and bore elliptical spores 6–8 μm long by 4–5 μm wide. The attached end was pointed. They took the Cotton Blue color in a Lactophenol-Cotton Blue stain.

The same *Lepiota* sporophores were grown by W. J. Robbins, A. Hervey and I. Wong (Leong) in 1965 from cultures of the fungus of *Myrmicocrypta buenzlii* from Trinidad (Weber 1966a). They were able to produce these sporophores in subsequent years (see Addenda).

Reactions to alien species of fungi. Colonies of various species were tested with free-living alien species of fungi (Weber 1945; 1972a). Fresh sporophores of Trinidad fungi, identified by the mycologists Briton-Jones and Wardlaw, were tested in 1934–1935. *Acromyrmex octospinosus* workers were immediately attracted to the stipe, pileus and velum of *Phallus* sp., lapped up the juices and cut off parts but soon discarded them. Workers of *Atta cephalotes* fed more extensively on the pileus and stipe of *Agaricus arvensis,* as well as the above *Phallus* sporophore. In a review of earlier records of fungi associated with ants in South America (Weber 1938), the various cases in the literature were found not to be definitely associated (Table 5.3). Generally, they were growing on the periphery of nests of *Atta* or *Acromyrmex* or on cast-out substrate.

Two tables (Weber 1972a) summarized a testing of identified species of fungi from the New York Botanical Garden with attines in the Swarthmore laboratory, through the collaboration again of Robbins, Hervey and Wong (Leong). There were eight ant species of five genera. The fungi were wild species of *Lepiota* and its relatives, *Lepiota* being chosen because of the identifications of the sporophores of this genus from cultures from *Cyphomyrmex costatus* and *Myrmicocrypta buenzlii.*

Workers were isolated for a day in petri dishes with moist filter paper and were thus deprived of their natural fungus. They were then tested by the introduction of samples of the known free-living fungi. There were three

Table 5.3 Attine Ant Fungi: A Chronological List.

Name	Source Ant	Remarks
Rozites gongylophora Möller, 1893 (Agaricaceae)	*Acromyrmex disciger*	true ant fungus
Xylaria micrura Spegazzini, 1899 (Xylariaceae)	*Acromyrmes lundi*	on nest debris; not proven an ant fungus
Bargellinia? Spegazzini, 1899 (Endomycetaceae)	*Acromyrmex lundi*	Hyphae over old nest debris
Rhizomorpha formicarum Spegazzini, 1899 (Xylariaceae)	*Acromyrmex lundi*	under piece of wood over old nest
Tyridiomyces formicarum Wheeler, 1907 (Cryptococcaceae)	*Cyphomyrmex rimosus*	yeast of fungus garden
Xylaria micrura Bruch, 1921 (Xylariaceae)	*Acromyrmex lundi*	discarded garden substrate; not proven an ant fungus
Locellina Mazzuchii Spegazzini, 1921 (Agaricaceae)	*Atta vollenweideri*	sporophore over nest; not proven an ant·fungus
Poroniopsis Bruchi Spegazzini, 1921 (Sphaeropsidaceae)	*Acromyrmex (Moellerius) heyeri*	sporophores from discarded garden substrate; not proven an ant fungus
Lentinus sp. Weber, 1938 (Agaricaceae)	*Atta cephalotes*	sporophores from ant nest; not proven an ant fungus
Rozites gongylophora various authors (Agaricaceae)	Various *Atta* sp.	identifications based on Möller's descriptions
Lepiota sp. Weber, 1957 (Agaricaceae)	*Cyphomyrmex costatus*	first sporophores to be reared in cultures
Lepiota sp. Weber, 1966 (Agaricaceae)	*Myrmicocrypta buenzlii*	sporophores reared in cultures, same fungus as above
Attamyces bromatificus Kreisel, 1972 (F. imperfecti)	*Atta insularis*	sterile mycelium of garden
Phialocladus zsoltii Kreisel, 1972 (F. imperfecti)	*Atta insularis*	contaminant?
Aspergillus Lehmann, 1975 (F. imperfecti)	Higher *Attini* and Macrotermitinae	contaminant of *Attini?*

major reactions: *positive rejection,* the ants moving their antennae about and avoiding the fungus; *neutral reaction,* the ants ignoring it; and *acceptance,* of three degrees: (1) a slight feeling of the fungus with the apices of the antennae; (2) a slight tasting of it with mouthparts after the favorable antennal reactions; and (3) a complete and rapid acceptance as the ants immediately ate the fungus.

Workers of *Myrmicocrypta buenzlii* showed neutral reactions to seven species of *Lepiota,* indicating that none was close to the ant fungus although known to be a species of *Lepiota.*

Workers of *Sericomyrmex urichi* (of the 10-year colony described below) tasted the fungus of three *Lepiota* species and explored a fourth, indicating that their own fungus may have been somewhat similar.

The northernmost *Trachymyrmex (T. septentrionalis)* and a wide-ranging tropical species, *T. urichi,* in general showed acceptance of eight species of *Lepiota,* mostly by eating them, and ate the fungus *Leucoagricus naucina* (Fr.) Singer, a relative of *Lepiota.* The former ant also ate the mycelium of *Cystoderima* and *T. urichi* slightly tasted *Pluteus namus,* both fungi being considered relatives of *Lepiota.*

The widespread *Acromyrmex octospinosus* responded similarly to *Trachymyrmex,* eating or tasting 10 species of *Lepiota,* and they accepted two of the above relatives. The southern South American *Acromyrmex lobicornis* tasted or ate six of the *Lepiota* and two of the relatives.

The responses of *Atta sexdens* and *A. cephalotes* were much less clear-cut. The *A. sexdens* workers were neutral to the species of *Lepiota,* except that they tasted *Lepiota excoriata* Schaeff. ex Fr. and also *Pluteus nanus* (Fr.) Kum. The *A. cephalotes* workers ate the fungus of the *Leucoagaricus* and tasted that of *Lepiota amantina* L. *naucina* and L. *molybdites* (Fr.) Sacc., being neutral towards the other species.

The results of this testing clearly showed the generally favorable response to wild *Lepiota* species by attines. Their response to species of other fungi in the wild is quite different from this, and the response to non-basidiomycetes in the laboratory, or in nature, by the higher attines is at the least negative. For example, *Penicillium* and *Aspergillus* mycelia with conidiophores, when brought near the attine garden, induce a hostile reaction, the ants removing them from the vicinity of the gardens.

This testing also indicated a discrimination between species of fungi. An attine ant species has evolved a special relationship with a particular strain of a fungus over the millenia, preventing it from sporulating by culturing practices. If the ant has evolved slightly different proportions of enzymes, excretions and other products, compared with other ant species, this may have influenced the formation of subtle differences in the particular strain of fungus. Both the ant and the fungus exist in isolation and ordinarily are not in contact with other ants and fungi. This would seem clearly to be a case of coevolution.

The ants showing the most plasticity in more or less acepting alien *Lepiota*-type fungi are those of the most widely ranging genera, *Acromyrmex* and *Trachymyrmex,* that occur in many habitats and climates. They belong to the higher group with the fungus showing well defined gongylidia. The highest genus, *Atta,* may be of comparatively recent evolution and shows little flexibility. Unfortunately colonies of the species of ants whose fungus produced the known *Lepiota* sporophores were not available for this testing.

Identifications; conflicting views. As in any other group of animals or plants, revisionary studies and additional material have caused changes in recognized names that may cause confusion.

The pioneer work on ant fungi by Möller (1893) is an example. The ants with which his fungal studies were associated were sent to Switzerland from Brazil for Forel to describe. The identifications then received were used in Möller's publication. Forel later (1929, v. 2, p. 247) changed the generic name of *Atta* to *Acromyrmex,* giving the species new names of *A. niger* (F. Smith) for *A. hystrix; A. subterraneus* Forel for *A. coronatus* E.; and *A. mölleri* Forel for *Atta* IV. For changes in other ant names see Kempf (1972) and Weber (1972a).

The ant fungus that Möller named *Rozites gongylophora* was widely quoted in the literature under this name as from *Atta.* Investigators in northern South America then assumed that the fungus in the nests of the widespread *Atta cephalotes* and *A. sexdens* was the same as that from the Möller species of *Acromyrmex.* Later mycologists, especially Heim (1957), Locquin and Singer (Weber 1966a; 1972a) came to other conclusions for the generic name *Rozites.* Also, fructifications were tentatively named from other attine ant genera (Weber 1957a; 1966a) as *Lepiota.* Möller's *Rozites gongylophora* and the *Lepiota* "n.sp." of Weber 1957a, then, have been more recently considered to be one species, *Leucocoprinus gongylphorus* (Möller) Heim, *Leucoagricus gongylophora* or *Agaricus gongylophora.* The experimental data above show that the well-known genus *Lepiota* or its segregates may be correct.

Conidiophores in fungus gardens. Möller was the first to describe conidiophores in abnormal ant gardens *(Acromyrmex disciger),* calling them the "strong form." The relationships to the fungus as cultured by the ants in normal gardens was unclear. Stahel and Geijskes, Weber and other authors also reported these from *Atta* nests, without giving these conidiophores a name. They were figured (Weber 1966a) from the nests of *Trachymyrmex septentrionalis* and noted in other nests of various attines as a sign of deterioration of such gardens. Kreisel (1972) described them from nests of the Cuban *Atta insularis* Guerin as *Phialocladus zsoltii* Kreisel, a new genus and species in Fungi Imperfecti, and the normal fungus as *Attamyces bromatificus.*

Lehmann (1975) suggested that the gongylidia or inflations of attines, and specifically of *Acromyrmex subterraneus* and *Atta sexdens,* are "primitive ascomycetes with some phycomycetic characters." He suggests that the above conidial form "conforms well with *Aspergillus*" and is also the fungus of fungus-associated beetles and termites. The evidence to the contrary is disregarded. Twice (Weber 1972a, p. 115), tentative identifications of an unusual *Aspergillus* in abnormal gardens of *Acromyrmex* and *Atta* colonies were received. These were strongly avoided by the ants. They had every sign of being contaminants developing on abnormally wet gardens from over-use of succulent flowers.

The test of the German culture would be to return them to the ants to see whether or not the ants eat and culture them. Yeasts were also reported from attine gardens (Craven et al. 1970) that appeared to be a contamination in abnormal laboratory gardens and not the well-known yeast of *Cyphomyrmex rimosus* gardens. These and bacteria may be present in senile gardens.

Colonies in the laboratory ten years. Attine colonies kept for ten years in the laboratory afford positive evidence for the longevity of the single queen responsible for each, for the amount of substrate needed for gardens of a given size, and for various biological cycles (Table 5.4). This evidence could not be obtained in nature and it is unique. The tropical colonies were isolated from each other and kept in a clean temperate zone laboratory (Figs. 5.10, 5.11, 5.12). This was free from possible contamination except for the remote possibility of air- or leaf-borne organisms in the room in which the relative humidity varied between 8 and 60 percent. The fresh green leaves and flowers for substrate were collected near the laboratory in an area not sprayed with insecticides. The fungus cultured by the ants was the lineal descendant of that originally brought by the queen to her incipient colony. It was morphologically indistinguishable throughout ten years with that repeatedly found with congeneric colonies in the tropics.

Common contaminants always possible on the leaves would presumably include *Fusarium* and others that must have been inactivated by the ant excretions and manipulations. No sporulating fungi developed in the ten years, and throughout there were the usual gongylidia clustered as staphylae. The cast-out exhausted substrate that in nature might have had proliferating alien fungi and other organisms was taken by the ants to unused chambers and dried as pale brown granules. This was periodically removed and could not be a source for contamination.

There was no opportunity for the ants from one colony to acquire the fungus or parts of the garden of the other colonies since each was permanently isolated from each other. Had contact been made, the hostility of each species would be quickly apparent.

The three colonies were *Sericomyrmex urichi* (Weber 1976a), *Acromyrmex octospinosus* (Weber 1977b) and *Atta cephalotes* (Weber 1976b), all from Trinidad. They were taken in July 1965 within 100 meters of each other. The former genus is confined to the tropical American mainland, and the species is known only from Trinidad but might be expected on the adjacent Venezuelan mainland. The *Acromyrmex* is a widespread species in northern South America and Central America; *Atta cephalotes* has a somewhat similar range.

Table 5.4 represents a comparison of the monthly and annual substrate intake in grams with the minimum and maximum monthly garden size in milliliters of the three colonies.

These three records represent species of the higher genera. Similar data are

Table 5.4 Comparison of Monthly and Annual Substrate Intake (gm) with Minimum and Maximum Monthly Garden Size in Estimated Milliliters of Three 10-Year Attine Colonies.

		Substrate used (gm)		Total	Garden Size (estimated ml)	
		Minimum	Maximum	for Year	Minimum	Maximum
Sericomyrmex urichi	1966	10	73	188	15	1100
	1967	42	102	828	900	2100
	1968	79	152	1093	2150	2550
	1969	8	177	1221	2150	2270
	1970	48	140	1052	2200	2300
	1971	44	161	981	1900	2250
	1972	31	158	1069	2250	2250
	1973	45	138	1146	2200	2620
	1974	85	126	586[a]	1400[b]	2250
	1975				1500[b]	1800[a]
Acromyrmex octospinosus	1966	20	145	500	12	480
	1967	46	325	2214	150	3900
	1968	190	530	4722	1950	6100
	1969	250	488	3929	3900	5300
	1970	160	300	2689	2500	5000
	1971	30	175	1496	800	3100
	1972	87	410	1001	1800	2700
	1973	56	460	2137	500	2900
	1974	45	130[a]	720[a]	300	2500[a]
	1975				650	1700[a]
Atta cephalotes	1965 (July -)	128	335	1513	220	3200
	1966	159	860	5718	3200	4350
	1967	270	735	6290	3800	7800
	1968	365	930	7008	3900	10,000
	1969	305	880	6915	8000	11,600
	1970	230	735	6142	7900	11,700
	1971	290	668	5800	5000	11,300
	1972	180	450	3949	4000	7100
	1973	110	965	5468	3800	6500
	1974	240	562	2189[a]	6200	7900
	1975				600	3600

[a]Through June.
[b]For explanation of marked changes in garden size of this and other species in the table, see original references in the text.

available for other species of *Atta (A. sexdens, A. colombica tonsipes)* in Weber (1977a) and, to a limited extent, for *A. cephalotes isthmicola* Weber (unpublished). No other species of *Sericomyrmex* has been investigated to this extent and no species of *Trachymyrmex* (the fourth of the higher attines) has shown similar growth except for the first two weeks of colony life.

The three 10-year records show a sigmoidal record of growth of the gardens

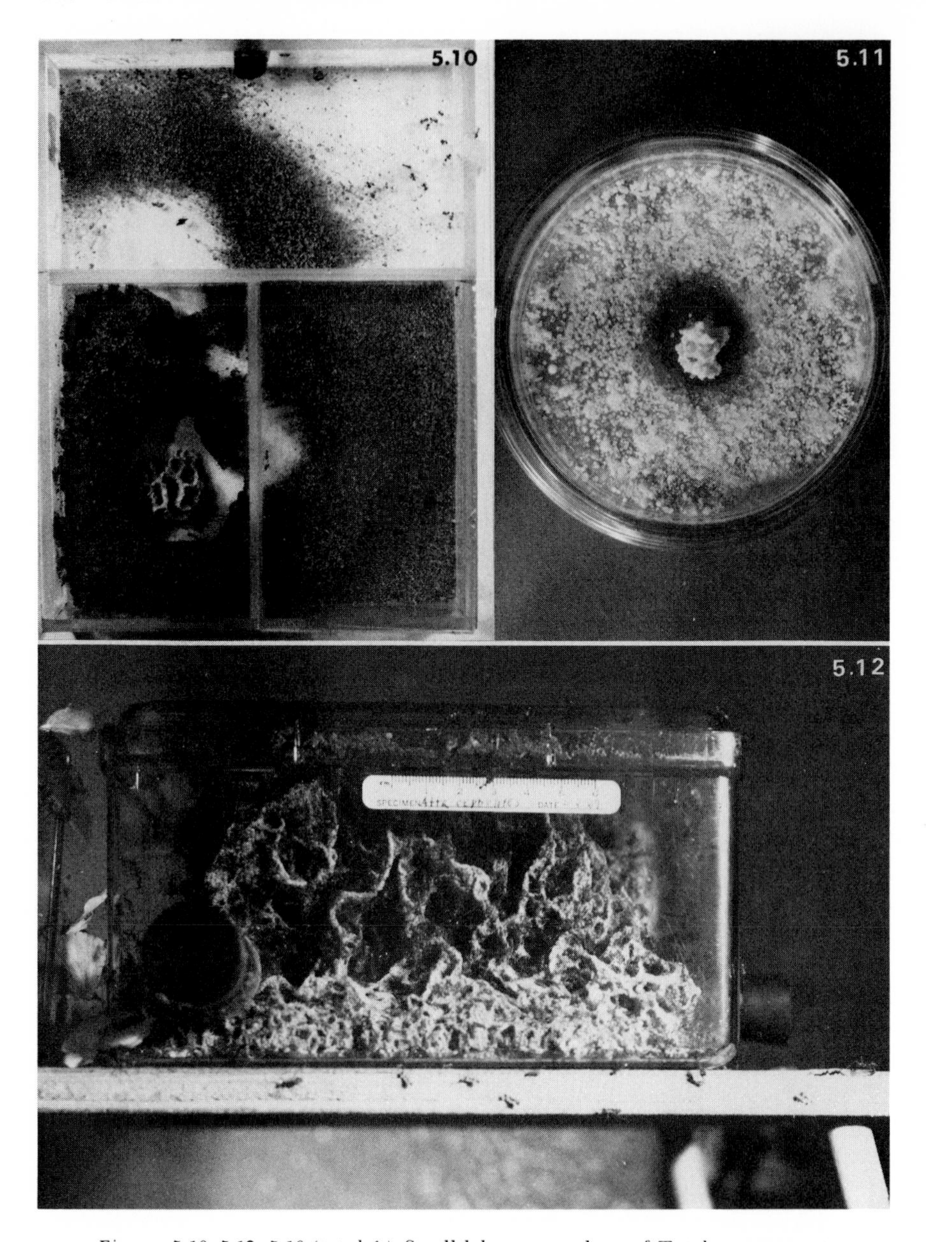

Figures 5.10–5.12. 5.10 (top left). Small laboratory colony of *Trachymyrmex septentrionalis,* showing semicircular crater and fungus garden. *5.11. (top right)* Small garden of *T. septentrionalis* on cassava granules. The ant-specific fungus was maintained even though the surrounding PDA substrate was contaminated by alien fungi and bacteria. *5.12 (bottom).* Fungus garden of a laboratory colony of *Atta cephalotes*; the lower portion contains the oldest substrate.

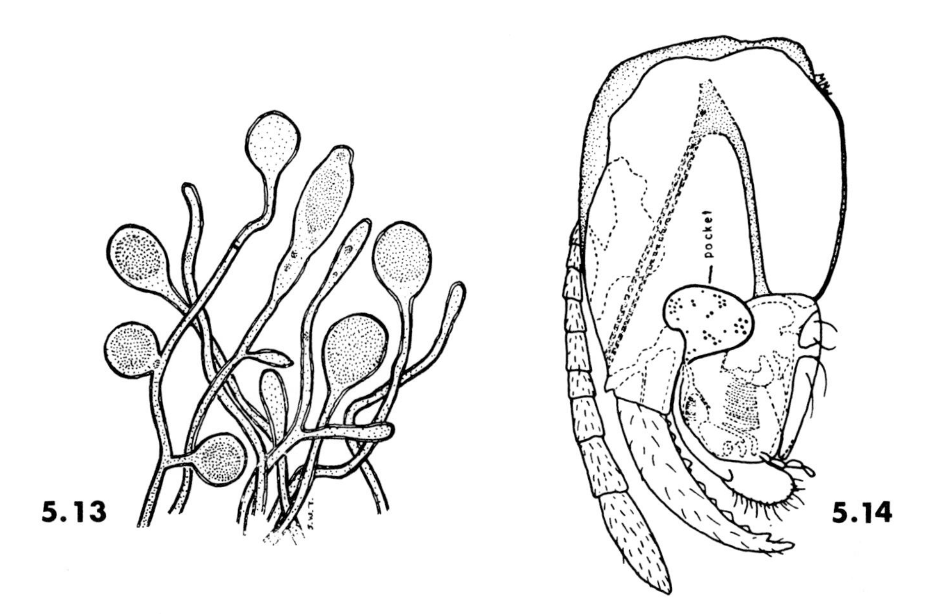

Figures 5.13–5.14. 5.13 (left). A cluster of inflated hyphae (a staphyla) from fungus garden of *Trachymyrmex septentrionalis. 5.14 (right).* Section of head of *T. septentrionalis* showing infrabuccal pocket emptying into mouth above the mouthparts. The pharyngeal tube leading to the thorax is not shown. The V-shaped structure is the internal skeleton. (Courtesy of the American Philosophical Society.) (Weber, 1972a.)

that would correspond with a sigmoidal growth of ant populations since the relationship is direct. A large garden of large size can only be developed and maintained by a correspondingly large ant population. Large gardens (and populations) were attained in the third year of colony life (1968), increasing in the fourth and fifth year in *Acromyrmex* and *Atta.*

The three show an ascending total growth (garden volumes, substrate totals and populations) corresponding to their evolutionary scale and size of nests in nature. The *Atta* colony used a total of 51.0 kg of substrate, the *Acromyrmex* colony 19.4 kg, and the *Sericomyrmex* 8.2 kg. Under conditions in nature the first two could have used several times as much substrate with correspondingly more gardens (or could have suffered adversities) and the last might have attained an even greater size, though unknown in nature. The ratio of garden weight to volume was 1:8 or 1:9. There are other records of the three that also indicate an evolutionary scale. The *Sericomyrmex* had broods of both sexes in each year after the second; the total deaths in nearly 11 years were 37,216 workers, 2,585 females and 16,321 males. Sister colonies of the *Acromyrmex* produced female broods, beginning late in the second colony year, and in the third year in the 10-year colony. The *Atta* colony produced no sexual brood though in nature the third year appears to be the time for several species.

For three classes of substrate, pith (albedo) of citrus, rose flowers, and green leaves, the much smaller *Sericomyrmex* workers were much slower in cutting and transporting the pieces in a given length of time, due to their much smaller mandibles and slower movements. They cut 20 g of pith at the rate of 1.3 g per hour, *Acromyrmex* cut at the rate of 2.0 to 4.4 g per hour, and *Atta* at 1.3 to 3.3 g per hour. Pith is not an important substrate for the latter and their mandibles are better for cutting leaves.

The ants reared the same type of mycelium and its gongylidia on leaves of temperate zone plants as on tropical foliage. On an annual basis, chosen for acceptance by the ants and for availability in Pennsylvania, the chief plants were, starting with January, rhododendron, forsythia (and flowers), miscellaneous new spring foliage, ash (*Fraxinus americana*) in late summer, then rose, rhdodendron and others up to December. In the early nest stages of these same colonies, rose flowers and young leaves were especially suitable. During the brief period in Massachusetts, *Acromyrmex* and *Atta,* especially, used lilac, followed in Florida by pecan, elderberry (*Sambucus canadensis),* and *Stachys floridana.* During the 10 years the ants took many other leaves and stems, including those of succulents.

The rate of growth of a new garden in the *Atta* colony was probably equal to that in nature and shows the suitability of the substrate used. For example, one of the gardens grew from 0 to 1,500 ml in 18 days in a new chamber. In the next month it then became the normal 1,800 ml size.

ADDENDA

Since the above was written, an important publication, received 19 August 1977, has added much useful information on the ant fungus collection deposited in the New York Botanical Garden. "Studies on Fungi Cultivated by Ants," by A. Hervey, C. T. Rogerson and I. Leong (1977), deals with their culturing results with the Weber Ant Fungus Collection (Table 5.1). They have used the original and permanent field note numbers for the ant colonies, since deposited at Harvard University (Museum of Comparative Zoology). Their data confirmed those in Weber 1972a, especially the utility of potato dextrose agar, lack of antibacterial activity of these fungi against *Staphylococcus aureus* and *Escherichia coli,* oatmeal (instead of whole oats) for producing fruiting bodies and the production of gongylidia. They have adopted Kreisel's name for the latter stage, *Attamyces bromatificus* Kreissel (Mycelia Sterilia of the Fungi Imperfecti). Mature *Lepiota* basidiocarps were produced from the fungus of *Myrmicocrypta auriculatum* Wheeler. They considered this fungus was not that cultured by other ant species and genera.

At this stage in investigations, considering all the evidence, the following conclusions may be made:

There are two major basidiomycete ant fungi: 1. *Lepiota* sp. These basidiocarps are definitely the fungus cultured by species of four primitive genera of ants (Weber 1957a; 1972a: Figs. 165, 166). The mycelium lacks the clear-cut typical gongylidia and their compact aggregates (staphylae) of the higher, leaf-cutting genera. 2. One or more other basidiomycete genera may be cultured by the *Trachymyrmex-Sericomyrmex-Acromyrmex-Atta* genera, including Möller's original *Rozites gongylophora* and Kreissel's *Attamyces bromatificus* (Fungi Imperfecti). Whether these are *Lepiota* or some segregate of it is a problem for mycologists to determine. It is difficult to believe that the *Attamyces* is not a vegetative form of a typical wood-rotting basidiomycete. In any event its biochemical characteristics are those described in the body of this manuscript.

Hervey et al. (1977) showed that the fungus of *Apterostigma mayri* (W15) is a basidiomycete (as also mentioned in Weber 1957a and 1972a). The Samuels data quoted in Weber (1972a) would seem to indicate on other evidence the "domestication" of fungi of different types. The apparent *Xylaria*-like fungus from primitive ants described in the earlier Weber studies, and of ascomycetous nature, emphasizes the complexity of the problem.

The requirement should be repeated, to return cultures of ant fungi to the original ant species to verify the identity of the culture. This was done repeatedly with those of the New York Botanical Garden list.

CONCLUSIONS

Success of the mutualistic relationship. The success of this example of mutualism is attested to objectively. The latest indirect evidence is the recent creation of an informal organization in Great Britain, with its own newsletter, devoted primarily to the control of the leaf-cutting species. Sums in the millions of dollars have been spent in many countries on chemical control. Various entomologists have made and are making their living in this manner. Until the recent action of the U. S. Government in forbidding the use of a chemical primarily used throughout the Americas, it was used in great quantities. Other countries are now engaged in reassessing the harm done by this chemical. The result is that the present efforts at control involve the use of pheromones and natural products produced by organisms that may not have untoward effects on nature in general.

This evidence of success is supported by the persistence of these ants in maintaining and spreading their populations, despite the efforts of man. Best examples are in those parts of South America where deforestation is proceeding on a large scale. In Brazil, for example, *Atta laevigata* F. Smith, *A. capiguara* (F. Smith) and others have moved into newly created grasslands or waste lands in unprecedented numbers. The same is true of the newly created grasslands of Venezuela where *Acromyrmex (Moellerius) landolti* has been in an unprecedented population explosion. We should recognize that man's exploitation of the environment changes direction from time to time. What is considered desirable control in one generation will be altered in another generation.

The basic fitness in nature of this example should be emphasized. Green plants are well understood to be indispensable to life on earth. But if they grow without control, they may be their own worst enemies and suffocate their own kind as well as other organisms. Saprophytes and animals are essential in the turnover of nutrients, making possible more green plants. Fungi play an essential role in converting the cellulose that would otherwise blanket the earth's surface. The fungi of the attine ants thrive because of the ant care and propagation. Biochemical evidence shows that the attine ants recycle nutrients originated by the ant fungi, and in effect speed up a normal and essential ecological cycle. The conclusion therefore is that this mutualistic association is overwhelmingly desirable in perpetuating nature. Any human interference should be rigorously controlled and localized.

ACKNOWLEDGMENT

I thank Dr. Suzanne Wellington Tubby Batra, a former student at Swarthmore College, for reviewing the manuscript.

LITERATURE CITED

Autuori, M. 1950. Contribuçao para o conhecimento da Saúva (*Atta* spp.). (V), Numero de formas aladas e reduçao dos sauveiros iniciais. Arq. Inst. Biol. Sao Paulo, 19: 325–331.

Boyd, N. D., and M. M. Martin. 1975. Faecal proteinases of the fungus-growing ant *Atta texana*: Properties, significance and possible origin. Insect Biochem. 5: 619–635.

Bonetto, A. A. 1959. Las hormigas "cortadoras" de la Provincia de Santa Fe, Argentina. pp. 1–79.

Cherrett, J. M. 1968. The foraging behavior of *Atta cephalotes* L. (Hymenoptera, Formicidae). Foraging patterns and plant species attacked in tropical rain forest. J. Anim. Ecol. 37: 387–403.

Craven, E. E., M. W. Dix, and G. E. Michaels. 1970. Attine fungus gardens contain yeast. Science 169: 184–186.

Forel, A. 1929. The social world of the ants compared with that of man. A. and C. Boni, New York. 2 vols.

Haines, B. 1973. Impact of leaf-cutting ants on vegetation development at Barro Colorado Island. Vol. II, Trop. Ecol. Springer-Verlag, New York.

Heim, R. 1957. A propos du *Rozites gongylophora* A. Möller. Rev. Mycol. 22: 293–299.

Hervey, A., C. T. Rogerson, and I. Leong. 1977. Studies on fungi cultivated by ants. Brittonia 29: 226–236.

Jonkman, J. C. M. 1976. Biology and ecology of the leaf-cutting ant *Atta vollenweideri* Forel, 1893. Z. Ang. Entomol. 81: 140-148.

———. 1977. Biology and ecology of the leaf-cutting ant *Atta vollenweideri* Forel 1893 (Hym. Formicidae) and its impact in Paraguayan pastures. Thesis, Univ. Leiden. 132 pp.

Kempf, W. W. 1972. Catalogo abreviado das formigas da região neotropical (Hym. Formicidae). Studia Entomol. Vol. 15, fasc. 1-4. 3-344 p.

Kreisel, H. 1972. Pilze aus pilzgärten von *Atta insularis* in Kuba. Allg. Mikrobiol. 12: 643-654.

Lehmann, J. 1975. Ist der nahrungopilz der pilzzuchtenden blattschneider-amerisen und termiten ein *Aspergillus*? Waldhygiene (1974) 10: 252-255.

Lewis, T., G. Pollard, and G. Dibley. 1974. Rhythmic foraging in the leaf-cutting ant *Atta cephalotes* (L.) (Formicidae-Attini). J. Anim. Ecol. 43: 129-141.

Lugo, A. E., E. G. Farnsworth, D. G. Pool, J. Jerez, and G. Kaufman. 1973. The impact of the leaf-cutter ant, *Atta colombica*, on the energy flow of a tropical wet forest. Ecol. 54: 1291-1301.

Martin, M. M. 1970. The biochemical basis of the fungus-attine ant symbiosis. Science 169: 16-20.

———. 1974. Biochemical ecology of the Attine ants. Acc. Chem. Res. 7: 1-5.

Martin M. M., G. A. Carls, R. F. N. Hutchins, J. G. MacConnell, J. S. Martin, and O. D. Steiner. 1967. Observations on *Atta colombica tonsipes* (Hymenoptera: Formicidae). Ann. Entomol. Soc. Amer. 60(6): 1329-1330.

Martin, M. M., R. M. Carman, and J. G. MacConnell. 1969. Nutrients derived from the fungus cultured by the fungus-growing ant *Atta colombica tonsipes*. Ann. Entomol. Soc. Am. 62: 11-13.

Martin, M. M., J. Geitselmann, and J. S. Martin. 1973. Rectal enzymes of attine ants. Amylase chitinase. J. Insect. Physiol. 19: 1409-1416.

Martin, M. M., and J. S. Martin. 1970. The biochemical basis for the symbiosis between the ant, *Atta colombica tonsipes*, and its food fungus. J. Insect Physiol. 16: 109-119.

Martin, M. M., and N. A. Weber. 1969. The cellulose-utilizing capability of the fungus cultured by the attine ant, *Atta colombica tonsipes*. Ann. Entomol. Soc. Amer. 62: 1386-1387.

Möller, A. 1893. Die pilzgarten eniger südamerikanischer Ameisen. Heft VI, Schimper's Botan. Mitth. aus d. Tropen, 127 pp.

Robbins, W. J. 1969. Current botanical research. Pages 71-73 in Current topics in plant science. Academic Press, New York.

Rockwood, L. L. 1975. Distribution, density, and dispersion of two species of *Atta* (Hymenoptera: Formicidae) in Guanacaste Province, Costa Rica. J. Anim. Ecol. 42: 803-817.

Weber, N. A. 1938. The biology of the fungus-growing ants. Part III. The sporophore of the fungus grown by *Atta cephalotes* and a review of reported sporophores. Rev. Entomol. (Rio de Janeiro) 8: 265-272.

———. 1941. The biology of the fungus-growing ants. Part VII. The Barro Colorado, Canal Zone species. Rev. Entomol. 12: 93-130.

———. 1943. Parasymbiosis in Neotropical "ant-gardens". Ecology 24: 400-404.

———. 1945. The biology of the fungus-growing ants. Part VIII. The Trinidad, B. W. I. species. Rev. Entomol. (Rio de Janeiro) 16: 1-88.

———. 1947. Lower Orinoco River fungus-growing ants (Hymenoptera: Formicidae, Attini). Biol. Entomol. Venezolana (Caracas) 6: 143-161.

———. 1956a. Symbiosis between fungus-growing ants and their fungus. Yb. Amer. Philos. Soc. 153-157.

———. 1956b. Treatment of substrate by fungus-growing ants. Anat. Rec. 125: 604-605.

———. 1957a. Fungus-growing ants and their fungi: *Cyphomyrmex costatus*. Ecology 38: 480-494.

———. 1957b. Dry season adaptations of fungus-growing ants and their fungi. Anat. Rec. 128: 638.

———. 1958. Evolution in fungus-growing ants. Proc. 10th Internat. Congr. Entomol. (Montreal) 2: 459-473.

———. 1959. Isothermal conditions in tropical soil. Ecology 40: 153-154.

———. 1966a. The fungus-growing ants. Science 153: 587-604.

———. 1966b. Fungus-growing ants and soil nutrition. Actas. Prim. Coloq. Latinoamer. Biol. Suelo, monogr. I. Centro Coop. Cien. Amer. Latina (UNESCO, Montevideo) 221-256.

———. 1972a. Gardening ants, the Attines. Mem. 92. Amer. Philos. Soc. (Philadelphia). 146 pp.
———. 1972b. The attines: the fungus-culturing ants. Amer. Scientist, 60: 448–456.
———. 1972c. The fungus-culturing behavior of ants. Amer. Zool. 12: 577–587.
———. 1976a. A ten-year colony of *Sericomyrmex urichi* (Hymenoptera: Formicidae). Ann. Entomol. Soc. Amer. 69: 815–819.
———. 1976b. A ten-year laboratory colony of *Atta cephalotes*. Ann. Entomol. Soc. Amer. 69: 825–829.
———. 1977a. Recurrence of *Atta* colonies at a Canal Zone site (Hymenoptera: Formicidae). Entomol. News 88 (3–4): 85–86.
———. 1977b. A ten-year colony of *Acromyrmex octospinosus* (Hymenoptera; Formicidae). Proc. Entomol Soc. Washington 79: 284–292.
Wheeler, W. M. 1907. The fungus-growing ants of North America. Bull. Amer. Mus. Nat. Hist. 23: 669–807.
Wilson, E. O. 1971. The insect societies. Belknap Press, Harvard, Cambridge, Mass. 548 pp.
———. 1975. Sociobiology, the new synthesis. Belknap Press, Harvard, Cambridge, Mass. 697 pp.

chapter 6

Termite-Fungus Mutualism

by L. R. BATRA* and S. W. T. BATRA†

ABSTRACT

Nests of *Odontotermes obesus* (Rambur) of India were studied in the field and termite behavior was observed in laboratory colonies. This provided data on division of labor, royal couple, feeding habits including trophallaxis and probable proctodeal exchange, soldier secretion and behavior, nutrition, activities regarding the mutualistic *Termitomyces* and *Xylaria* fungi and regulation of growth of these fungi in the comb, microhabitat of termitaria, and chemical composition of the comb and its role in conserving and recycling nutrients. These and other data are compared with information on Indian species of *Microtermes* and *O. gurdaspurensis* (Holm. & Holm.). *Termitomyces albuminosus* (Berk.) Heim and *Xylaria nigripes* (Klotzsch) Cooke, both cellulolytic, are consistently together associated with the termites and two species of cellulolytic bacteria in the Punjab. Simultaneous investigations of termite behavior and their two fungal symbionts revealed true mutualism between them. Chemical analysis of eight fungus gardens of various ages showed that the nitrogen content of the comb was severalfold that of the associated raw materials. *Termitomyces* and *Xylaria* are heterotrophic for a mixture of vitamins. The habitat of the mutualistic symbionts is briefly described and the development and systematics of the following significant fungi are discussed: *T. albuminosus, T. microcarpus* (Berk. & Br.) Heim, *X. nigripes* and *X. furcata* Fr.

INTRODUCTION

"Should anyone wish to improve on my observations, I shall be glad to acknowledge this and beg him to authenticate any corrections under his

*U.S. Department of Agriculture, Agricultural Research Service, Plant Protection Institute, Mycology Laboratory, BARC, Beltsville, Maryland 20705.
†U.S. Department of Agriculture, Beneficial Insect Introduction Laboratory, Insect Identification and Beneficial Insect Introduction Institute, BARC, Beltsville, Maryland 20705.

INSECT-FUNGUS SYMBIOSIS /Batra (ed.) / Allanheld, Osmun, Montclair, NJ

own name. My circumstances have hindered me greatly in observing with precision; and, if I myself should find any errors, I shall be ready to point these out with all sincerity in the future" (König 1779). The subject to be discussed in this paper is still little understood 200 years later. Due to the difficulty of working with these tropical insects, information regarding their biology remains largely anecdotal and fragmentary.

There are about 1700 species of termites assigned to some 168 genera (Coaton 1961). They are all social insects and live in communities composed of a pair of long-lived reproductive adults, one or more kinds of workers, and soldiers of both sexes. Immatures are nymphs, young sexuals are alates. Adult sexuals, the royal couple or king and queen, remove their wings after nuptial flight before beginning a new nest. Termites feed on material rich in cellulose and lignin. For its decomposition they must depend either on bacteria and protozoa in their guts, or on mutualistic fungi and bacteria primarily outside their bodies. Many species of termites are attracted to wood that is being decayed by various fungi.

Many coincidental saprophytic fungi of all classes that attack wood play a significant role in the life of various termites. The Termopsidae and Rhinotermitidae usually infest wood decayed by fungi. Some species are clearly beneficial nutritionally to termites; others are essential. A critical review of termite ethology by Sands (1969), and recent data by Becker and Lenz (1976), indicate that termite attraction to or repulsion from the affected wood depends on the insect, the fungus, the wood species and on the period of interaction between the last two. The significance of the coincidental fungal associates of termites lies in the fact that: (a) the attractants may be used as baits to control insects with suitable chemicals more effectively than is now possible; and (b) it may be possible to control termites with fungal toxins from harmful species. Our work, however, deals only with the mutualistic saprobionts peculiar to Macrotermitinae, and coincidental commensals (Batra and Batra, 1966); species-specific parasites are excluded here.

Macrotermitinae. The fungus growers are exclusively in the Old World termitid subfamily Macrotermitinae, with 12 genera. Of these, seven are from the Ethiopian region only, two (*Hypotermes, Euscaiotermes*) from the Oriental region only, and three (*Macrotermes, Odontotermes, Microtermes*) from both of these regions (Sen-Sarma 1974). They are mutualistic with fungi and they do not have the protozoa found in most other termites. Figure 6.1 shows the Paleotropical distribution of fungus-growing termites; the fungus-growing attine ants occupy a similar ecological niche in the Neotropical region. The fungus-growing adaptation in Macrotermitinae seemingly evolved in Africa and then spread to other areas where further speciation of both the insects and the fungi took place.

The habitat of these termites is variable and ranges from African savanna (Coaton 1946; 1961) to South Asian rain forest (Matsumoto 1976) and desert

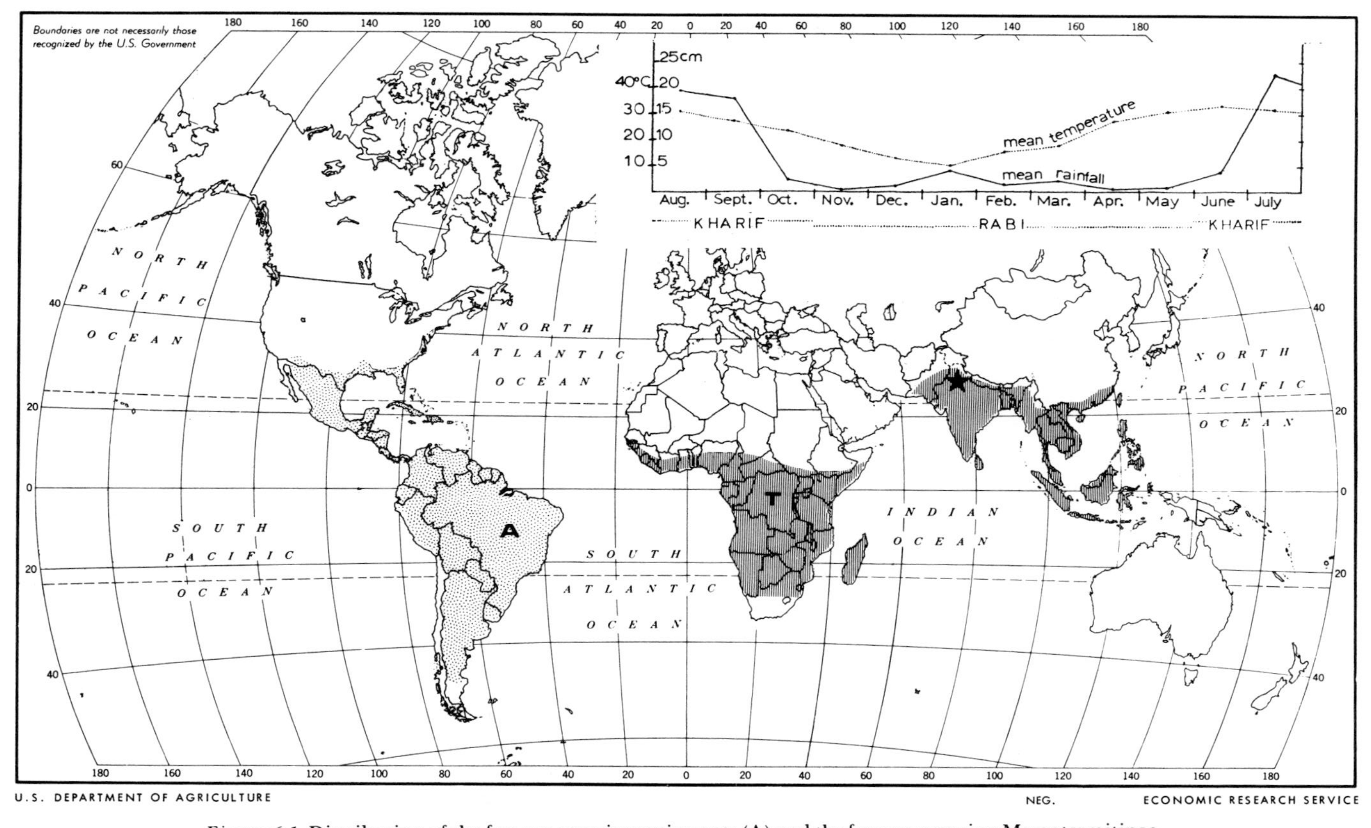

Figure 6.1. Distribution of the fungus-growing attine ants (A) and the fungus-growing Macrotermitinae (T) in the tropics of the world. Star indicates location of Punjab, India; inset shows crop seasons and data for Ludhiana, Punjab.

(Akhtar 1974). All fungus growers build nests or termitaria in the soil. When the nest system is concentrated at one site, a mound of earth is generally built over the nest. As in the case of some fungus-growing, Neotropical attine ants (Fig. 5.7), termite mounds are often a conspicuous feature of the landscape. These mounds may become nuclei of woodland in a treeless grassland, and under suitable conditions, such mound vegetation has eventually produced a closed canopy forest.

Each colony is founded by a winged couple (alates) which mate and seal themselves for a life of many years in a subterranean vault of hard clay (royal chamber). The wings are discarded immediately after the nuptial flight, before a nest is established. The physogastric queen's abdomen becomes enlarged and sausage-like; the king remains much smaller and relatively unmodified (Figs. 6.5, 6.6). Because they are unable to leave the royal chamber, food is brought to them, and eggs are removed by the numerous workers soon produced. Young termites (nymphs) stay in the fungus garden (comb), which usually surrounds the royal chamber. Irregular pillars or partitions of clay give the cavities containing the comb the appearance of a cave, complete with neatly carved stalactites and passageways leading into smaller chambers.

The nests of mound-building termites may be divided into two parts: the nest itself which is primarily underground, and the above-ground part called the mound. However, some fungus growers are entirely subterranean and lack mounds. In Africa, the individual mounds may be up to 6 meters high and 3 meters across at the base and may contain over a million termites; in South Asia the mounds are smaller.

The mound generally is a conical structure of hard clay, with or without buttresses. It usually contains air-filled shafts leading to the fungus garden, which may be a large, central mass, with or without accessory combs, or it may be chains of distinct, small combs in individual cavities. Normally there are no openings to the outside air, but inside the upper wall, there are thin-walled cupulate pits through which air may diffuse with minimal water loss. These shafts ordinarily are sealed at the surface except during the nuptial flight of alates in the rainy season, or in the inactive nests where they have been damaged by human or animal activity.

Economic importance. The fungus-growing termites have a profound and pervasive effect on agriculture, soil structure and human habitations or artifacts in the Paleotropical region. Organic debris removed from the soil surface by the workers is carried to the fungus gardens (combs), thus impoverishing the soil and concentrating nutrients beneath their mounds; these nutrients may persist there for decades. These termites are major pests of forests, orchards, and crops such as wheat and sugarcane (Sands, 1977; Usher, 1975; Varma et al. 1974). Their mounds interfere with cultivation and they are extremely destructive to wooden parts of buildings and to their organic contents (Harris, 1961; Krishna and Weesner 1969; Lee and Wood

1971). In India, they directly compete with the cattle because the *bhusa* or wheat straw fodder stored outdoors is carried away into their nests and incorporated into the combs.

Although *Termitomyces* and *Xylaria* decompose plant debris in the combs and provide food for the destructive termites, *Termitomyces* mushrooms are among the most prized of edible tropical species (Alasoadura 1966, 1967; Cheo 1948; Heim 1958; Oso 1975, 1977; Petch 1906; Zoberi 1972, 1973) and their cultivation has been attempted by us and others (Petch, 1906). Because vernacular terms or names for several mushrooms, both edible and inedible, may be the same, it is important that the reputation of each *Termitomyces* species as to its edibility must be authenticated by comparison with voucher specimens.

HABITAT, MATERIALS AND METHODS

Habitat. At Chohla Sahib village, Amritsar district, Punjab, a field laboratory was established for studying the division of labor and fungus-culturing behavior of the locally abundant species *Odontotermes obesus* (Rambur) (det. P. K. Sen-Sarma). Fields surrounding this village (pop. 6000), four miles from the Beas River were of cleared, leveled, irrigated former semi-arid scrub forest planted mainly in sugarcane (*Saccharum officinarum* L.), wheat (*Triticum* spp.), toria (*Brassica napus* L. var. *dichotoma* Prain) and gram (*Cicer arietinum* L.). The fine, light loam, loesslike soil was slightly alkaline (pH 7.2–8.5). Due to intense agricultural disturbance and wet irrigated soils in the fields, termite mounds were found primarily in elevated areas such as along the roadsides, near wells and gurdwaras, in vacant lots and, in the village, inside the brick or adobe buildings. Termite tunnels penetrated a 30 cm-thick cement floor, eroded mortar between bricks, and termite chambers formed beneath floors weakened these structures. Between Chohla Sahib and Sarhali (4.2 km), 67 mounds were counted along the roadside among trees of *Acacia nilotica* (L.) Delile, *Capparis aphylla* Roth, *Ziziphus nummularia* (Burm.) Wight & Arn., and *Dalbergia sissoo* Roxb. All mounds in this area were low (Figs. 6.3, 6.4), unlike those of *Odontotermes obesus* in the forests of South India (Fig. 6.2). This was probably due to constant trampling and erosion caused by numerous grazing animals herded along the roadsides and in other areas where mounds are common. Undisturbed mounds, where protected by buildings, were taller and of a more conical shape. In the desert of Rajasthan, this species does not form mounds (Roonwal 1975).

Materials and methods. Six nests were systematically and completely excavated, and their contents were measured and analyzed, and 19 others were sampled. The careful excavation of a termitarium requires about three man-days. Because small accessory combs usually extended beneath the level soil at each side of the mound, a 1 m deep trench was started about 0.6

Figure 6.2. Nest of *Odontotermes obesus* at Bangalore, Mysore, South India. Much of the central fungus garden below the mound has been removed.

Figures 6.3–6.4. 6.3 (top). Nest of *Odontotermes gurdaspurensis* at Ludhiana, Punjab, northwest India. The arrow indicates the top of this low, vegetation-covered mound at the edge of an irrigated field. *6.4 (bottom).* Nest 4 of *O. obesus* at Chohla Sahib, Punjab, India. This type of nest morphology is prevalent in Punjab. Accessory fungus gardens extended beneath the level ground surrounding the mound.

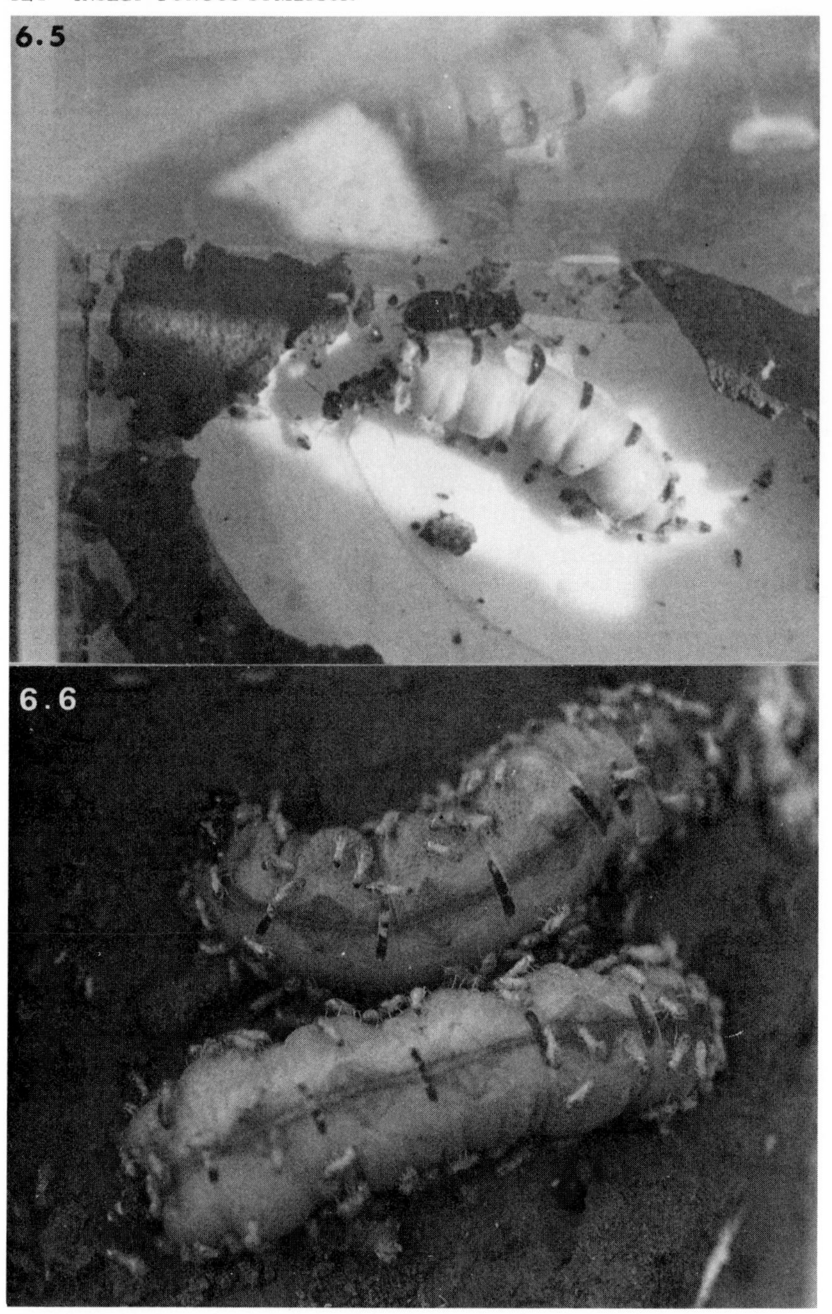

Figures 6.5–6.6. 6.5 (top). Small queen and king from *O. obesus* nest 3 at Chohla Sahib. The queen rests on a filter paper disc in the plastic observation chamber; the royal cell is being reconstructed around the couple. *6.6 (bottom).* Two queens from *O. obesus* nest 4; the single king is hidden between them. Major and minor workers are crawling over the queens and licking them as they do in nature.

m away from the edge of each mound. This trench was gradually widened so that sequential longitudinal sections of the mound and surrounding area were made. Excavations were continued until no more combs were found and the royal chamber was located. Samples of freshly collected workers, soldiers, nymphs, fungus garden and termitophiles were preserved in 70% ethanol for later study and dissection. Portions of combs, termites and surrounding soil were immediately placed in three acrylic plastic rearing chambers or in sterile petri dishes, for more detailed laboratory study (Fig. 6.9). These were incubated in darkness at 25 ± 2°C.

Microbiological investigations were carried on simultaneously with ethological observations. Unfortunately this has not been done by earlier workers, leading to considerable confusion and speculation in the literature and many misidentifications or nonidentifications of fungi or of termites; for example, Lehmann (1975); Sannasi (1969). Misuse of mycological (for example, Sannasi 1969) or entomological (Otieno 1964; 1968) terminology is also frequent.

Samples of soil or wood substrata for the isolation of fungi and bacteria were collected aseptically and they were plated within a few hours on two percent water agar and on potato dextrose or yeast extract malt extract agars with or without 6 ppm streptomycin sulphate and 10–15 units of penicillin per ml of medium (all media from Difco).* Spherules of *Termitomyces* and apical meristems or hyphae of other macroscopic fungus structures were also picked up aseptically from fungus gardens and directly plated on nutrient media with the antibiotics. All cultures were characterized at 25°C. Fungus garden and soil samples were measured in the field for moisture and pH, air dried and shipped to Beltsville for further analyses. The temperatures of nest, fungus garden and field soil were recorded by battery operated telethermometers.

Deposit of voucher specimens. For micromorphological investigations the fungi and termites were fixed in formalin acetic alcohol or air dried. Microscopic preparations were made using routine mounting media but all measurements, including estimated volume of gut contents, were made in water. Unless indicated otherwise, such material and cultures are on deposit with the National Fungus Collections at Beltsville. Voucher specimens of termites and associates are at the National Museum of Natural History, Washington, D.C. *Xylaria nigripes* (Klotzsch) Cooke (used in all nutritional studies) and *Termitomyces* spp. were, respectively, identified (by L. R. B.) using monographs by Dennis (1961) and by Heim (1958, 1977) and the identifications were further confirmed by comparing our material with voucher specimens from the herbaria at New Delhi, Kew and

*Trade names are used solely for the purpose of providing specific information. Mention of a trade name does not constitute a guarantee or a warranty of the product by the U.S. Department of Agriculture or an endorsement by the Department over other products not mentioned above.

Beltsville. Termites and other associated fauna were determined by specialists.

ETHOLOGY OF MACROTERMITINAE IN INDIA

Field observations. The morphology and contents of several Indian mounds or nests of *Odontotermes* and *Microtermes* have been summarized in Table 6.1. Generally, findings agree well with those of earlier workers such as König (1779), Petch (1906), Kemner (1934), Mukerji and Mitra (1949), Roonwal (1962) and Sen-Sarma (1974), and therefore will not be discussed in detail here. The most unusual finding was the occurrence of normal comb in a nest lacking the royal pair and having only about 100 worker and soldier termites (Table 6.1, *O. obesus* nest 1 at Chohla Sahib). Nests of this species usually contain up to 90,900 individuals (Roonwal 1962).

Laboratory observations. Maintaining cultures of Macrotermitinae for detailed behavioral study is difficult, primarily due to rampant growth of *Xylaria* on the combs when these are removed from termitaria and exposed to air. Termites of all castes, including the royal couple and comb, ordinarily live for only a few days when taken from large termitaria (Kemner 1934; Sands 1956; Ausat et al. 1960). Laboratory colonies started from paired alates are more successful (Ausat et al. 1960). There has been very little study of communication, division of labor and feeding habits of fungus-growing termites (Krishna and Weesner 1969), probably due to the difficulty of maintaining laboratory colonies and the lack of techniques to observe their life inside undisturbed, intact nests.

The royal couple (or trio), workers, soldiers, nymphs, soil and comb of *O. obesus* were removed from two mounds at Chohla Sahib and immediately placed in each of two acrylic plastic rearing chambers (Fig. 6.9). These, and petri dish cultures of soldiers, workers, nymphs and comb were used for behavioral observations, which are summarized in Table 6.2.

Queens. The three queens lived for two weeks. Queens taken from nests in previous years and kept in petri dishes with access to only a few workers and small amounts of comb shrank rapidly, deteriorated, developed a rancid odor and died within three days, probably due to lack of food to support oviposition. The royal cells, when opened for inspection, were sealed with wet earth by the workers within an hour. Queens are capable of peristaltic abdominal movements and have functional legs. They changed positions considerably within the cells, sometimes reversing direction by 180°. They continually laid eggs (about one every two seconds) which were carried away by a series of minor workers, to be deposited beneath neighboring fungus gardens. Major and minor workers constantly licked the queen's abdomen, removing the copious secretions (Figs. 6.6, 6.7) and keeping

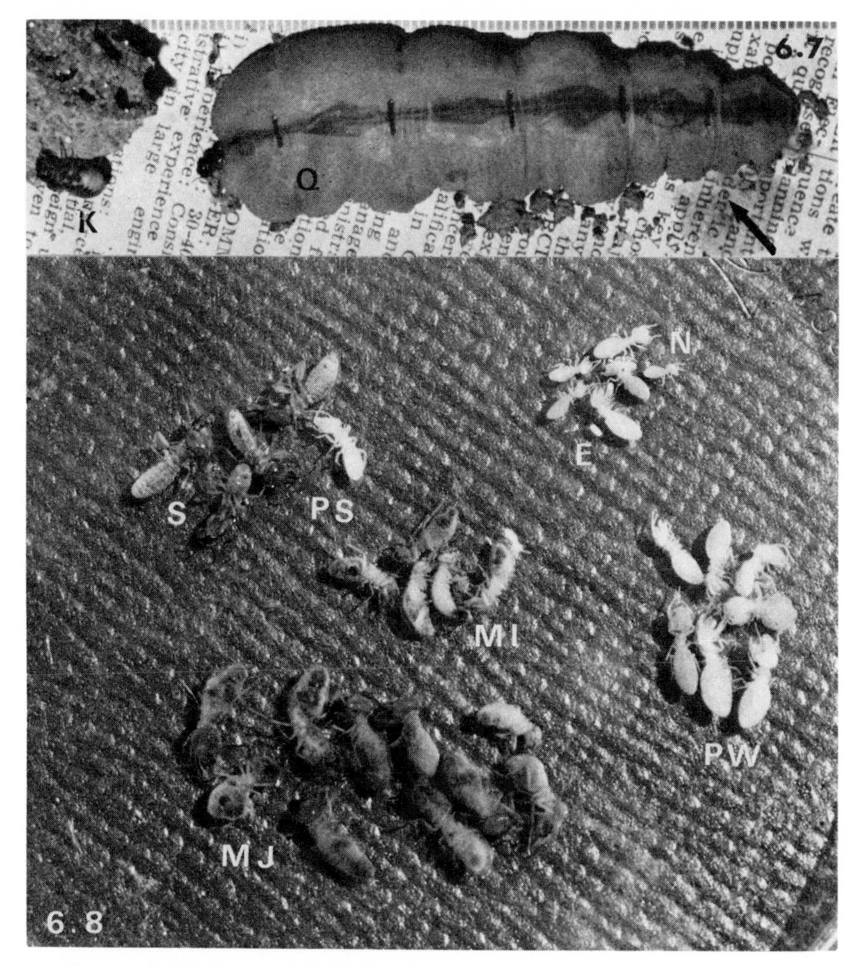

Figures 6.7–6.8. 6.7 (top). King (K) and queen (Q) of *Odontotermes obesus.* The queen's exudate (arrow) has saturated the newspaper; natural size. *6.8 (bottom).* Stages of *O. obesus*: E=egg; N=nymphs; PW=preworkers; MJ=major workers; MI=minor workers; S=soldiers, including a white presoldier (PS) in the group (not to same scale as *6.7*).

them dry. When fluorescein or methylene blue powder was placed on the queen's abdominal terga it was immediately licked off by workers. Some was swallowed (readily visible in crop and midgut, fluorescein becoming green), the rest was regurgitated and it was soon incorporated into the wet soil being used to reconstruct the royal chambers (Fig. 6.10). A little dye was also found in soil plastered around the adjacent fungus gardens but no dye was seen in the gardens. Major and minor workers and soldiers, when in the royal chambers, frequently vibrated (back and forth horizontal reciprocating movement of about 60 cycles/second lasting perhaps two seconds). When the queen from nest 3 died, one of the two queens from nest 4 was

Table 6.1 Six Completely Excavated Nests of *Odontotermes* from India.

	O. gurdaspurensis Ludhiana, Pb. May-June, 1965	*O. obesus* Bangalore, Mys. Jan. 1965	*O. obesus* Chandraprabha, U.P. Nov. 1971
Habitat	On bund at edge of irrigated cultivated field	In mixed tropical hardwood forest	In scrub forest, beneath *Zizyphus* bush
Soils	Sandy clay, poorly drained, water table 1-2m, pH 8.1	Stony lateritic red soil, pH unavailable	Red lateritic clay, pH 5.6
Age of mound	8 years	10 years	Unknown
Total length, maximum height above soil	4.0m long x 0.6m high	1.4m diameter x 1.5m high	3m long x 0.6m high
Shape of mound	Long, low, narrow (Fig. 6.3) *Calotropis,* grasses on it	Tall, conical, with vertical ridges (Fig. 6.2), bare, of hard clay	Low, oblong, with irregular pinnacles, bare
Number of chimneys or pinnacles	None	None, but has 8 vertical ridges instead	15, distinct, closed at surface
Ventilation shafts	Numerous, closed at surface, vertical, circular cross section	Numerous, vertical, anastomosing inside the vertical ridges	Wide, vertical, irregular, end beneath pinnacles, max. 20cm wide x 75cm deep
Shape or location of main fungus garden or chamber	None, large empty central cavity below soil level	Large central mass, 1.5 x 1.4m, partly divided by irregular clay partitions	Comb in accessory chambers, no central comb
Number and size of accessory fungus gardens or chambers	About 120; 80% with fungus combs, chambers 1-40cm diam. ($\bar{x}$ 4cm)	About 20; at periphery of nest, 2-10cm diam.	About 40; 2-21cm diam.
Condition of fungus garden	Empty chambers near mound surface; combs hemispherical or spherical, lamellar	Comb with irregular lamellar lobes; moisture content 47.2%	Often hemispherical, hollowed out below, moisture unknown
Total wet weight	Not weighed	28.5 kg	Not weighed
Temperature variations from ambient	0-12°C warmer than soil of same depth Dec. to May; 0-3°C cooler June	Not measured	Not measured
Location and size of royal chamber	In soil 15cm above ground level, 14cm long x 5cm wide x 2cm high	In mass of hard clay, 29cm x 15cm, 30cm below ground, inside central fungus garden, 13cm x 11cm x 3cm	In hard, moist subsoil, 1m below surface, below fungus garden, 13cm x 7cm x 2cm
Royal couples (all primary reproductives)	1 queen, 7.8 x 1.4cm 1 king, 1cm long	1 queen, 7.3 x 2.2cm, 1 king (Fig. 6.7)	1 queen, 7cm x 1.5cm, 1 king
Other animals in mound	Ants in exterior wall, *Microtermes* sp. with own fungus combs and alates inside mound; some galleries entered *O. gurdaspurensis* fungus chambers. *Gastrotheus* thysanuran *Haplothrips* sp.	Lepismatid Blattid Calliphorid	*Microtermes* sp. with alates and own fungus combs, mites (?*Odontoxenus*) on head and legs of soldiers and workers *Gryllacris* sp. nymph Blattelid adults and nymphs Tineid moth adult Isopods *Steatoda* spider *Termitodiscus* staphylinid adults *Cyphoderus* collembola *Reduvius* bugs adults *Chlaenius* and *Callistomimus* carabids Centipedes

O. obesus No. 1 Chohla Sahib, Pb. Jan. 1978	*O. obesus* No. 3 Chohla Sahib, Pb. Jan. 1978	*O. obesus* No. 4 Chohla Sahib, Pb. Jan. 1978
Edge of irrigated field near building and *Melia* sp.	In mango grove at edge of irrigated *Cicer* field	At roadside, among *Acacia* trees and *Saccharum bengalensis*
Loess, pH 7.8	Loess, pH 7.5	Loess, pH 7.5
Old, few termites (+ 100 in entire nest)	Probably young	Unknown
3.4m long x 0.6m high	0.5m long x 0.4m high	2.0m long x 1.0m high
Low, oblong with irregular pinnacles, bare	Conical, small, with 1 pinnacle, *Achyranthes aspera* on it	Low, oblong with irregular pinnacles, bare (Fig. 6.4)
8, small, some open at surface	1, closed	7, all closed
20, irregular, vertical, some beneath pinnacles, with minute pores, some with white salt deposits, max. 6cm wide x 1m deep	One, max. 14cm wide x 55cm deep, closed at surface, branching near top	About 40-50, vertical, irregular, 7-35cm diam.
None; fungus gardens in accessory chambers	One central garden 30 x 25 x 25cm (Fig. 6.13)	Large central mass, 1.2 x 1.0m
About 40 chambers, only 15 with gardens in them. Chambers 7-30cm diam.	About 15, at nest periphery, small, 5-8cm diam.	Several, at periphery, 3-8cm diam.
Laminated, hollow hemispheres to spheroidal. Spherules normal	Spheroidal, with concentric laminations, moisture 45.0% (Fig. 6.15)	Central garden divided by irregular clay partitions, moisture 43.2%
Not weighed	3.8kg	31.3kg
Not measured	18°C, or 3° warmer than soil	27°C, or 7° warmer than soil
In earth among tree roots, 45cm to surface, 9cm x 3cm x 3cm	In lower central part of main fungus garden inside thin-walled hard clay cell 5cm x 3cm x 2cm (Fig. 6.13)	In clay cell inside lower center of main fungus garden, 45cm below soil surface, 15cm x 7cm x 3cm
No royal pair, cell empty	1 queen, 3.15cm long 1 king (Fig. 6.5)	2 queens, each 6.5cm x 1.5cm, 1 king (Fig. 6.6)
Blattid nymphs and adults Isopoda Gryllids *Ctenonomia* and anthophorid bees nesting in chimneys Collembola	*Termitodiscus* Phorid (?*Termitoxenia*) *Collembola*	*Termitodiscus* Phorid (?*Termitoxenia*) Tineid moth adults Blattids Isopods Collembola Carabids

Table 6.2 Observed Division of Labor in *Odontotermes obesus*.

Major workers	Minor workers	Soldiers	Larvae or Nymphs (all instars)
Eat wood and debris outside nest, carry it to nest in gut	Eat wood and debris outside nest, carry it to nest in gut	Protect foraging workers	
Construct and repair mound and runways with wet soil	Construct and repair mound and runways with wet soil carried orally	Guard construction workers	
Carry nymphs	Carry nymphs	Guard mound interior	Carried by workers
Groom nymphs Feed nymphs orally Eat nymph excreta (proctodeal feeding)	Groom nymphs Eat nymph excreta Pick up eggs from queen, carry them to fungus garden, transfer eggs to other minor workers	Respond to drafts by raising head and opening mandibles; respond to human breath by biting and spitting	Trophallaxis with major workers Proctodeal feeding of major and minor workers Stay in interior crevices of fungus gardens near royal chambers
Lick queen caudal exudate and surface Eat dead queen Feed queen and king orally Lick king caudally Repair royal chamber	Lick eggs Lick queen caudal exudate and surface Eat dead queen Repair royal chamber	Walk on queen but do not lick her Guard royal chamber	Usually absent from royal chamber
	Lick soldiers at grooming station	When licked by minor workers vibrate in response to grooming	
	Eat soldier feces	Proctodeal feeding of minor workers; feces brown or clear	
Trophallaxis with minor workers	Licks major workers Trophallaxis with major workers		
Feed soldiers orally		Fed orally by major workers	
Vibrate in presence of nymphs			Vibrations by all instars when workers are active nearby
	Lick condensation water droplets Bury dead termites under soil Repair broken fungus garden, carrying pieces in mouth Antennate new fungus garden when provided		
Carry *Termitomyces* spherules Eat spherules Transfer spherules to other major workers *Termitomyces* cell germination enhanced in gut	Eat spherules		
Bite *Xylaria* mycelium Remove *Xylaria* from fungus garden and bury it under soil	Bite *Xylaria* mycelium Remove *Xylaria* from fungus garden and bury it under soil		

taken from the royal cell and introduced into the nest 3 royal chamber, along with about 15 workers and soldiers from nest 3 and about 20 from nest 4. After 30 minutes, workers from nest 3 (recipient) had completely sealed off the royal chamber, but after three hours, access was restored by the termites, and workers from nest 3 (some dyed) were crawling over the new queen, licking her and removing eggs in the normal way. Mukerji and Mitra (1949) also successfully exchanged queens between nests in the field.

When the queens died, major and minor workers swarmed over their bodies, licking their abdomens as with living queens, but they also chewed deep holes in their integuments and devoured oocytes and abdominal fat deposits. Antennae, legs, and palpi were amputated. *Xylaria* was growing on soil beneath one dead queen, but no queens contained internal mycoses. Evidently dead queens are quickly eaten, thus explaining the empty royal chamber found in nest 1 from Chohla Sahib (Table 6.1), which seems to have lost its queen 3–6 months before excavation (see Mukerji and Mitra 1949). Dead nymphs, soldiers, and workers, however, were usually not eaten but were heaped up and buried by workers, under wet soil, or were rarely incorporated into the fungus garden. According to Kemner (1934), in *Macrotermes gilvus* Hag., the dead queens and other castes are buried.

Division of labor. Observed division of labor in *O. obesus* is summarized in Table 6.2, which should be consulted together with Table 6.3, concerning feeding habits of the castes. No distinction was observed in the activities of various instars of nymphs, which were often picked up by their thoraces (facing forward) and carried by major or minor workers, usually to be deposited in the interior crevices of the combs among dense clusters of spherules. Nymphs sometimes vibrated when workers came nearby and they were fed orally. It has been generally assumed for many years that the nymphs of Macrotermitinae are fed saliva; however, there is no evidence for this, just as there is no proof for the widely held assumption that worker saliva is used in nest construction.

Termites bring fine, wet subsoil particles from near the water table for mound construction (Hesse 1955; Lee and Wood 1971; Watson 1975). In the laboratory they were not able to build with dry or merely moist soil and it was not noticeably moistened by oral secretions. When wet soil was provided, there was much construction, however. Workers were also seen licking water that had condensed on the plastic. It would be necessary to biochemically analyze worker saliva and then see if this substance appears in mound soil and in the guts of nymphs.

Table 6.3 indicates that small nymphs ingest at least some *Termitomyces* conidia and hyphae; large nymphs are fed pulverized wood and *Termitomyces* conidia and hyphae. Workers licked the anal regions of nymphs, king, queen and soldiers and were probably involved in a form of proctodeal feeding. This is common in other termites but is not previously reported in Macrotermitinae (Harris and Sands 1965).

Table 6.3 Contents of the Alimentary Canals of the Castes of *Odontotermes* and *Microtermes*.

Stage	*O. obesus* Punjab, Jan. 1978 Nest 3 & 4	*O. obesus* U.P., Jan. 1971	*O. obesus* Mysore, Jan. 1965	*O. gurdaspurensis* Punjab, Sept.-June, 1965	*Microtermes* Punjab, Nov. 1965
Nymphs (small)	No *Termitomyces* spores, hyphae or plant debris	---	Hyphal fragments only	Occasional *Termitomyces* conidia only	Occasional *Termitomyces* conidia
Nymphs (large)	Wood elements to 32x10x10 μm, not degraded by fungi, pulverized plant debris, *Termitomyces* conidia	Hyphal fragments, conidia, wood elements	Wood elements, pulverized plant tissue, *Termitomyces* conidia	Abundant *Termitomyces* conidia, plant fragments	Some *Termitomyces* conidia and plant fragments
Minor workers	Some with abundant conidia and hyphal elements; others with only wood elements. (? foragers) Gut content volume to 0.0013 mm^3	Wood and spherule elements, conidia	Wood elements, *Termitomyces* and other fungi, including yeasts	Abundant *Termitomyces* conidia, hyphal fragments, complete spherules	Predominantly (80–90%) plant debris, few conidia of *Termitomyces;* other fungi present
Major workers	Wood pieces to 240x240x50 μm (foragers), abundant conidia in midgut and paunch, entire spherules present; wood volume to 0.063 mm^3, fungus-wood ratio 20=80				
Soldiers	Rectum of some without particles; others contain up to 0.016 mm^3 wood fragments (to 100x80x25 μm) and *Termitomyces* conidia. In culture soldiers ate *Xylaria* rind cell fragments (to 50x20x20 μm)	Occasional bacteria	---	No fungi or plant elements; bacteria	---
Alates	(none)	Finely pulverized fungi and occasional conidia	Mostly pulverized *Termitomyces*, some entire spherules, occasional conidia and plant fragments	60–80% pulverized *Termitomyces* spherules, 15–40% plant elements to 175–225 x 11–15 μm	Finely pulverized fungal and plant material, some viable *Termitomyces* conidia
King	No fungi or plant elements; bacteria not observed	Packed with *Streptomyces* sp. only	Rod-shaped bacteria only	Rod-shaped bacteria only	---
Queen	No fungi or plant elements, bacteria not observed	Packed with *Streptomyces* sp. only	Rod-shaped bacteria only	Rod-shaped bacteria, amorphous material only	---

Figures 6.9–6.11. 6.9. Plastic, sectioned observation chamber used for behavioral observations with lid open, showing queen on filter paper (Q), comb (C), and pieces of wood (W) provided as food. Small holes in partitions and lid provide access while lid is normally closed and sealed. (×1/6). *6.10.* Major (MJ) and minor (MI) workers walking on queen (Q) as they repair a hole made in the royal cell. (×5.) *6.11A,B.* Soldiers biting and spitting labial gland secretion (arrow) onto filter paper strips. *(11A, ×6; 11B, ×2.)*

By measuring head capsules of nymphs it was possible to distinguish three white instars as follows: (1) head capsule width 0.43–0.45 mm, overall body length 1.3–1.4 mm; (2) head capsule 0.55–0.65 mm, length 1.8–2.0 mm; (3) head capsule 0.70–0.75 mm, length 2.5–2.7 mm. Later instars differentiated into white preworkers and presoldiers with sclerotizing mandibles as follows: (1) presoldiers, head capsule 0.9 mm, length 4.2 mm; (2) minor preworkers, head capsule 0.9 mm, length 3.0 mm; (3) major preworkers, head capsule 1.2–1.4 mm, length 3.5–4.8 mm. Mature castes were: (1) soldiers, head capsule 1.2–1.3 mm, length variable, to 5.5 mm; (2) minor worker, head 1.0–1.2 mm, length 3.5–4.0 mm; (3) major worker, head 1.5–1.7 mm, length 3.7–4.7 mm. These sclerotized insects, including soldiers, contained vascular plant material and *Termitomyces* (Table 6.3).

Soldiers. Soldiers of *O. obesus* were induced to bite sterilized strips of filter paper and to discharge the contents of their large abdominal labial glands for future analysis (Figs. 6.10, 6.11, 6.12). They responded readily even without biting, when gently breathed on, but currents of ambient air or forceps manipulation did not cause labial gland discharge, although soldiers were attracted toward air currents. This reaction to mammalian breath may be an adaptation to defend the nest against vertebrate predators such as the pangolin, *Manis crassicaudata* Gray and the sloth bear, *Melursus ursinus* (Shaw) which regularly eat termites in India. Sheppe (1970) found that soldiers of *Odontotermes latericus* (Haviland) are ineffective against ants and other invertebrate predators. After discharge, the soldiers' abdomens shrank by about half. The sticky transparent labial gland contents, when exposed to air, immediately became an opaque brownish pink and rapidly solidified. The secretion had the pungent, acrid odor of *p*-benzoquinone (a sample of which was taken into the field for comparison). This chemical is secreted by soldiers of *Odontotermes badius* (Haviland) in Africa (Wood et al. 1977). Maschwitz and Tho (1974) found that soldiers of the oriental *O. redemanni* (Wasmann) and *O. praevalens* (John) secreted toluquinone; benzoquinone was secreted by species of

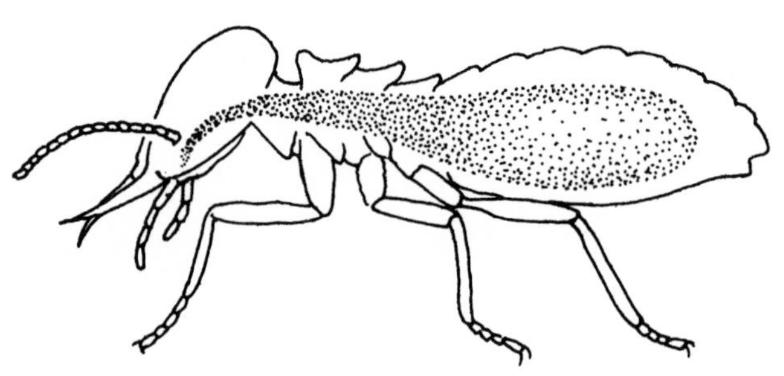

Figure 6.12. Soldier of *Odontotermes obesus,* showing extent of large, paired labial glands (stippled) that produce defensive secretion. (Diagrammatic, ×12.)

Figures 6.13–6.15. Nest of *Odontotermes obesus*. *6.13 (top left).* Entire fungus garden from nest 3 with small royal cell (arrow) in center. Note concentric layers of garden. (×¼.) *6.14 (top right).* Alert soldiers (arrow) on the surface of a fungus garden. (×2.) *6.15 (bottom).* Detail showing *Termitomyces albuminosus* spherules growing densely in the layered crevices of fungus garden. These crevices are usually occupied by many nymphs and eggs. (×2.)

Hypotermes and *Macrotermes*. Quinones are toxic to fungi (Rich 1969), and secretions of *O. gurdaspurensis* (Holm. and Holm.) soldiers have some fungistatic effect (Batra and Batra 1966). Whether soldiers' secretion influences the growth of fungi in the comb is, however, unknown. Soldiers evidently attacked or were fed *Xylaria* stromata emerging from combs in culture (Fig. 6.34), since *Xylaria* rind cells were found in their guts (Table 6.3).

Comb construction. There has been considerable controversy regarding the method used in construction of the fungus comb by termites (see reviews in Sands 1969; Sen-Sarma 1974). Some scientists speculate that it is made of masticated plant material (see Martin and Martin 1978) and many others believe that it is of fecal origin. These structures were first studied and described by König (1779) and have frequently been morphologically redescribed. Sands (1960), who made the only direct observations of comb construction, found that the combs are entirely made of aggregations of feces, and that combs are periodically reingested. Workers of *O. obesus*, when kept in petri dishes with combs but without soil, were seen to carry pieces of comb orally, and used them to repair broken portions (Fig. 6.24), or to bury dead termites and developing *Xylaria* stromata. Such structural use of fungus gardens may have contributed to the belief that masticated material is used in normal fungus garden construction.

The workers of *O. obesus* often carry, transfer and lick the spherules of *Termitomyces* (Fig. 6.16; Batra 1975). This behavior is reminiscent of egg- and nymph-carrying and grooming behavior which leads to the tentative hypothesis that *Termitomyces* spherules may physically or chemically mimic eggs or nymphs. Eggs, nymphs and spherules are normally kept in the inner crevices or chambers of the fungus comb. When nests are opened, all three are soon removed by workers from exposed areas and placed in

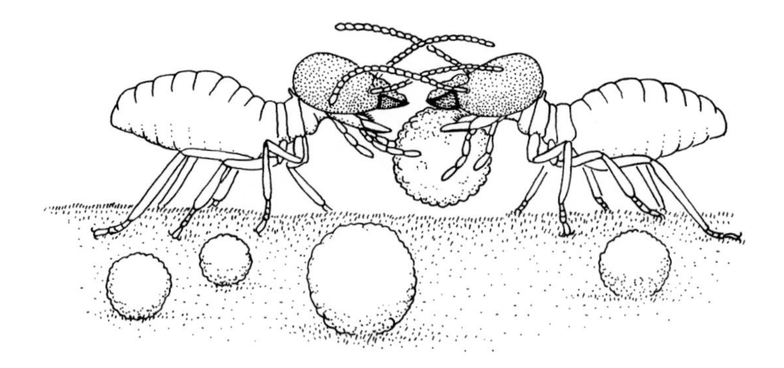

Figure 6.16. Two workers of *Odontotermes obesus* transferring a *Termitomyces* spherule between them. Eggs and larvae are similarly carried by the maxillae. (Diagrammatic, reprinted from Batra, 1975.)

protected niches. Eggs, nymphs and spherules are usually most numerous in portions of the fungus garden nearest the royal chamber in the center of the nest.

MICROBIOLOGY AND ECOLOGY

Soil moisture and minerals in termitaria. Airspaces in soils that can support vegetation are saturated with water vapor (Buckman and Brady 1969); therefore it may be expected that air surrounding fungus gardens is also saturated. In Punjab, we found that during the winter rains, although the mound and adjacent soil gained free water, there was no increase in water content of the combs, indicating saturation.

Except for Weir's (1975) work, few critical data are available for soil moisture movement within termite nests and the effect of temperature gradients that exist within their nests, as compared with the surrounding soils. He and Watson (1972) demonstrated that a large volume of water, some of it of metabolic origin, was evaporated from openings in mounds of *Macrotermes.* However, as noted by Petch (1906), many Macrotermitinae have subterranean nests or mounds that lack the so-called ventilation shafts, and nests of most lack openings to the outside air.

One reason for the high pH of mound soil in *O. obesus* and *O. gurdaspurensis* may be due to mineral deposits caused by evapotranspiration, often visible as a thin white layer inside ventilation shafts. It is also known that these species and others (Lee and Wood, 1971) bring soil and moisture from several meters' depth. Due to irrigation, salt concentrations in subsoil are high in the Punjab plains. The nature of the association of fungus-growing, mound-building termites and trees in part may be explained in terms of water and minerals brought to the surface. Mukerji and Mitra (1949) did not find any significant differences in pH between the termitarium of *O. redemanii* and surrounding soil.

All combs that we excavated were acidic in pH (4.50–6.85), but the majority of them came from nests in the Punjab districts of Amritsar and Ludhiana where most soils are alkaline; of the 1,730 Ludhiana soil samples 60% were pH 6.0 to 8.5 and 40% were above pH 8.5; of 2,295 Amritsar samples 85.6%, pH 6.0 to 8.5; 14.4% above pH 8.5 (Kanwar and Raychaudhuri 1971). Organic carbon of such soils in the upper 45 cm varies from 0.2–0.4%; and according to Grewal and Kanwar (1967) nitrogen varies from 229 to 1,008 ppm. We found that the fine clayey soil that lined the chambers containing combs was always alkaline, often with a pH 8.5 in the ventilation shafts where white $CaCO_3$ deposits were visible, as was reported by Watson (1974, 1975) in Africa. *Xylaria nigripes* and *Termitomyces albuminosus* (Berk.) Heim apparently derive few nutrients, if any, from the nest or neighboring soil, for they did not grow on basal medium supplemented with it, either as presumed source of carbon or of nitrogen.

Temperature in the nest. Movement of water under the influence of a temperature gradient is a major factor in the water regime of soils (Rode 1965). Heat is generated by the termites and fungal metabolism in the combs, which may create a temperature gradient that assists termite activity in bringing up subsoil water. Nests with large, centralized fungus gardens (nest 4, Table 6.1) may become warmer relative to their surroundings than small nests (nest 3, Table 6.1), or nests having many small combs in numerous scattered chambers (*Microtermes, O. gurdaspurensis,* Table 6.1). Heat is produced by the fungus gardens of various termites (Escherich 1911; Geyer 1951; Lüscher 1951; Sands 1969; Rohrmann 1977) and also by large masses of ants (Geyer 1951) and non-fungus growers (Krishna and Weesner 1969).

We removed the entire 31.3 kg of fungus garden, including termites, from nest 4 (Table 6.1), placed it at once in five large polyethylene bags; for insulation wrapped them in aluminum foil, two blankets and two sleeping bags and placed them on a *charpai* (cot). Temperatures were measured by a telethermometer probe kept in the center of the pile. Within 10 minutes of wrapping up the fungus garden, the temperature of the pile was 18 °C, 1 above ambient (17 °C); after one hour, the temperature was 16 °C, or 2 above ambient (14 °C); after 14 hours, it was 20 °C, or 10 above ambient (10 °C); and after 25 and 27 hours it rose to 22 °C, or 7 above ambient (15 °C). Such heating resembles the heating of compost piles by microbial respiration (Gerrits 1969; Singer 1961). It is realized that thermal conductivity of soil and of insulation used in the aforementioned *charpai* experiment may be quite different.

Structure of the fungus garden. Fungus combs, collectively called the fungus garden, are firm, moist, sponge-shaped fecal masses. They are lodged in miniature caves or chambers, solitary or confluent with the neighboring combs. They are built of finely divided plant tissues and proportionately a very small amount of insect remains and sometimes also insoluble inorganic particulate matter (see chapter 5, this volume). The entire mass is permeated with the mycelia of *Termitomyces* and *Xylaria,* and at places contains large masses of ungerminated conidia of the former. With respect to the nature of the substrate in the fungus comb of *Ancistrotermes guineensis* (Silvestri), Sands (1960) observed that "The fungus combs in the [glass] plate nests were constructed entirely from balls of faeces, partially moulded by the rectum. In the course of many hours under observation no chewed wood was ever added to the comb by the termites. On the other hand, the process of deposition of faecal pellets, resulting in the gradual building up of new combs, was observed many times." Although we did not observe comb construction, our comparative histological observations of the rectal contents of actively foraging workers of *Odontotermes obesus* and of the fungus comb support his conclusions: (1) the mean particle size of plant debris in rectum and comb is the same; (2) of the 53 such workers examined, 42 had abundant, ungerminated *Ter-*

mitomyces conidia within their guts and similar conidia were found within comb pellets, presumably of recent origin. An essentially similar situation prevails in *O. gurdaspurensis* and *Microtermes obesi* Holm. which we also examined (Table 6.3).

Chemical composition of the comb. Fungus gardens from eight active nests of *Odontotermes obesus* and two active nests of *O. gurdaspurensis* were analyzed. Although the soil of the termitarium and lining of the chambers surrounding the garden may be alkaline, the fungus garden itself is acidic (pH 4.50 to 6.85, 17 samples). König, with whom we concur, in 1779 noted the bitter taste of pieces of fungus garden. Combs of *O. redemanii* had a pH of 3.9–4.3 (Mukerji and Mitra 1949).

The percentage chemical composition of several samples of air-dried fungus garden (termites removed) and of selected raw materials being used in its construction at the time of excavation in one case are summarized in Table 6.4. Portions of the garden from vigorous nests of various ages were collected during September through January, broken into small pieces,

Table 6.4 Chemical Composition of Fungus Comb $\overline{X}$ and (R) for Eight Nests of *Odontotermes obesus* and Raw Materials ($\overline{X}$ of 2 samples).

	Fungus Comb	Wheat Straw	Paddy Straw	Mango Bark
		Percentage of air-dried samples		
Moisture	7.41 (8.46 - 6.09)	7.70	10.10	7.44
Dry Matter	92.59 (91.54 - 93.91)	92.30	89.90	92.56
		Percentage of oven-dried samples		
Total Ash	31.42 (27.00 - 42.95)	8.60	18.19	24.48
Acid Insol. Ash	16.54 (10.10 - 23.14)	---	---	9.60
Nitrogen	2.01[a] (1.80 - 2.12)	0.23	0.44	0.02
Cellulose	18.03[b] (9.68 - 25.60)	34.20	33.25	22.10
Lignin	12.00 (9.97 - 16.62)	17.00	12.90	34.90
Cellulose : lignin ratio	1.50 (0.97 - 2.25)	2.01	2.57	0.63

[a]Samples from 4 nests only.

[b]Combs with *Termitomyces* basidiocarps consistently had less cellulose and more lignin than those without them.

quickly air-dried, and shipped to Beltsville for analysis. They were oven-dried at 60°C, crushed into a fine powder and this material was used for various chemical determinations or extractions. Nitrogen determination was either by micro-Kjeldahl method or by F and M 1815 Scientific Corporation carbon-hydrogen-nitrogen analyzer, and cellulose, lignin, moisture, total ash and insoluble ash were determined by methods routinely used for forage fiber analyses (Goering and Van Soest 1975). In some samples, the lower one-third, concave portion and the upper one-third, convex area were analyzed separately. They were, respectively, presumed to be old and new construction in the fungus comb, as demonstrated by Josens (1971). This author showed that in *Odontotermes pauperans* (Silvestri) and *Ancistrotermes cavithorax* (Sjostedt), two African termites, the comb is removed from below and fresh material is added at the top.

We found that the nitrogen content of *O. obesus* comb was consistently higher and the cellulose correspondingly lower than that of presumed raw materials. The lowest cellulose was found in combs which had borne a flush of *Termitomyces* sporocarps or *Xylaria* stromata. This parallels similar cellulose depletion in commercial compost at the time of mushroom harvest (Gerrits 1969). *Odontotermes obesus* comb was drier, had a somewhat higher total ash content and lower acid insoluble ash content than that of the African *O. badius;* nitrogen content was higher and lignin was about half as great in *O. badius* comb (see Cmelik and Douglas 1970). The cellulose and lignin content did not correlate with the presumed ages of the combs, however pH of the upper portions was consistently higher than those of lower, presumably older parts. An old nest (nest 1, Table 6.1) contained combs with highest lignin and lowest cellulose and pH.

Parasymbiosis or mutualism between Termitomyces and Xylaria. In an active, normal termitarium, *Termitomyces albuminosus* and *Xylaria nigripes* (or *X. furcata* Fr.) occupy fungus combs of *Odontotermes obesus* and *O. gurdaspurensis* in a balanced way throughout the year. One without the other was not observed in well developed combs obtained from any part of intact nests in our work area. Only twice (Table 6.1, nest *O. gurdaspurensis,* early June) two small combs (? incomplete or young—with the highest cellulose content for this species) situated about 30 cm above the ground level yielded *Xylaria* only. The apparently sterile combs of young colonies of *Ancistrotermes* reported by Sands (1960) were not microbiologically analyzed; these may have contained *Xylaria.* It is possible that *Xylaria* is the colonizer of new combs, preparing the substrate for *Termitomyces.* Although *Xylaria* can be a rapidly growing fungus, its growth is suppressed on normal combs, and *Termitomyces* derives nutrients from it (see Table 6.5 and section on nutritional requirements). The fungi may be assisted by two species of cellulolytic bacteria isolated from the comb and worker excreta.

Combs, when removed from the nest, exposed to air, and incubated for

Table 6.5 Growth of *Termitomyces*, 28 Days, and *Xylaria*, 14 Days, on Selected Media.

Medium	T. albuminosus	X. nigripes	T. albuminosus	X. nigripes
	Relative growth on agar		Mycelium dry wt., mg.	
Oat meal, Sabouraud glucose, or yeast extract malt extract	+++++[a]	+++++	---	---
Potato dextrose	+++	++	---	---
Fungus garden filtrate	++	++	15.5	48
Fungus garden residue	++	++	---	---
BM[b] plus carboxymethyl cellulose[c]	++	++	10.5	27.0
BM plus vit. mixture	+++++	+++++	36.5	95.4
BM minus vit. mixture	-	-	0.0	0.0
BM minus pyridoxine	+++++	+++++	25.0	86.0
BM minus riboflavin	+++++	+++++	35.5	82.5
BM minus biotin	+++++	-	22.0	0
BM minus thiamine	±	-	0.0	0
BM filtrate, after *Xylaria* harvest	+++++	+++	38.4	12.0
BM filtrate, after *Termitomyces* harvest	+++	++	12.5	35.0

[a]+++++, maximum relative growth; -, no growth.
[b]BM, basal medium.
[c]See text for explanation, p. 147.

about 48 hours at 25 to 30°C, are within two days covered with a dense mat and stromata of *Xylaria*, which soon smothers *Termitomyces* completely and the termites die. However, the termites in the laboratory were attracted to small pieces of effused or upright *Xylaria* stromata placed on the surface of a normal comb (seven tests). One to five major and minor workers immediately began chewing the stromata; within five minutes the stromata were removed and buried under earth (Table 6.2). In the field, a small comb of *O. gurdaspurensis* overgrown by *Xylaria* mycelium and stromata in a glass beaker was replaced in the termitarium. This comb was soon removed and the beaker was plastered with wet earth.

Xylaria and *Termitomyces* can coexist in a nest, even when termites are essentially absent, provided that they are not exposed to outside air. In nest 1 (Table 6.1), normal combs with *Termitomyces* spherules were present and *Xylaria* had not formed rhizomorphs and stromata even though the royal couple were absent (estimated to have died three to six months previously) and only about 100 termites (mature workers and soldiers) were found in the entire nest. *Xylaria* stromata developed as usual when comb was taken from this nest.

It thus seems apparent that normally high soil CO_2 (up to 5.0% in soil vs. 0.03% atmospheric) or other factors related to the subterranean habitat, and not always direct mechanical or chemical stimuli by the termites are responsible for regulating the growth of *Xylaria* (Batra and Batra 1966). The termites may also indirectly control fungus growth by regulating their environment. In nest 1 and nest 3, broad, thin (1–2 mm thick) sheets of clay protruded from the chamber walls above some fungus combs, separating the combs from the roofs of their chambers and associated ventilation shafts or channels. These evidently blocked air currents.

Observations of pure cultures reveal that the two fungi can be distinguished on the basis of micromorphology alone. *Xylaria* thrives only as mycelium in the intact nests and its older hyphae are characteristically gnarled or thick-walled, often with highly refractive septa. Moreover, the wall around septa in most hyphae is conspicuously distended and reminiscent of a node (Fig. 6.36). In *Termitomyces* the hyphae, conidiophores, and conidia are rather thin-walled and the latter two are organized in a distinctive structure called a spherule (Fig. 6.20). They usually stain intensely in lactophenol-cotton blue. The septa are nonrefringent, less numerous, and thick-walled hyphae are absent.

Development and cultural characters of Termitomyces. Most of the growth of *T. albuminosus* is on the surface of the comb in the form of effused mycelium and raised spherules (Figs. 6.15, 6.25). The latter are globose to subglobose, short-stipitate or subsessile, pearl white, 0.5–1.5 (-2.5) mm wide sporodochia and are by far the most distinctive feature of a fungus garden. When fully mature they are detached readily (Fig. 6.20), the cells of anchor hyphae now having lost chromaticity.

The spherules originate as upright wefts of 8–15 intertwining hyphae which repeatedly branch sympodially and stain intensely towards their distal, enlarged ends (Figs. 6.17-6.18). When the initials are about 300–500 μm tall, the terminal 3 to 5 or more, cells begin to swell up (Fig. 6.19), become catenulate or monilioid, sprout repeatedly and measure up to 15 μm in diam. Further growth is apparently slow since tagged spherules on the fungus garden as well as on agar media did not form conidia for up to three weeks. The conidia are narrowly ellipsoid, usually catenulate, thin walled, and borne in verticils on the distal swollen cells of the spherule. Spherules are readily formed on agar media with sufficient nutrients (Table 6.5). Except for minor variation in size, the micromorphology and ontogeny of the spherule, sphaerocytes and conidia in all species of *Termitomyces* from India, Pakistan and Thailand is similar. Heim (1942 et seq.) reported similar uniformity among the species from Africa. Spherules were observed to give rise to rudiments of basidiocarps in *T. albuminosus*, *T. microcarpus* (Br. and Br.) Heim and *T.* taxonomic sp., thus confirming earlier reports (Bathellier 1927; Batra and Batra 1966; Cheo 1948; Heim 1942).

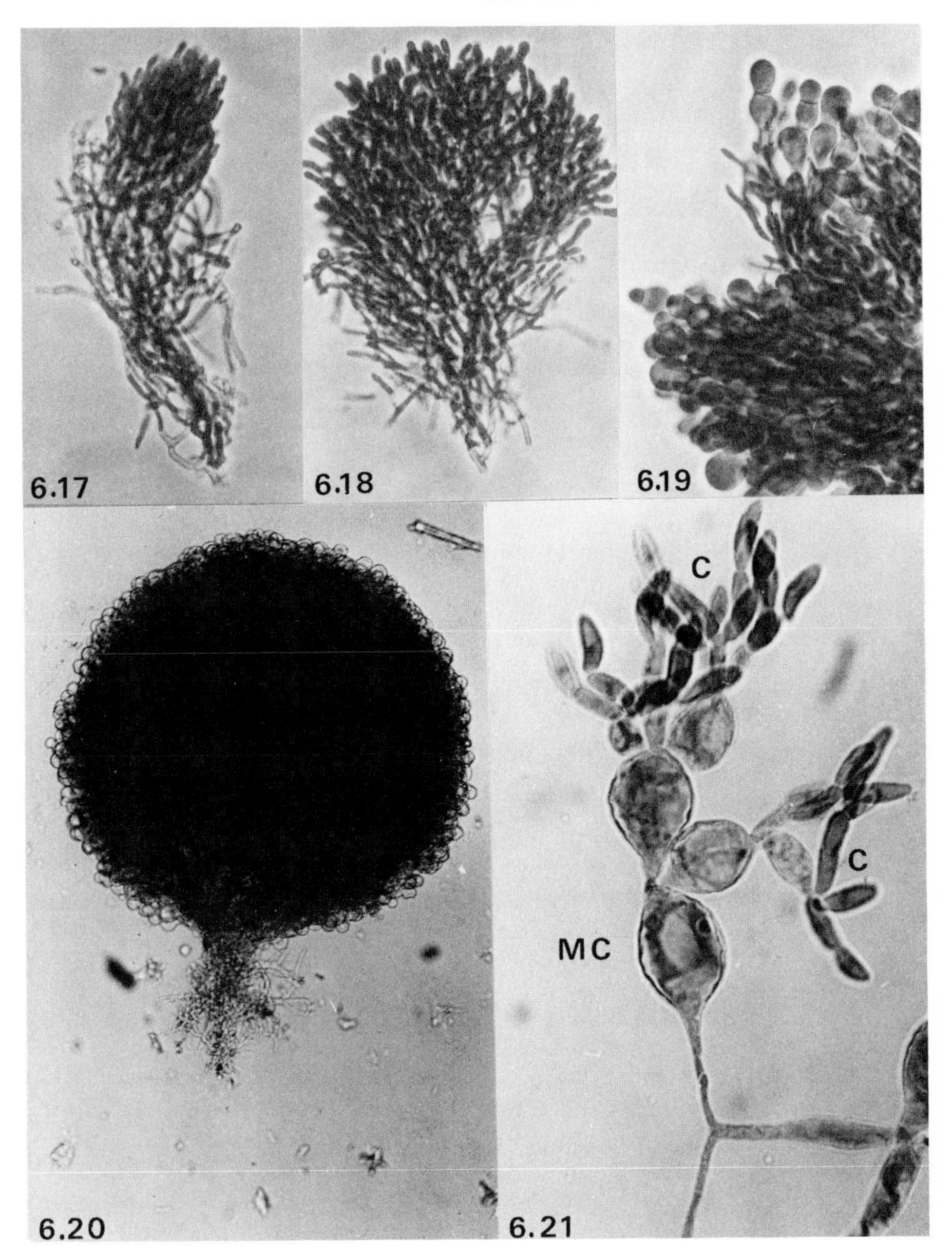

Figures 6.17–6.21. Termitomyces albuminosus—Ontogeny of a spherule or a sporodochium. *6.17.* Deeply staining upright generative hyphae. (×450.) *6.18–6.19.* Abstriction and enlargement of conidiogenous cells on conidiophores, some monilioid. (×450.) *6.20.* Conidiophores aggregated into a nearly ripe spherule or sporodochium. (×60.) *6.21.* Monilioid conidiophores, (MC) bearing branching chains of ellipsoid conidia (C). (×450.)

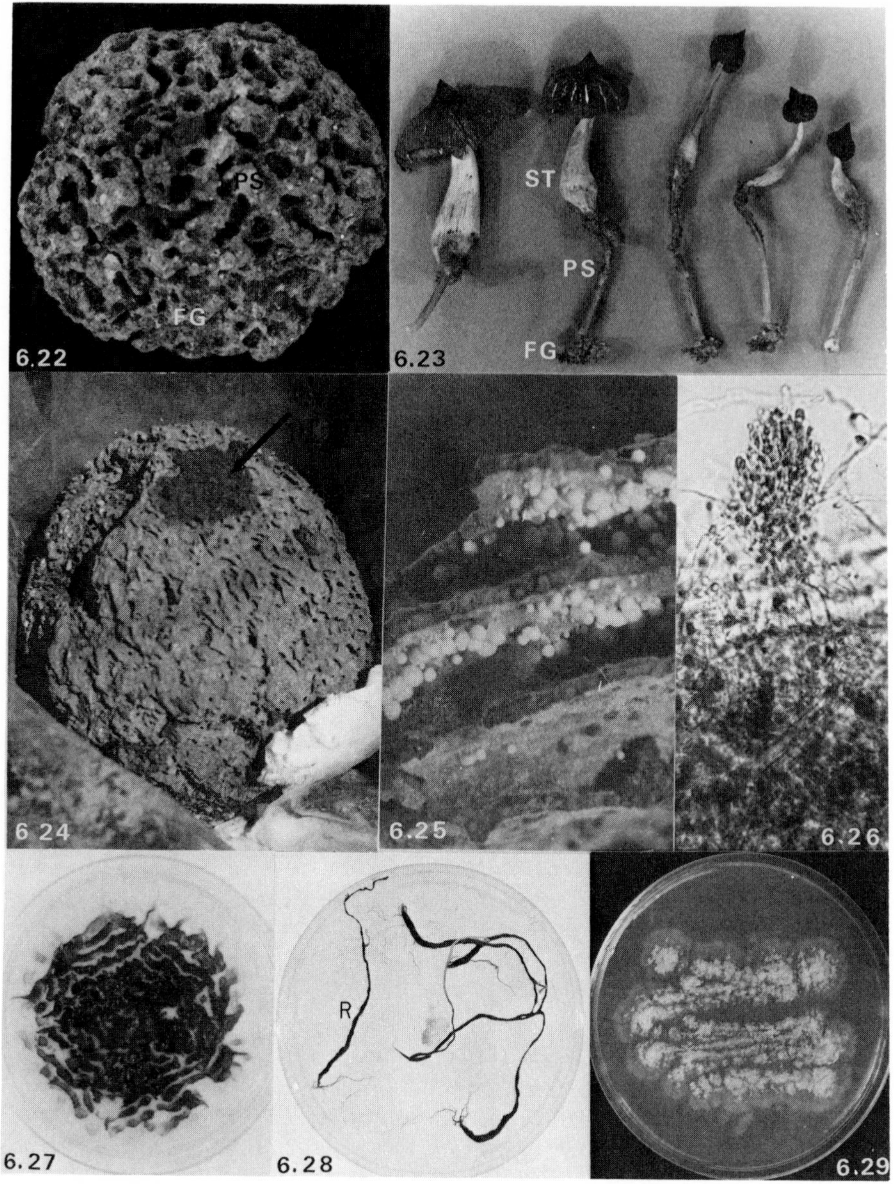

Figures 6.22–6.29. Termitomyces taxonomic sp., 1, *T. albuminosus,* and *Xylaria nigripes* on fungus comb and in culture. *6.22–6.23. T.* taxonomic sp. 1. *6.22.* Spherules and remnants of pseudorhizae on comb. (×½ of dried sample.) *6.23.* Basidiocarps (dried). (FG. = remains of fungus garden; PS = pseudorhiza; ST = stipe) (×½.) *6.24–6.27. T. albuminosus* on comb of *Odontotermes obesus. 6.24.* Fungus garden stored in a plastic bag overnight. The termites have begun to enclose the exposed upper surface with soil mixed with comb debris (arrow). (×1/10.) *6.25.* Spherules on the comb (×1.5.) *6.26.* Spherule initial on the comb. (×425.) *6.27.* Streak cultures on yeast extract malt extract agar. Note farinose sporodochial masses. (About ×½.) *6.28–6.29. Xylaria nigripes. 6.28.* Typical upright stromata on potato dextrose agar, note rhizomorphs (R). (About ×½.) *6.29.* Effused stromata mostly intramatrical cultures from similar stromata from the comb. (About ×½.)

Depending upon their age at the time of transfer from the garden onto media listed in Table 6.5, the spherules of *Termitomyces albuminosus* begin to enlarge, remain white, bear conidia, and in about seven days at 25°C, produce towards the margin a limited amount of mycelium, mostly intramatrical; or they do not grow further, become opalescent and translucent, soft and somewhat deliquescent (also normal for ripe spherules on the comb). Bacterial contaminants are uncommon on spherules thus transferred and *Xylaria* never appeared in any of the 400 cultures thus obtained from fungus gardens of diverse termites. The ellipsoid-allantoid conidia did not germinate but other blastic cells did so readily, usually with a polar germ tube, but often with two, three or more tubes from various sites (Fig. 6.21).

Termitomyces albuminosus grows slowly, as do all other species investigated by Heim (1977), and newly isolated cultures are generally slower than their subisolates. Micromorphologically the cultures consist of, from the margin inward: (1) an effuse, sparsely septate mycelium with cells 3–5 (–8) μm wide; (2) loose aggregates of repeatedly dichotomously branched erect mycelium with enlarged, ellipsoid or ovoid blastic cells, 10–15 (–20) μm wide (Fig. 6.21); and (3) initially discrete, but soon coalescing, rather compact aggregates of globose to subglobose sprout cells, 30–40 (–55) μm wide and bearing ellipsoid-allantoid, 15–26 × 6–13 μm, conidia (Fig. 6.21). These aggregate, may be several millimeters across, and are similar to ambrosial masses of symbiotic fungi of other insects (Batra and Batra 1967). They also resemble spherules on the comb in nature but here the cells, though similar in shape and size, are less compactly arranged. Subisolates of cultures maintained over a period of two to four years were predominantly mycelial but upon plating on Sabouraud or nutrient agar they produce conidia bearing aggregates of sprout cells.

Colonies at one week on potato dextrose agar from field-collected spherules measure 5 mm, with a margin of effuse intramatrical mycelium; at four weeks 2.5 cm, mycelium sometimes with concentric areas, aerial and intramatrical, white, effuse, and often furfuraceous, bearing 0.50 to 0.75 mm wide, discrete aggregates of sprout cells bearing conidia, but soon coalescing; secondary and subsequent colonies about twice as large under similar growth conditions, otherwise indistinguishable from the primary ones. The best growth is on Sabouraud glucose agar, primarily consisting of yeast-like masses of sprout cells and conidia, and on malt agar it is equally good but with abundant mycelium interspersed with sprout cell aggregates. Most isolates are white but some are pinkish or pale cream.

On cellulose agar the colonies are 1 cm in diameter at one week, surrounded by a 1–1.5 cm wide clear area; at four weeks they are 5 cm in diameter, bear spherules and conidia, and are otherwise similar to those on potato dextrose agar. As pointed out by Heim (1940, 1977), culturally *Termitomyces* species are quite similar and we confirm his findings with respect to three Indian species.

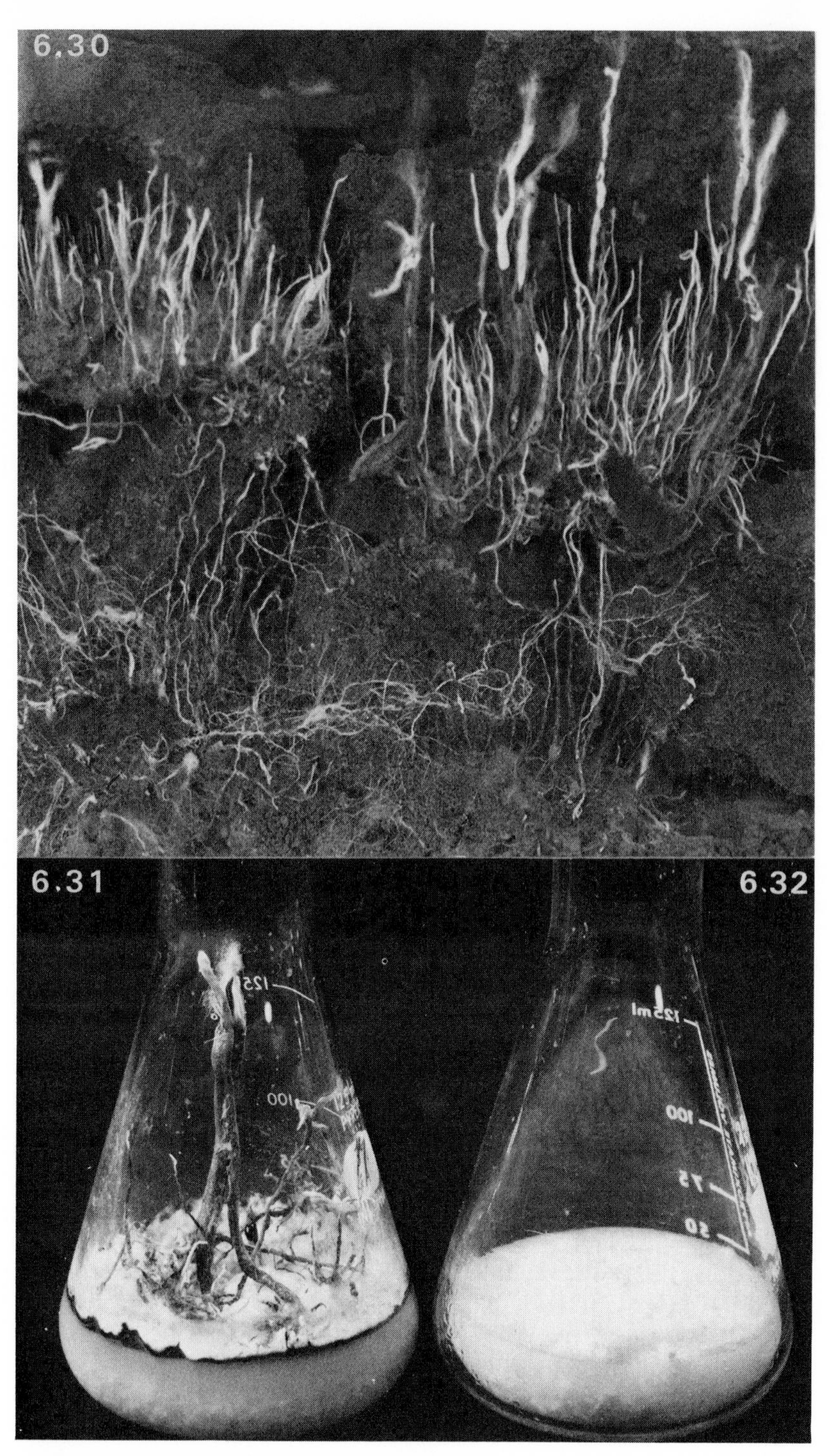

Figures 6.30–6.32. Xylaria nigripes and *Termitomyces albuminosus. 6.30–6.31. X. nigripes* upright white stromata, respectively on comb in an observation nest and on oat agar. (Both $\times\frac{2}{3}$.) *6.32. T. albuminosus* on oat agar. ($\times\frac{2}{3}$.)

Nutritional requirements of T. albuminosus and X. nigripes. Preliminary growth studies on Difco media, with and without vitamins, revealed that *T. albuminosus* and *X. nigripes* required a mixture of vitamins. Similarly, it was determined that they could use NH_4NO_3 as the sole source of nitrogen even though they grew somewhat better on media with organic nitrogen. Our basal medium had the following composition: NH_4NO_3, 1 g (except in case of testing for other nitrogen sources); KH_2PO_4, 0.90 g; $K_2H\ PO_4$, 0.70 g; $MgSO_4$ 7-H_2O, 0.75 g; glucose, 20 g (except in case of testing for other carbon sources); Difco yeast extract powder, 0.3 g; distilled water, 1 liter. The pH of the medium after autoclaving is 6.4. Except for the addition of yeast extract and glucose in place of cellulose powder or cotton fibers, it is the same medium as used by Marsh (1953) in studies on fiber decomposition by microorganisms. All experiments were performed at least twice and in each case, three replicates of each treatment were set up. Vitamin requirement experiments were conducted with the basal medium solidified with 2% agar in acid cleaned glassware, with usual precautions. It received a vitamin mixture to yield the following final concentration (μg/1): biotin, 5; pyridoxine, 100; riboflavin, 100; thiamine, 100. Experiments were performed at room temperature, 25 ± 2° C, since preliminary observations at 5° to 35° C, with 5° C interval, gave good growth at 25° C. Dry weight of mycelium (Table 6.5) is the average of three flasks. In the case of carboxymethyl cellulose or oats only aerial mycelium was weighed.

Termitomyces albuminosus is a relatively slow growing fungus on the basal medium, as compared with *Xylaria nigripes* (Table 6.5). It grows just as well on cell free filtrate of basal medium (or comb water extract), from which 14-day-old *Xylaria* mycelium has been harvested, indicating that some nutrients, growth factors, or both are available to it. *Xylaria* on the other hand did not grow so well on similar filtrates of *Termitomyces* cultures. *In vitro Xylaria* is heterotrophic for biotin and thiamine; *Termitomyces* is heterotrophic for thiamine. Both fungi grew well on basal medium with carboxylmethyl cellulose or Walseth cellulose as carbon sources (Batra and Batra 1977), with a distinct clearing zone in the latter medium. To a limited extent *Xylaria* grew on Marsh's medium (but *Termitomyces* did not) with cotton fiber as the carbon source. Moreover, *Xylaria* penetrated the fibers and the damage could be microscopically detected. Cellulolytic activity of *Xylaria* and *Termitomyces* was positive according to Rautela and Cowling's method (1966), each clearing the medium to a depth of about 8 mm in 15 days.

TAXONOMY OF TERMITOMYCES AND XYLARIA

Genus *Termitomyces* Heim (Agaricales: Agaricaceae) Arch. Mus. National Hist. Nat. Paris Ser. 6, 18: 147. 1941a.

=*Rajapa* Singer, Lloydia 8: 142. 1945.

Termitomyces was established by Heim in 1941 although several species had been known for about 100 years as *Flammula, Collybia, Lepiota,*

Armillaria, etc. Heim (1977) recently updated his prior work (1941a, 1941b, 1942, 1948, 1952, 1958) but several recent species described by Natarajan (1975) and Otieno (1964, 1968) were not treated. We follow Heim (1977) for generic concept but comments on trama are adapted from Singer (1975). Singer treats *T. microcarpus* (Berk. & Br.) Heim as the type of his *Podobrella,* a genus related to *Termitomyces.* He also raises the possibility that all Asiatic species are "identical with each other" (1975:278), a view difficult to accept in light of sociobiological work with these species and the termites.

Basidiocarps originating from fungus garden of macrotermitid termites, usually fleshy, angiocarpic, without latex; pileus umbonate, umbo often well developed and sharp-pointed, cuticle hardly separable, smooth or punctate in the center, usually dry, epicutis consisting of repent, hyaline hyphae; gills free to almost adnate but emarginate or with decurrent tooth, pale, spore print pink; hymenophoral trama bilateral in young condition, often remaining so for a relatively long time, but in adult specimens regular and consisting of parallel, thin, filamentous hyphae, inamyloid, without clamp connections (Singer 1975); stipe central, often very long, variable in thickness, with or without a pseudorhiza, fleshy, compact, and readily separable from the pileus; spores hyaline, inamyloid, or hardly amyloid, ellipsoid, smooth, with a germ pore; cystidia present.

Type species. *T. striatus* (Beeli) Heim, Arch. Mus. Nat. Hist. Ser. 6, 18: 147. 1942.

Termitomyces species, distribution and associated termites. In addition to the type species, the following are known. Authentic edibility (indicated by an asterisk) and species distribution are primarily from Heim (1958, 1977) or from Alasoadura (1966, 1967), Oso (1975), Petch (1907), and Zoberi (1972, 1973). Earlier references, recorded by Sands (1969), on associated termites are omitted.

**Termitomyces albuminosus* (Berk.) Heim (=*T. cartilagineus* and *T. eurhizus),* Termites et Champignons, p. 100, figs. 32-34, 1977; Ceylon, India, Indonesia; the record by Coaton (1961) from Africa may be a misidentification. Termites: *Odontotermes horni* (Wasmann), *O. gurdaspurensis* Holm, and Holm., *O. obesus* (Rambur), *O. obscuriceps* (Wasmann), *O. redemanni* (Wasmann), *O. sundaicus* Kemner, and *Microtermes insperatus* Kemner.

**T. aurantiacus* Heim, also described as *T. striatus* var. *aurantiacus,* "var. nov." Termites et Champignons p. 56, 1977; subtropical West Africa; The Congo. Termite: *Pseudoacanthotermes militaris* (Hagen) (Heim 1977).

T. badius Otieno, Sydowia 22: 162, pl. 1, fig. 3. 1968; Kenya.

T. biyii Otieno, Proc. E. Afr. Acad. 2: 114, pl. 3 (sic) fig. 1-5, 1964; Kenya.

**T. cartilagineus* (Berkeley) Heim, Arch. Mus. Nation. Hist. Nat. Paris Ser. 6, 18: 116, fig. 8, 1942; a later name for *T. albuminosus.*

T. citriophyllus Heim, ibid., p. 148, fig. 15; redescribed as "sp. nov." (Heim 1977: 89); Ivory Coast.

**T. clypeatus* Heim, Bull. Jard. Bot. Bruxelles 21: 207. 1951, pl. 5, C, redescribed as "sp. nov." (Heim 1977: 95); The Congo, Nigeria, Kenya.

T. congolensis (Beeli) Singer, Papers Mich. Acad. Sci. 32(1946): 134, 1948; The Congo.

**T. eurhizus* Heim, Arch. Mus. Nation. Hist. Nat. Paris Ser. 6, 18: 146, 1942; now treated by Heim (1977) synonymous with *T. albuminosus.*

**T. entolomoides* Heim, Denskschr. Schweiz. Naturf. Gesellsch. 80: 23, pl. 2, a,á and fig. 2. 1952; Congo-Brazzaville. Termite: "*Macrotermes* sp." (Heim 1952), probably misidentified (Sands 1969).

**T. fuliginosus* Heim, Arch. Mus. Nation. Hist. Nat. Paris Ser. 6, 18: 147. pl. 9, fig. 1; pl. 10, fig. 1-3; and figs. 9-14. 1942; redescribed "sp. nov." (Heim 1977: 83); The Congo.

**T. giganteus* Heim, nomen nud. (see *T. schimperi* below).

**T. globulus* Heim & Goossens, Bull. Jard. Bot. Bruxelles 21: 216, pl. 6, A; fig. 47, 1951; The Congo.

T. indicus Natarajan, Kavaka 3: 63, figs. 1A-1C, 1975; India.

T. lanatus Heim, Termites et Champignons, p. 97, figs. 30, 31. 1977; Central African Republic.

**T. le-testui* (Pat.) Heim. Arch. Mus. Nation. Hist. Nat. Paris Ser. 6, 18: 110, pl. 9, C, pl. 10, fig. 4; text fig. 2, 4; 1942; The Congo, subtropical West Africa, Nigeria. Termite: *Macrotermes natalensis* (Haviland).

T. magoyensis Otieno, Proc. E. Afr. Acad. 2: 112, pl. 2, fig. 1-6. 1964.

**T. mammiformis* Heim, Mem. Acad. Sci. Inst. France Ser. 2, 64: 53, pl. 9, fig. 6; pl. 10, E, 1941 (validated with Latin diagnosis, Arch. Mus. Nation. Hist. Nat. Paris Ser. 6, 18: 147, 1942); but described "sp. nov." (Heim 1977: 61); The Congo, Nigeria. Termite: *Pseudoacanthotermes* sp., probably *P. militaris* according to Sands (1969).

T. medius Heim & Grassee, Rev. Sci. 88: 8, fig. 14, 1950; Central Africa Republic. Termite: *Ancistrotermes latinotus* (Holmgren).

**T. microcarpus* (Berk. & Br.) Heim, Compte Rendu Acad. Sci. Paris 213: 147. 1941. Ceylon, India. "Indo-China", subtropical Africa, South Africa, Nigeria, Uganda. Termites: *Odont. badius. O. transvaalensis* (Sjostedt), and *O. vulgaris* (Haviland).

T. nairobiensis Otieno (as *Narobiensis*) Proc. E. Afr. Acad. 2: 110, pl. 1, figs. 1-15, 1964.

T. orientalis Heim nomen nudum, without latin diagnosis (Heim 1977: 125).

T. perforans Heim, Termites et Champignons, p. 138, pl. 4, figs. 3, e.f., 1977; Central African Republic.

T. rabouri Otieno, Proc. E. Afr. Acad. 2: 1, 1964; Kenya.

**T. robustus* (Beeli) Heim, Bull. Jard. Bot. Nat. Belg. 21: 210, 1951; The Congo, Kenya, Nigeria. Termite: *Acanthotermes acanthothorax* (Sjostedt) Heim (1958), "but comb photo not this genus." (Sands, 1969: 508).

**T. schimperi* (Pat.) Heim, sensu Heim, Arch. Mus. Nation. Hist. Nat. Paris Ser. 6, 18: 114, 1942, renamed as *T. giganteus,* without designating type, Termites et Champignons, p. 67, 1977; The Congo, Cameroon, Ivory Coast. Termite: *Macrotermes natalensis.*

T. spiniformis Heim, Termites et Champignons, p. 65, fig. 4, sporocarps 6, 7, 1977; Central African Republic.

**T. striatus* (Beeli) Heim, Mem. Acad. Sci. Inst. France 64: 47, 1940; The Congo, Nigeria. Termite: *Pseudoacanthotermes militaris* and probably with *Macrotermes* spp. (Heim 1977).

**T. tylerianus* Otieno (as *tyleriana*), Proc. E. Afr. Acad. 2: 116, pl. 4, figs. 1–6, 1964.

Termitomyces is a common genus during the rainy season in South and Southeast Asia. Except for two minor taxonomic papers from India (Natarajan 1975; Purkayastha and Chandra 1975) and incidental comments by Heim (1977) and Petch (1913b), it remains to be investigated in Asia. We could identify the following species from this area: *T. albuminosus, T. clypeatus, T. indicus, T. letestui, T. microcarpus, T. striatus,* and two undescribed small species that are likely to be confused with *T. microcarpus.* Pending a taxonomic revision and investigations into their biological relationships with the Asian Macrotermitinae, we cite only a few collections each of the two distinctive species described below.

Termitomyces albuminosus (Berk. & Br.) Heim.

Armillaria eurhiza Berk., *A. termitigena* Berk.; *Collybia albuminosa* (Berk.) Petch; *C. sparsibarbis* Berk. & Br.; *Lentinus cartilagineus* Berk. & Br.; *Lepiota albuminosa* Berk.; *Termitomyces cartilagineus* (Berk.) Heim; *T. eurhizus* Heim. (all on the basis of material from India or Sri Lanka).

An illustrated description of *T. albuminosus* is given by Purkayastha and Chandra (1975). Except for minor details, our account, based on specimens cited, matches well with their observations. Basidiocarps of our specimens originated from comb of nests inhabited by *Odontotermes obesus* or *O. gurdaspurensis.* It is an edible species and in Hindi it is called *bhoin phur* or *kukker mutta.*

Pseudorhizae are 5–8 mm wide, length depending on the depth at which the comb is located, solid; stipe is white, regular or twisted, 10–30 cm long, 5–10 mm wide, enlarged three to four times near the annulus in our collections, solid or fistular, fibrous and does not break clean even when dry; pileus is conical, white, glabrous, smooth, distinctly umbonate (oval and without an umbo when young), faintly striate halfway towards the margin which is inrolled when young, lacerating towards maturity; umbo is obtuse, remains smooth when dried (in contrast to wrinkled appearance of the remaining pileus); gills are thin, crowded, free, 5–6 mm wide, pink; annulus is ascending, irregular, permanent, margin lacerate; basidia are clavate, 4-spored, 25–32 × 6–9 μm, including sterigmata, which are 4 × 2 μm;

basidiospores are hyaline, smooth, thin walled, ellipsoid, 6.6–8.8 × 4–6 μm; and cheilocystidia and pleurocystidia are present.

Specimens examined (all collected August, September). India: National Botanical Garden, Lucknow, 163–668; L. R. Batra 1635 (immature, material used in experiments by Batra and Batra (1966); 3386–1 (nest 1, Chohla Sahib, Table 1), 3392; Pakistan: 2703, 2705A, both coll. Sultan Ahmad.

Termitomyces microcarpus (Berk. & Br.) Heim.

Syn: *Entoloma microcarpum* Berk. & Br., *Mycena microcarpa* (Berk. & Br.) Pat.

The most distinctive habitat characteristic of *T. microcarpus* is its origin from the fungus comb deposited outside the nest, usually contiguous with the nest. This occurs just before, or soon after, the first rains. This is a small, gregarious species, its basidiocarps measure up to 5 cm high, with pileus up to 2 cm in diam. It lacks a pseudorhiza, the stipe being an extension of a loose weft of hyphae, cemented together by dirt particles and comb remnants. Based on comb structure and the size of its pellets, a circumstantial evidence at best, *T. microcarpus* in the Punjab is probably associated with *Microtermes obesi,* a species lacking mounds. Occasionally the termite dwells in mounds of *Odontotermes* spp. (Table 6.1), with which *T. albuminosus* is associated, but the two *Termitomyces* spp. were not collected from the same mounds.

Stipe is clavate, twisted or regular, up to 4 cm long and 3 mm at the base but gradually tapered to 1 mm towards the apex, solid, breaking clean, i.e. not fibrillose, without an annulus; pileus is conical, dirty white to pale tan, glabrous, smooth, umbonate, striate, margin inrolled when young, pileus up to 1.5 cm in diam.; umbo is pointed or obtuse, sometimes quite distinct and set apart from the pileus by a circular depression, darker than pileus; gills are free, thin, crowded, 1.5 to 2.0 mm wide, white or pink; trama is compact, up to 200 μm thick, of parallel hyphae up to 30 μm wide, cells sausage-shaped, subhymenial hyphae 5–10 μm wide; basidia are clavate, 4-spored, 24–30 × 6–8 μm including sterigmata, which are up to 2 μm high and often blunt; basidiospores are hyaline, smooth, thin walled, ellipsoid, uniguttate, 7–8 × 4.4–5.5 μm, mostly 8 × 5 μ; pleurocystidia are thin walled, club shaped, sometimes greatly enlarged, up to 80 × 15 μm.

Specimens examined (all collected late August) *India:* L. R. Batra 3394, coll. Pushpa Arora; *Pakistan,* 2704, 2705, collected Sultan Ahmad.

Genus Xylaria Hill ex Greville (Euascomycetidae: Spaeriales). There are about 100 species of *Xylaria,* a cosmopolitan genus common on wood, plant debris and similar substrata. The perithecia are carbonaceous, superficial or immersed to varying degrees in upright clava. Two species, *X. nigripes* and *X. furcata,* are recorded from termitaria of the Macrotermitinae. Heim (1977) synonymizes the two but we confirm observations by Petch (1906, 1907, 1913a, 1924) that the two species are distinct. *Xylaria*

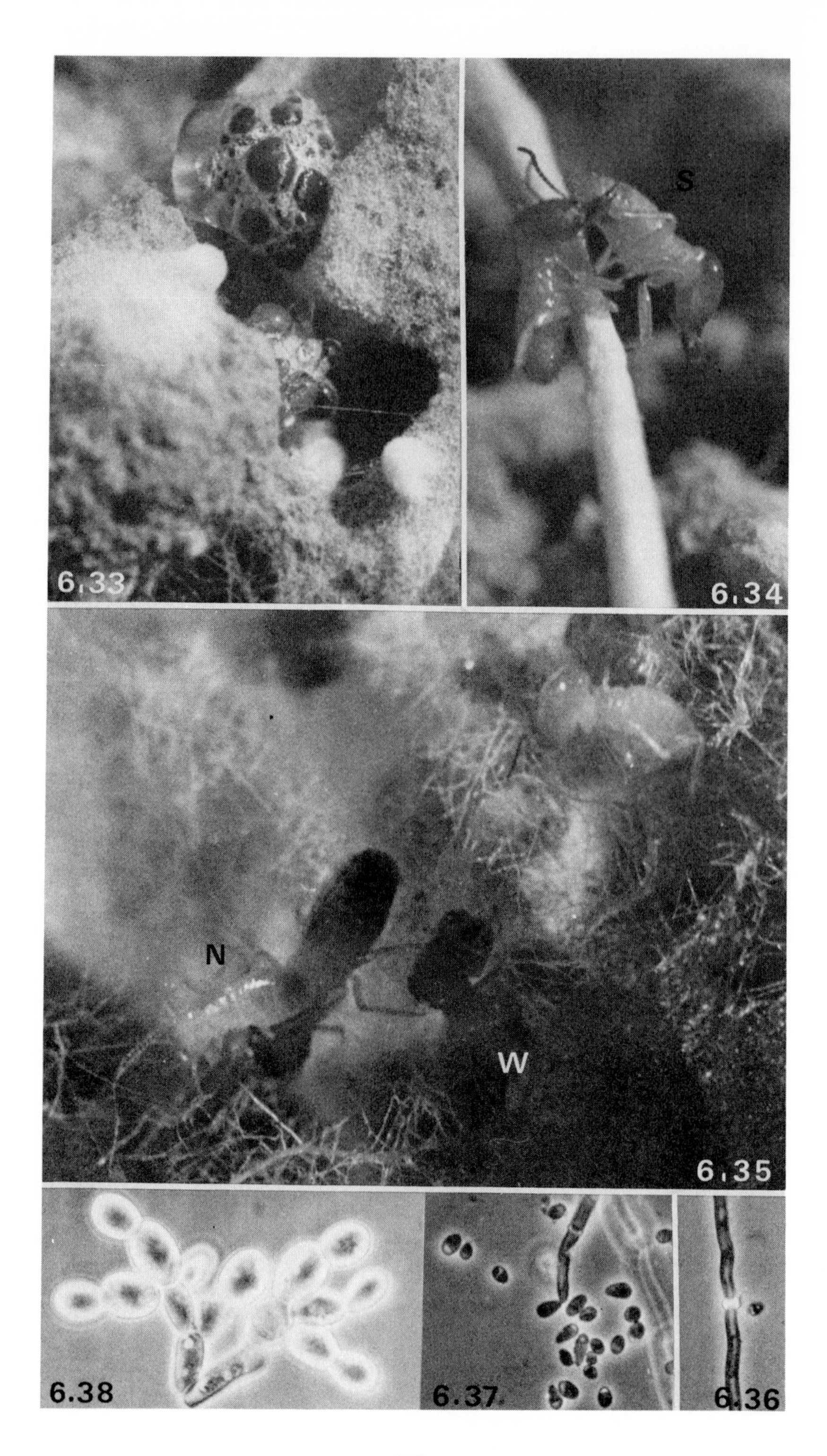
S
6.33
6.34
N
W
6.35
6.38
6.37
6.36

micrura Speg., another species with small ascospores, occurs in subterranean ant nests in South America (Dennis 1961), as occasionally does *X. brasiliensis* (Theissen) Lloyd (specimens from ants' nest at Beltsville). For the first time from the Old World, Dennis (1958) reports the latter species from the nests of *Macrotermes* in Sierra Leone. New World records of *X. termitum,* a synonym of *X. nigripes,* are apparently misidentifications; Beltsville material of most such records is *X. brasiliensis.*

We confirm observations by Petch (1906, 1907, 1913a) that *X. nigripes* and *X. furcata* are associated with combs of *Odontotermes,* however the two were not found together in the same nest as observed by him (1913a). In the Bangalore nest (Table 6.1) *X. furcata* was present; in other nests of this genus only *X. nigripes* was recovered. We also confirm most of Petch's micromorphological conclusions on the two *Xylaria* species. He states, "But the chief difference between *Xylaria nigripes* and *X. furcata* is in the conidiophore. The conidia of the former are borne singly on short parallel conidiophores (or basidia) closely arranged side by side along the clava in the typical *Xylaria* fashion. But the conidial stage of *X. furcata* is not typical; its ultimate components arranged along the clava consist of somewhat flattened spheres, each sphere being formed by a compound conidiophore, which terminates in a lobed head, on which are borne flask-shaped basidia with catenulate spores, 4–5 μ diameter" (Petch, 1913a: 332). He further states, p. 332, that, "it has occurred to me, . . . that the *furcata* conidial stromata may really be *nigripes* stromata parasitized by a hyphomycete, the conidiophores observed being those of the parasite, but I have not been able to carry out experiments to test that suggestion." We obtained cultures from single hyphal tips and single conidia and remove his doubts.

In contrast to our observations of *X. furcata,* conidiophores of this species from the combs of *Odontotermes natalensis* from Ghana are dichotomously branched at the apex and produce two types of conidia which develop in pairs, one upon the other. They are violently discharged, as a unit, and this movement apparently separates the globose, 10–12 μm upper conidia, with a basal collar, from the lower, apiculate 10–15 μm long conidium with a basal ring 6.0 μm across (Dixon, 1965). Dixon's illustrations are the type of the hyphomycetous genus *Padixonia* Subramanian (1972). What remains to be investigated is to relate the respective asexual stages with ascigerous stromata, beginning with ascospores of *X. nigripes* and *X. furcata.* Petch (1906, 1913a), beginning with combs, obtained both stages of each species.

Figures 6.33–6.38. Xylaria nigripes and termites and *Termitomyces. 6.33–6.37. X. nigripes* and termites. *6.33.* Effused stromata on comb; note exudate droplets which are eaten by termites. (×3.) *6.34–6.35.* Dead soldier(s) (S), nymphs (N) and workers (W) entangled in rapidly growing *Xylaria* mycelium on garden removed from nest. (×3.) *6.36.* Swollen refractive septum of a hypha from agar culture. (×450.) *6.37.* Conidia and a conidiogenous cell from agar culture. (×450.) *6.38.* Blastic cells of *T. albuminosus* from agar culture. (×450.)

A key to Xylaria from termitaria.

Ascigerous stromata usually simple, clavate or cylindrical, perithecia immersed and crowded, ascospores mostly 4.0 × 3.0 μm; conidia acropleurogenous, on cylindrical, unbranched conidiophores: *X. nigripes*
Ascigerous stromata branched, often repeatedly, narrow cylindric, perithecia superficial, ascospores mostly 5.0 × 2.5 μm; conidia acrogenous, on repeatedly branched, flask-shaped conidiophores: *X. furcata*

Xylaria nigripes (Klotzsch) Cooke.

Syn. *Sclerotium stipitatum* Berk. & Currey (*puttu manga,* or white ant mango); *X. escharoides* (Berk.) Fr.; *X. flagelliformis* Currey; *X. gardneri* (Berk.) Berk.; *X. piperiformis* Berk.

Xylaria nigripes is a distinctive ascomycete, highly variable but seldom confused with lignicolous species, probably because of its characteristic habitat (the termitaria in the Old World tropics), and its rhizomorphs, sclerotia and minute ascospores. Occasionally it has been reported from wood, as is *X. furcata,* another termiticolous sp., but this is dubious; voucher specimens relating to such reports lack substratum.

In contrast to superficial growth of *Termitomyces,* the growth of *Xylaria* is within the comb, as determined by paraffin sectioning, and thus is invisible to the unaided eye. The hyphae are generally thick-walled, 2–3.5 (-5) μm wide, often swollen near the rather thick and refractive septa (Fig. 6.36) and spores of any kind are usually lacking. In *Termitomyces* the hyphae are thin-walled, slightly constricted near the septa, often moniloid (Fig. 6.38), and enlarged into spherules or sporodochia (Fig. 6.21).

Development of stromata and cultural characteristics: White cottony mycelium, soon turning gray to gray brown, develops rapidly on combs removed from active nests and forms upright initials of stromata within 24 hours at 25° C. An initial consists of a white fascicle of parallel hyphae and forms a white strand, 0.5 mm in diam. Its growth is apical where hyphae are coarse and loosely arranged, but some distance posteriorly they are compact, gray to dark grayish brown and may bear conidiophores. Stromata are repeatedly dichotomously branched and may grow as much as 2.5 cm per 8 hours for the next 2–3 days (Fig. 6.30).

On combs of *Odontotermes obesus* nest 1 (Table 6.1), *Xylaria* formed loaf-shaped, white, spherical, stromatic initials that may be confused with large spherules of *Termitomyces.* However, these masses are cottony, devoid of conidia, often bear large drops of brownish exudate (Fig. 6.33) and eventually become hard stromata with a black rind and white medulla. Subcultures from such initials, or from mature stromata, are very slow growing (Fig. 6.29), predominantly intramatrical, effuse-stromatic, but eventually give rise to typical upright, branched stromata. Effused stromata are also occasionally formed on yeast extract malt extract agar.

Colonies at 48 hours on potato dextrose or on yeast extract malt extract agar are white and effuse; hyphae are dichotomously branched, 7–8 μm in diam., rather thick-walled, with conspicuous septa (Fig. 6.36); at places white, upright and stringy, or effuse, stromatic initials develop after 60 hours. Stromatic areas soon turn tan at the surface and dark brown to black underneath, growth rate of upright stromata from this point through 96 hours is 2.5 cm per 8 hours, growth slower but otherwise similar on cellulose agar; best growth, with stromata, is in flask cultures on oatmeal agar, or on crushed oats alone (Fig. 6.31), where average weight per flask (six flasks, each with 3 g oats and 25 ml H_2O) of superficial mycelium and upright stromata was 0.461 g, an unusually large quantity; stromata simple or branched, grayish to grayish brown below but white toward the acute apex, with an inner white medulla of pseudoplectenchyma and an outer rind of thicker-walled, brown hyphae, stromata anchored into agar, as on the comb, with branched, black, 250–500 μm thick rhizomorphs (Fig. 6.28).

Conidial and ascigerous stages: Conidia are sympodial, subglobose to broad ellipsoid, smooth, hyaline, acropleurogenous, produced blastogenously from apices of conidiophores, but eventually forming small heads, with a basal refractile scar corresponding to a similar one on the conidiophore denticle at the point of attachment, (4-) 7.5–11 (-14) × 4–5 (-6.6) μm; conidiophores are cylindrical or club-shaped, septate, hyaline above, light brown below, 8–11 μm wide, arranged in a palisade originating from stromatal plectenchyma or on sporodochia formed directly on agar plates with 6% (air dry wt.) fungus comb as source of nutrients, and conidiophore denticles are up to 1.5 μm high.

Ascigerous portion of stroma is narrow-cylindric, up to 8 cm × 5 mm, apex obtuse, sometimes acute, tan to gray brown, almost black when rind cells peel off, thus exposing ostioles and underlying black stromatic tissue; ascigerous stroma borne on a similarly shaped, somewhat slender sterile stalk, with or without subglobose to spindle-like bulbous base situated on fungus garden; perithecia cylindrical, densely aggregated, up to 1 mm long, completely immersed except for the ostioles; ascospores brown, ellipsoid or subcymbiform, with broad ends, 4–5 (-6.5) × 2.5–3 (-4) μm; conidia often present on ascigerous clava but they are somewhat smaller than those formed in cultures mentioned above.

We identify *Xylaria nigripes* with Rehm's Ascomyceten No. 1810 (National Fungus Collection Set) collected on termite nest soil, Bogor (Buitenzorg), Indonesia, "1907-08." Both the conidial and ascigerous stages are represented in this collection and match our material well. Other collections examined at Beltsville are: *China*: S. C. Teng 737 (one of the two boxes had a mixture of two species); H. N. Shen 319, S. C. Deng 440, 3517; *India:* Lloyd 292, 10497, 12215, 12223, the latter three collections by Bose (1923); Butler 1160, 1168, 1173 (= Herbarium Indiae Orientalis, New Delhi, 185, 184, and 183, respectively); L. R. Batra 1633, 1635 (cultures used by Batra and Batra [1966] in experiments with *Odontotermes gurdaspurensis* and referred to as "*Xylaria*-like fungus"); 2586, 2587-3 (used by Batra [1975]), 3158-2, 3386-2, 3388-2, 3388-3, 3389-2 (3386–2 used in present experiments); *Indonesia:* Lloyd 10076, 11783; *The Philippines:* Lloyd 12196–12199, 12225, Ferrer 9668; *Singapore:* Lloyd 12202, 12224; *Sri Lanka:* Lloyd 12203–12205 (all coll. and det. by Petch); *Taiwan:* K. Swada, 19 VI 1931.

Xylaria furcata Fries.

Syn. *Podosordaria furcata* (Fr.) Martin (1970), *Xylosphaera furcata* (Fr.) Dennis.

We have not investigated *X. furcata* in detail but the form and growth rate on comb and on agar media is generally similar to that of *X. nigripes*. However it is a slenderer species and it lacks rhizoids and sclerotia. Asexual stages of the two species are comparatively discussed above. The following description is based on the specimens cited below.

Stromata are tan, gray or grayish brown, usually many times dichotomously branched, at maturity often conidiferous towards the apex, distinctly stipitate when emerging from combs *in situ*, stipe up to 2 mm thick and variable in length; conidia are blastic, acrogenous, subglobose to broad ellipsoid, 6–8 μm or (4–) 6 × 8 (–10) μm, with a minute refractive projection corresponding to a minute refractive scar on the usually conical conidiogenous cell; conidiophores solitary or arranged in sporodochia, repeatedly dichotomously or sympodially branched towards the apex and thus forming a head, subhyaline or pale brown, conidiogenous cell conical or clavate, up to 20 μm long, often with several conidial loci and scars but no measurable growth after formation of successive conidia.

Perithecia are borne intermixed with sporodochia or on exclusively ascigerous stromata which are repeatedly branched or forked, tips tapering to a fine point, and surface is covered with a tomentum of flexuous, brown hyphae throughout. Perithecia are superficial, subglobose, 300–400 μm thick and of about the same height, with a well developed, papillate ostiole. Ascospores are cymbiform or ellipsoid, with gradually tapering ends, dark brown, and 5.0–6.5 × 2.5–3.0 (–3.5) μm.

Collections examined (all from combs of Macrotermitinae): *India:* Bangalore, L. R. Batra 1639, from comb of *Odontotermes obesus; Indonesia,* Rehm Ascomyceten 1812, Bogor (Buitenzorg), Coll. V. Höhnel, 1908; *Malaysia* Lloyd 10425, coll. M. Noor, 23.1. 1920 (=Reinking 5625; =Malay Peninsula Collection 15366); *Sri Lanka*: Lloyd 11858, Coll. T. Petch.

CONCLUSIONS

The nature of the termite-fungus mutualism is complex. Two major questions remain to be answered: (1) what is the nutritional contribution of each of the partners: *Termitomyces, Xylaria,* and cellulolytic bacteria and termites; and (2) how is the growth and development of *Termitomyces* and *Xylaria* regulated by the termites?

Nutritional cycle and termite behavior. The nutrient cycle in *Odontotermes obesus* appears to be as follows: major and minor workers forage, bringing coarse fragments of wood and other plant debris to the nest in their guts (see Tables 6.2 and 6.3). The gut contains various carbohydrases (Singh 1975, 1976) including cellulase and cellobiase (Misra and Ranganathan 1954) which may be produced by bacteria (Misra and Ranganathan 1954; French 1975; Zhuzhikov et al. 1971), or by previously ingested viable or germinating *Termitomyces* sprout cells (Batra and Batra 1966; Table 6.3) or by spherules in workers of *Macrotermes natalensis* (Martin and Martin 1978). This material is either defecated onto the comb, as in *Ancistrotermes guineensis* (Sands 1960), or it may be stomodeally regurgitated to soldiers, large nymphs, and perhaps to other workers; some proctodeal exchange also may occur (Tables 6.2, 6.3). Cellulose and other material inside the comb is digested by *X. nigripes* and by the superficially growing mycelium of *T. albuminosus.* The two fungi live in a balanced way inside the nest. *In vitro* the two are heterotrophic for some vitamins and one may satisfy the needs of the other for a deficient vitamin. The worker termites and alates ingest spherules (Petch 1906; Batra and Batra 1966; Batra 1975; Table 6.3) as well as pulverized, degraded comb substrate (Sands 1969; Table 6.3). Both materials evidently are fed to some soldiers (Petch 1906; Table 6.3) and to large nymphs; *Termitomyces* conidia alone are fed to some small nymphs (Table 6.3). The food of other small nymphs, soldiers and the royal couple, which contained cotton blue-staining amorphous material and bacteria, is unknown; however it may consist of deliquesced cells abraded by workers from mature spherules (Batra 1975).

The comb is usually continually eaten away from below (Josens 1971); and feces derived either from ingested comb or plant material brought from outside the nest are added above. In this way, nutrients are continually recycled in the comb, greatly increasing the nitrogen and ash content above that of foraged vegetation, while cellulose and lignin are decreased by the activity of mutualistic fungi and bacteria (Batra and Batra 1966; Table 6.4). During the dry season when foragers are less active, the comb is a food reserve (Sands 1977). The Macrotermitinae seem to be more efficient utilizers of substrate than attine ants, because material in the comb is continually recycled (Fig. 6.39), but the spent fungus gardens of the ants are dumped after fungi are produced (see Chapter 5, this volume).

Decomposition of cellulose and lignin. Hundreds of forest fungi decompose cellulose, lignin, or both. This has been repeatedly demonstrated by weight loss tests, by the detection of various cellulases and polyphenol oxidases, or by microscopic examinations (Duncan and Lombard 1965; Nilsson 1973; 1974a,b). Some species use cellulose, leaving lignin residue, and thus cause the brown rots; others may use more lignin than cellulose, or use both more or less to the same extent, and thus cause the white rots.

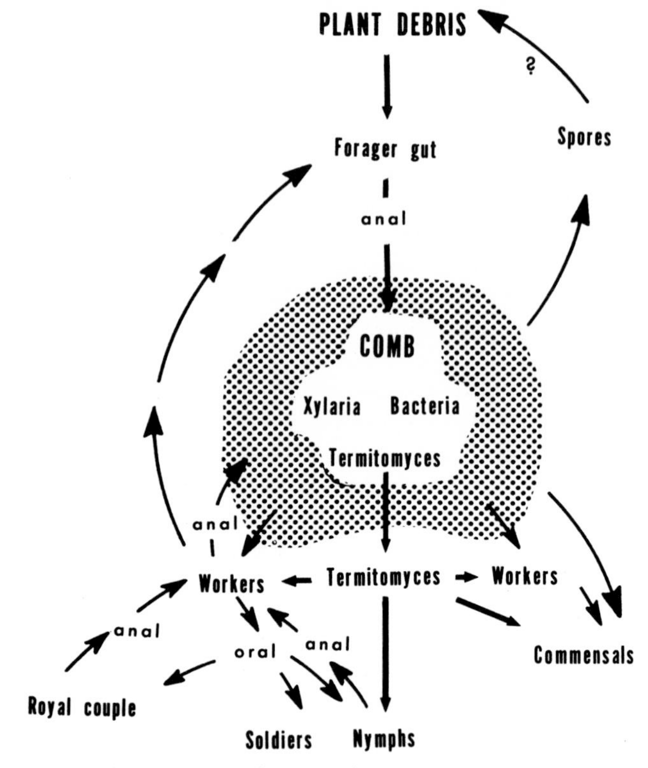

Figure 6.39. Nutrient cycle in *Odontotermes obesus.*

Fungus combs of Indian *Odontotermes* spp. are consistently brown to dark brown, indicating a proportionately greater loss of cellulose than lignin.

Macrotermitinae generally use materials similar to wood in various stages of decomposition but some also harvest other plant parts richer in nitrogen. All combs consistently have less cellulose than any of the raw materials presumably used in the comb (Table 6.4). Both *X. nigripes* and *T. albuminosus* grow on Walseth cellulose or on carboxylmethyl cellulose agars and produce a clear zone ahead of their growth (Table 6.5). Both also produce polyphenol oxidase(s), as indicated by Bavendamm test (1928) and *Termitomyces* cultures were positive for laccase using syringaldazine as an indicator, as recommended by Harkin et al. (1974). Thus, they may also use lignin. Increased cellulose decomposition during sporocarp production of *T. albuminosus* and during stromata production of *X. nigripes* parallels similar decomposition of compost during commercial mushroom growing. In both cases there is considerable drop in residual cellulose during a short period of time. (Table 6.4, and Gerrits 1969).

Termitomyces autolysis and comb carbon-nitrogen ratio. Autolysis of *Termitomyces* spherules on the comb and in cultures on media with low carbon-nitrogen ratio (C/N ratio) is a conspicuous phenomenon. A maximum of 14% of the intact, field-collected spherules transferred onto nutrient agar media gave rise to vegetative growth, the remainder became translucent and pale tan. This change in pigmentation was also noted by Heim (1977). The slightest disturbance, such as touching the spherule with an inoculation loop, collapses it and releases a copious amount of protoplasm but with a minimal amount of wall material. Whether handling of spherules by worker termites (Batra 1975) initiates, enhances, or expedites release of autolytic and other enzymes and cell contents, and whether these are imbibed, is worth investigating. Perhaps it is the autolysed protoplasm that is most often fed to such individuals as young nymphs, soldiers and the royal couple, for the stainability of their gut contents is similar to that of autolysed fungus protoplasm.

Apart from congenial growth conditions and enzyme systems, the results of microbial decomposition of plant material also depend on a suitable C/N ratio. Fungus comb has several times more nitrogen than most of the raw materials and some of it is undoubtedly in readily assimilable form derived from the autolysis of mycelia and other living matter. This is common among mutualistic fungi (Batra 1966, 1967). Agar media containing only 6% of fungus comb (dry wt.) of *O. obesus* as the sole source of nutrients could support good growth of 21 fungi, including *X. nigripes* and *T. albuminosus* (Table 6.5). Growth of the latter two species was slow, poor, and subnormal on powdered *Liriodendron* sap wood, or wheat or rice straw. But when supplemented with NH_4NO_3, or with water extract of intact comb, their growth was normal and vigorous, thus indicating recycling of comb nitrogen (and carbon) in the utilization of cellulose and lignin. Benemann (1973) reports nitrogen fixation by termite gut of non-fungus growers; similar possibilities for the Macrotermitinae may be worth investigating.

Future research. Many fundamental questions about termite-fungus mutualism remain unanswered, particularly with respect to the extent of species-specificity of the symbionts. Such information is necessary to reduce effectively economic losses due to them. Specifically, (1) proper controls, adequate number of replicates and known prior nutritional condition of the symbionts is essential in experiments to study their interaction; (2) asexual and sexual stages of *Termitomyces* spp. and *Xylaria* spp. must be related and their role in dissemination and establishment of the fungi ascertained; (3) some other questions that require answers, or confirmation thereof, are: Do the termites and the fungi get together each generation accidentally, as suggested by Sands (1969)? Or are there morphological structures or adaptations similar to mycangia or fungal pouches in termites to facilitate their mutualism from generation to generation? (See Chapters 2 to 5, this volume.) Do these fungi exist apart from termites? What materials, if any, or environmental factors, biotic or abiotic, in the nest regulate cohabitation of *Xylaria* and *Termitomyces*, while at the same time inhibiting growth of alien fungi? Fungus and bacterial cellulases occur commonly but to what extent do they act on cellulose in comb or within the termite gut? Martin and Martin (1978) found cellulases in *Macrotermes natalensis* gut and based on this single example, broadly theorize that, "Because of the ability of fungi to produce stable digestive enzymes active against a wide spectrum of natural substrates, such as cellulose, chitin, lignin, we predict that acquired digestive enzymes will be found to play a particularly important role in the biology of fungus-feeding invertebrates and in food chains based upon litter, detritus, and dead wood." If proven correct, such a generalization would require emendation of concepts about the nutrition of many xylophagous and xylomycetophagous arthropods such as the rest of the Macrotermitinae, ambrosia beetles, siricid woodwasps, and others. Brown rots and white rots of wood, caused by external digestion by fungi, are common major substrata for thousands of such organisms (Buchner 1965; see Chapters 2 to 5, this volume).

ACKNOWLEDGMENTS

We thank the head granthi, Sri Kartarpur gurudwara, Chohla Sahib, India for providing facilities. We also thank Swaran Singh alias Panna, Swaran Singh alias Gangarama, Ashok Kumar Singh, Ved, Mira, and Persa Batra who cooperated in the digging of nests and for logistics. Marion Simpson provided technical leadership and facilities to perform chemical analyses and T. L. Erwin, R. C. Froeschner, A. B. Gurney, J. M. Kingsolver, K. Krishna, W. B. Peck, P. K. Sen-Sarma, D. L. Wray, and P. Wygodzinsky identified Arthropoda.

LITERATURE CITED

Akhtar, M. S. 1974. Zoogeography of the termites of Pakistan. Pakistan J. Zool. 6: 85–104.

Alasoadura, S. O. 1966. Studies in the higher fungi of Nigeria. II. Macrofungi associated with termite nests. Nova Hedwigia 11: 387–393.

———. 1967. Studies in the higher fungi of Nigeria. I. The genus *Termitomyces* Heim. J. W. African Sci. Assoc. 12: 139–146.

Ausat, A., P. S. Cheema, T. Koshi, S. L. Perti, and S. K. Ranganathan. 1960. Laboratory culturing of termites. Pages 121–125 *in* UNESCO Proc. New Delhi Symposium, Termites in the humid tropics. UNESCO, Paris. 259 pp.

Bathellier, J. 1927. Contribution à l'etude systématique et biologique des termites de l'Indo Chine. Faune Colonies France 1: 125–365.

Batra, L. R. 1966. Ambrosia fungi: extent of specificity to ambrosia beetles. Science 153: 193–195.

———. 1967. Ambrosia fungi: a taxonomic revision and nutritional studies of some species. Mycologia 59: 976–1017.

Batra, L. R., and S. W. T. Batra. 1966. Fungus-growing termites of tropical India and associated fungi. J. Kansas Entomol. Soc. 39: 725–738.

———. 1967. The fungus gardens of insects. Sci. Amer. 217: 112–120.

———. 1977. Termite-fungus mutualism. Page 43 *in* H. E. Bigelow and E. G. Simmons, eds. Second Int. Mycol. Cong. Abstr. Univ. of South Florida, Tampa.

Batra, S. W. T. 1975. Termites (*Isoptera*) eat and manipulate symbiotic fungi. J. Kansas Entomol. Soc. 48: 89–92.

Bavendamm, W. 1928. Uber das Vorkommen und den Nachweis von Oxydasen bei holzzerstörenden Pilzen. Z. Pflanzenkrank. Pflanzenschutz 38: 257–276.

Becker, G., and K. Seifert. 1962. Ueber die Chemische Zusammensetzung des nest- und Galerie-materials von Termiten. Insectes Sociaux 9: 273–289.

Becker, G., and M. Lenz. 1976. Einfluss von Moderfaulepilzen in Holz auf Frasstatigkeit, Galeriebau und Entwicklung einiger Termiten-Arten. Angewandte Entomol. 80: 232–261.

Benemann, J. R. 1973. Nitrogen fixation in termites. Science 181: 164–165.

Bose, S. R. 1923. The fungi cultivated by the termites of Barkuda. Rec. Indian Mus. 25: 253–258.

Buchner, P. 1965. Endosymbiosis of animals with plant microorganisms. John Wiley & Sons, Inc., New York. 909 pp.

Buckman, H. O., and N. C. Brady. 1969. The nature and properties of soils. Macmillan Co., New York. 653 pp.

Cheo, C. C. 1948. Notes of fungus-growing termites in Yunnan, China. Lloydia 2: 139–147.

Cmelik, S. H. W., and C. C. Douglas. 1970. Chemical composition of "fungus gardens" from two species of termites. Compar. Biochem. and Physiol. 36: 493–502.

Coaton, W. G. H. 1946. The Pienaars river complex of wood-eating termites. J. Entomol. Soc. S. Africa 9: 130–176.

———. 1961. Association of termites and fungi. African Wild Life 15: 39–54.

Dennis, R. W. G. 1958. Some *Xylosphaeras* of Tropical Africa. Rev. Biol. 1: 175–208.

———. 1961. *Xylarioides* and *Thamnomycetoideae* of Congo. Bull. Jard. Bot. Bruxelles 31: 109–154.

Dixon, P. A. 1965. The development and liberation of the conidia of *Xylosphaera furcata*. Tr. Brit. Mycol. Soc. 48: 211–217.

Duncan, C. G., and F. F. Lombard. 1965. Fungi associated with principal decays in wood products in the United States. U.S. Forest Serv. Res. Paper WO-4. Washington, D.C. 31 pp.

Escherich, K. 1911. Termitenleben auf Ceylon. Gustav Fischer, Jena. 262 pp.

French, R. J. 1975. The role of termite hindgut bacteria in wood decomposition. Material und Organismen 10: 1–13.

Gerrits, J. P. G. 1969. Organic compost constituents and water utilized by the cultivated mushroom during spawn run and cropping. Mushroom Sci. 7: 111–126.

Geyer, J. W. 1951. A comparison between the temperatures in a termite supplementary fungus garden and in the soil at equal depths. J. Entomol. Soc. S. Afr. 14: 36–43.

Goering, H. K., and P. J. Van Soest. 1975. Forage fiber analyses (apparatus, reagents, procedures, and some applications). U.S. Dep. Agr. Handbook 379. 22 pp.

Grewal, J. S., and J. S. Kanwar. 1967. Forms of nitrogen in Panjab soils. J. Res. Punjab Agr. Univ. 4: 477–480.

Harkin, J. M., M. J. Larsen, and J. R. Obst. 1974. Use of syringaldazine for detection of laccase in sporophores of wood rotting fungi. Mycologia 66: 469–476.

Harris, W. V. 1961. Termites, their recognition and control. Longmans Green & Co., London. 186 pp.

Harris, W. V., and W. A. Sands. 1965. The social organization of termite colonies. Symp. Zool. Soc. London. 14: 113–131.

Heim, R. 1940. Culture artificielle des mycotetes d'un agaric termitophile d'Africain. Acad. Sci. 11 Mars 1940: 410–412.

———. 1941a. Nouvelles études descriptives sûr les agarics termitophiles d'Afrique tropicale. Mus. national d'histoire nat. (Paris) Arch. VI. 18: 107–166.

———. 1941b. Les *Termitomyces* dans leurs rapports avec les termites pretendus champignonnistes. Mém. Acad. Sci. Inst. France 64: 146–148.

———. 1942. Les Champignons des termitières. Rev. Sci. 80: 69–86.

———. 1948. Nouvelles reussites culturales sûr les *Termitomyces*. Acad. Sci. Seance 10 Mai 1948: 1488–1491.

———. 1951. Les *Termitomyces* du Congo Belge recueillis par Madame Goosens-Fontana. Bull. Jard. Bot. Etat Bruxelles 21: 207–222.

———. 1952. Les *Termitomyces* du Cameroun et du Congo francais. Mém. Soc. Helvét. Sci. Nat. 130: 1–29.

———. 1958. *Termitomyces*. Pages 129–151 *in* W. Robyns. Flore iconographique des champignons du Congo. Ministry of Agriculture, Brussels.

———. 1977. Termites et champignons. Soc. Nouvelle des Edit. Boubee, Paris. 207 pp.
Hesse, P. R. 1955. A chemical and physical study of the soils of termite mounds in East Africa. J. Ecol. 43: 449–460.
Josens, M. G. 1971. Le renouvellement des meules à champignons construites par quatre *Macrotermitinae (Isoptera)* des savanes de Lamto-Pacobo (Cote-d'Ivoire). Compt. Rend. Acad. Sci. 272: 3329–3332.
Kanwar, J. S., and S. P. Raychaudhuri. 1971. Review of soil research in India. Indian Council of Agr. Res. New Delhi. 229 pp.
Kemner, N. A. 1934. Systematische und biologische Studien über die termiten Javas und Celebes', Kungl. Svenski Vetenskapsakademiens Handlingar III. 13(4): 1–241.
König, J. G. 1779. Natural history of so-called white ants. Translation by T. B. Fletcher. 1921. *In* Proc. Fourth Entomol. Meeting, Pusa, India, pp. 312–333.
Krishna, K., and F. M. Weesner, eds. 1969. Biology of termites. Academic Press, N.Y. Vol. 1, 598 pp.; Vol. 2, 643 pp.
Lee, K. E., and T. G. Wood. 1971. Termites and soils. Academic Press, New York, 251 pp.
Lehmann, J. 1975. Ansatz zu einer allgemeinen Losung des "Ambrosiapilz"—Problem. Waldhygiene 11: 41–47.
Lüscher, M. 1951. Significance of "fungus-gardens" in termite nests. Nature 167: 34–35.
Marsh, P. B. 1953. A test for detecting the effects of microorganisms and of a microbial enzyme on cotton fiber. Plant Dis. Reptr. 37: 71–76.
Martin, M. M., and J. S. Martin. 1978. Cellulose digestion in the midgut of the fungus-growing termite *Macrotermes natalensis.* The role of acquired digestive enzymes. Science 199: 1453–1455.
Maschwitz, U., and Y. P. Tho. 1974. Chinone als Wehrsubstanzen bei einigen orientalische Macrotermitinen. Insectes Sociaux 21: 231–233.
Matsumoto, T. 1976. Role of termites in an equatorial rain forest ecosystem of west Malaysia. Oecologia (Berlin) 22: 153–178.
Misra, J. N., and V. Ranganathan. 1954. Digestion of cellulose by the mound building termite, *Termes (Cyclotermes) obesus* (Rambur). Proc. Indian Acad. Sci. 31B: 100–113.
Mukerji, D., and P. K. Mitra. 1949. Ecology of the mound-building termite, *Odontotermes redemanni* (Wasmann) in relation to measure of control. Proc. Zool. Soc. Bengal. 2: 9–25.
Natarajan, K. 1975. South Indian Agaricales I—*Termitomyces.* Kavaka 3: 63–66.
Nilsson, T. 1973. Studies on wood degradation and cellulolytic activity of microfungi. Studia Forestalia Suecica. 104: 1–40.
———. 1974a. Microscopic studies on the degradation of cellophane and various cellulosic fibres by wood-attacking microfungi. Studia Forestalia Suecica. 117: 1–32.
———. 1974b. Comparative study on the cellulolytic activity of white-rot and brown-rot fungi. Material und Organismen. 9: 173–198.
Oso, B. A. 1975. Mushrooms and the Yoruba people of Nigeria. Mycologia. 67: 311–319.
———. 1977. Mushrooms, myths, and traditional medicine in Nigeria. Page 499 *in* H. E. Bigelow and E. G. Simmons. Abstracts. Second Intern. Mycol. Cong. Tampa, Fla.
Otieno, N. C. 1964. Contributions to a knowledge of termite fungi in East Africa; the genus *Termitomyces* Heim. Proc. E. Afr. Acad. 2: 108–120.
———. 1968. Further contributions to a knowledge of termite fungi in East Africa: the genus *Termitomyces* Heim. Sydowia Ann. Mycol. 22: 160–165.
Petch, T. 1906. The fungi of certain termite nests. Ann. Roy. Bot. Gard. Peradeniya. 3: 185–270.
———. 1907. *Sclerotium stipitatum* Berk. & Curr. Ann. Mycol. 5: 401–403.
———. 1913a. Termite fungi: a résumé. Ann. Roy. Bot. Gard. Peradeniya. 5: 303–341.
———. 1913b. White ants and fungi. Ann. Roy. Bot. Gard. Peradeniya. 5: 389–393.
———. 1924. *Xylariaceae Zeylanicae.* Ann. Roy. Bot. Gard. Peradeniya. 8: 119–166.
Purkayastha, R. P., and A. Chandra. 1975. *Termitomyces eurhizus,* new Indian edible mushroom. Trans. Br. Mycol. Soc. 64: 168–170.
Rautela, G. S., and E. B. Cowling. 1966. Simple cultural test for relative cellulolytic activity of fungi. Appl. Microbiol. 14: 892–898.
Rich, S. 1969. Quinones. Pages 647–648 *in* D. C. Torgeson, ed. Fungicides, Vol. 2. Academic Press, New York.
Roonwal, M. L. 1962. Biology and ecology of oriental termites. 5. Mound-structure, nest and moisture-content of fungus combs in *Odontotermes obesus,* with a discussion on the association of fungi with termites. Records Indian Museum 58: 131–222.

———. 1975. A new mode of egg-laying, in ribbons, and the rate of laying in the termite *Odontotermes obesus* (Termitidae). Zool. Anz., Jena, 195: 351–354.

Rode, A. A. 1965. Theory of soil moisture. Translation from Russian by U.S. Dep. Agr. Washington, D.C. 560 pp.

Rohrmann, G. F. 1977. Biomass, distribution and respiration of colony components of *Macrotermes ukuzii* Fuller (Isoptera: Termitidae: Macrotermitinae). Sociobiology 2: 283–295.

Sands, W. A. 1956. Some factors affecting the survival of *Odontotermes badius*. Insectes Sociaux 3: 531–536.

———. 1960. The initiation of fungus comb construction in laboratory colonies of *Ancistrotermes guineensis* (Silvestri). Insectes Sociaux 7: 251–259.

———. 1969. The association of termites and fungi. Pages 495–524 *in* K. Krishna, and F. Weesner, eds. Biology of Termites, Academic Press, New York.

———. 1977. The role of termites in tropical agriculture. Outlook Agric. 9: 136–143.

Sannasi, A. 1969. Possible factor responsible for the specific growth of *Xylaria nigripes* in the "fungus gardens" of the mounds of termite *Odontotermes redemanni*. Entomol. Exp. and Appl. 12: 183–190.

Sen-Sarma, P. K. 1974. Ecology and biogeography of the termites in India. Pages 421–472 *in* M. S. Mani, ed. Ecology and biogeography in India. W. Junk Publ., The Hague.

Sheppe, W. 1970. Invertebrate predation on termites of the African savanna. Insectes Sociaux 17: 205–218.

Singer, R. 1961. Mushrooms and truffles. Leonard Hill Ltd., London. 271 pp.

———. 1975. The Agaricales in modern taxonomy. J. Cramer, Lehre. 912 pp.

Singh, N. B. 1975. Digestion and absorption of carbohydrates in *Odontotermes obesus* (Isoptera: Termitidae). Entomol. Exp. and Appl. 18: 357–366.

———. 1976. Studies on certain digestive enzymes in the alimentary canal of *Odontotermes obesus* (Isoptera: Termitidae). Entomol. Exp. and Appl. 20: 113–122.

Subramanian, C. V. 1972. *Padixonia*, a new genus of Hyphomycetes. Current Sci. 41: 282–283.

Usher, M. B. 1975. Studies on a wood-feeding termite community in Ghana, West Africa. Biotropica 7: 217–233.

Varma, A. N., N. D. Varma, and C. B. Tiwari. 1974. Effect of BHC and Aldrin on the termite damage in irrigated wheat crop when insecticides were applied by different methods. Indian J. Entomol. 36: 221–225.

Watson, J. P. 1972. Some observations on the water relation of mounds of *Macrotermes natalensis* (Haviland) Fuller. Insectes Sociaux 19: 87–93.

———. 1974. Calcium carbonate in termite mounds. Nature 247: 74.

———. 1975. The composition of termite (*Macrotermes* spp.) mounds on soil derived from basic rock in three rainfall zones of Rhodesia. Geoderma 14: 147–158.

Weir, J. S. 1975. Air flow, evaporation and mineral accumulation in mounds of *Macrotermes subhyalinus* (Rambur). J. Animal Ecol. 42: 509–520.

Wood, W. F., W. Truckenbrodt, and J. Meinwald. 1977. Chemistry of the defensive secretion from the African termite *Odontotermes badius*. Ann. Entomol. Soc. Amer. 68: 359–360.

Zhuzhikov, D. P., E. Zolotarev, and E. A. Orlova. 1971. Cellulolytic bacteria in the intestines of termites (Rus). Nauch Doil. Vysshei Shkoly Biol. Nauk 5: 96–100.

Zoberi, M. H. 1972. Tropical macrofungi. Macmillan Press Ltd., London. 158 pp.

———. 1973. Some edible mushrooms from Nigeria. Nigerian Field 38: 81–90.

chapter 7

The Role of Fungi in the Biology and Ecology of Woodwasps (Hymenoptera: Siricidae)

by J. L. MADDEN* and M. P. COUTTS†

ABSTRACT

A mutualism exists between woodwasps and specific fungi which results in colonization of host trees, insect development and subsequent dispersal of the fungi. Mycangia on female larvae and adults ensure the continuity of the relationship although one siricid genus has evolved from symbiosis to be a parasitic generalist exploiting timbers already infested by other woodwasp species and their fungi.

Woodwasps generally invade moribund or dead trees; however, arthropod mucus and fungus may combine to kill living trees in the *Sirex noctilio–Amylostereum areolatum* association. The ecology of this phenomenon is described from Australian studies on the dynamics of host-tree susceptibility and resistance to attack.

Moisture levels and aeration determine fungal growth and invasion of stems while resistance to attack is by resinosis and the formation of fungiostatic polyphenols. Larvae derive their nutrition from the fungus and ultimate size is related to fungal growth.

The symbiotic fungus also operates as a key factor in the detection of hosts by siricid parasitoids and the life cycle of certain entomophagous nematodes.

*Faculty of Agricultural Science, University of Tasmania, Hobart, Tasmania.
†Forestry Commission, Northern Research Station, Roslin, Midlothian, Scotland.

INSECT-FUNGUS SYMBIOSIS /Batra (ed.) / Allanheld, Osmun, Montclair, NJ

INTRODUCTION*

It has been known for some time that insect species belonging to many orders are capable of utilizing highly refractive materials in their diet. This ability, in the case of many phytophagous forms, would not be possible without the involvement of microorganisms, viz. fungi, protozoa, bacteria and yeasts. These associations of insects and microorganisms are reciprocal, beneficial relationships termed symbioses, or more precisely, mutualisms. Mutualism is a form of symbiosis in which both parties derive advantage without sustaining injury (Henderson and Henderson 1963).

In the case of lignicolous or wood-feeding insects the involvement of a third organism, the tree, adds a further dimension in which the microorganisms may exert a toxic effect. Graham (1967), reviewing insect-fungus mutualisms in forest trees, concludes that the predominant gains derived from such associations are the maximal conversion of indigestible materials, e.g. celluloses, lignins, etc., into forms readily assimilable by insects, and the provision of otherwise unobtainable vitamins, while the insect acts as a vector in the microorganisms' dispersal.

Buchner (1965), in a treatise devoted to endosymbiosis of animals with plant microorganisms, described many morphologically distinct storage organs within many insect groups, which developed internally (mycetomes) or externally (mycangia) to the symbiont partner. These organs provided storage and sustenance to the microorganism and facilitated dispersal into new host material, usually at the time of oviposition. The contributions by Kok (Chapter 2) and Norris (Chapter 3), in this volume, emphasize the importance and complexities of scolytid symbioses.

SIRICID WOODWASPS AND ASSOCIATED FUNGI

Siricid woodwasps infest a variety of both soft and hard woods in the northern hemisphere (Cameron 1965; Wolf 1967), and they provide an interesting example of a symbiotic association with microorganisms. Most of the insect life cycle is spent in the wood of a tree. The female insect has an ovipositor of a complex structure and 1.0–2.5 cm in length, with which it drills through the bark of the tree stem into the xylem, where eggs are deposited. The larvae feed by gnawing galleries through the wood and eventually pupate inside the tree. The adult insect then emerges and chews its way to the outside.

Buchner (1928) was the first to draw attention to two pear-shaped organs at the proximal ends of the valvulae of the ovipositor of female siricids. These organs, now known as intersegmental organs, were invariably filled with the arthrospores of a basidiomycetous fungus and a mutualistic role

*See also P. H. B. Talbot, 1977. The *Sirex-Amylostereum-Pinus* association. Ann. Rev. Phytopathol. 15: 41–54—Editor.

was suggested because spores were found to be implanted during oviposition and the mycangia of developing females became recontaminated before the insect emerged from the tree (Cartwright 1929).

Parkin (1942) demonstrated the existence of paired mycangia in larvae and thereby emphasized the close affinity of the insect-fungus association. These organs, known as hypopleural organs, were situated on the lower dorsal surface of the first abdominal segment. The hypopleural organ, which consisted of symmetrically arranged deep crypts embedded in a glandular, secretory hypodermis, maintained arthrospores in a dormant condition by embedding them in brittle wax platelets.

Rawlings (1951) reported that hypopleural organs only occurred in female larvae and Francke-Grosmann (1957) demonstrated that fungal release from the wax platelets only occurred through mechanical dislodgement or shattering at the time of eclosion of the adult when its reflex movements break up the exuviae containing the hypopleural organs. Growth of the fungus at this time resulted in the invasion of the intersegmental sacs. The form of the mycangia varies from species to species. The genus *Xeris* appears to be a degenerate form in which hypopleural organs are wanting and the intersegmental sacs are reduced and contain a variety of wood-inhabiting fungi as contaminants rather than a specific association. Species of this genus occur in woody material already infested with other siricids, so at the expense of the loss of any symbiotic association this genus exploits a wider variety of wood types infested with different species of basidiomycetes, but they forfeit the ability to attack previously uninfected material.

A variety of fungi have been found to be consistently associated with certain species of woodwasps. Cartwright (1938) showed that the mycangia of *Sirex gigas* and the wood infested by this woodwasp contained *Stereum sanguinolentum* Alb. and Schw. ex Fr. A similar but different fungus was found associated within *Sirex cyaneus*. Francke-Grosmann (1939) concluded from the examination of five siricid species occurring in both soft and hard woods that a basidiomycete occurred with each species and that the fungus of any one species was closely related but dissimilar to that associated with the other species. *Stereum sanguinolentum* and *Amylostereum chailletii* were identified from different species (Francke-Grosmann 1939, 1957). The majority of these recovery experiments involved sampling the mycangia of adult females and wood infested with siricid larvae. Rawlings (1949) reported a *Stereum sanguinolentum*-like fungus associated with *Sirex noctilio* F. Talbot (1964) and Gaut (1969) concluded respectively that the true generic and species identity was *Amylostereum areolatum* (Fr.) Boidin.

The associated fungi appear to play a complex role in the insects' biology and ecology. Some insight has been obtained from observations on fungus-free females. By removing the exuviae from the pupation chamber it was possible to prevent transfer of the fungus to the adult female (Rawlings

1951; Boros 1968), and Stillwell (1966) employed this method to demonstrate the dependence of larvae on the fungus. In a comparison of the oviposition of fungus-infested and fungus-free *Sirex juvencus* females on *Abies balsamea,* successful development occurred only in those billets in which oviposition had been carried out by fungus-infected females. Rawlings (1951, 1953) reported that the fungus provided a suitable environment for egg incubation and that it regulated moisture content, while Francke-Grosmann (1939) had proposed a nutritive role. In *S. noctilio,* as discussed below, the associated fungus, with a mucus, was also involved in reducing the intensity of the host tree's reaction to attack, which, if not moderated, would be harmful to the insect's developmental stages.

INVESTIGATIONS OF *SIREX NOCTILIO* IN AUSTRALIA

Sirex noctilio was first discovered in Tasmania in 1952 (Gilbert and Miller 1952) and since that time has caused extensive damage to *Pinus radiata* plantations in the states of Victoria and Tasmania. The National Sirex Fund was established in 1960, and full-scale investigations of the insect, the fungus, and the tree commenced; and it is from these studies that further discussion was prepared.

Host trees were attacked during the summer months and successful attack resulted in the death of trees and invasion of the stem with fungus. The number of females attacking an individual tree varied with the size and availability of the field population; thus small trees (less than 2.5 cm d.b.h.) were sometimes attacked and killed by one female whereas dominant trees were observed with as many as 50 females drilling with their ovipositors into their stems. In many instances the attacked tree became chlorotic in the apical region 10–14 days after attack and this chlorosis was progressive throughout the foliage depending on attack intensity and the tree's susceptibility to attack. Irreversible wilting of crown needles was an early indication of successful attack (Coutts 1969a).

The chlorotic phenomenon was shown to be due to the effects of a mucus inoculated into the sapstream of the attacked tree by the female wasp during drilling (Coutts 1968; 1969a and b). Spradbery (1973) showed that this response was more likely to occur where mucus of *S. noctilio* was involved rather than with that from other siricid species and that the foliage of *P. radiata* was more susceptible than that of other tree species tested. The emplantment and successful germination of arthrospores resulted in localized drying in the vicinity of the drilling lesion and this drying effect was aided by the presence of mucus (Kile and Turnbull 1974a). The biochemical nature of the mucus was reported by Fong and Crowden (1973).

Individual *P. radiata* trees were rendered attractive to *S. noctilio* by a number of surgical treatments which resulted in different, yet characteristic patterns of attractiveness (Madden 1971). The basis of this attraction was shown, in a general way, to be the increased rate of release of monoterpene volatiles through the bark following alterations in its permeability by changes in its osmotic and respiratory relationships (Madden 1968a, 1977).

Drilling was accompanied by the deposition of arthrospores and mucus, and the kind of drill made through a single hole in the bark varied from a single up to multiples of five or six. In each instance, the drill made prior to the removal of the ovipositor contained the arthrospores and mucus while the other drills contained eggs (Coutts and Dolezal 1969; Madden 1974; and Table 7.1). Attacked trees with high osmotic pressures of the phloem sap ($> 18.0 \times 10^5$ Pascals) contained only single drills initially, i.e. only mucus and arthrospores were inoculated, and the combined effect of the contents of these drills acting with the initial stress treatment was to progressively reduce osmotic pressure so that increasingly higher proportions of multiple drills were made in time and, in consequence, greater numbers of eggs were deposited. The *S. noctilio* female therefore displayed an economy in egg output by regulating release with respect to host tree physiology.

Coincident with the suppression of host tree physiology by edaphic, climatological or biological factors, was the suppression of its defensive reactions to attack. Coutts and Dolezal (1966) reported that resinosis and polyphenol formation at the drilling lesion site resulted in egg and larval mortality and the containment of the fungus respectively. Coutts and Dolezal (1966) also showed that polyphenol formation in the xylem was reduced by isolating the bark containing drills from translocation. Hillis and Inoue (1968) identified the defensive polyphenols as pinosylvin and pinosylvin monomethyl ether, two of the four naturally occurring heartwood polyphenols of *P. radiata* and the only two possessing appreciable fungistatic activity. Kile and Turnbull (1974b) found that monoterpene volatiles suppress *A. areolatum* growth. The reductions of turgor and

Table 7.1 Frequency of Eggs in Single and Multiple Drills of *S. noctilio*.

	Single drills	Double drills	Treble drills	Quadruple drills	Total Total
Total drills dissected	749	392	172	55	1368
Number with no eggs	718	152	27	2	899
Number with one egg	31	208	56	7	302
Number with two eggs	0	32	70	27	129
Number with three eggs	0	0	18	15	33
Number with four eggs	0	0	1	4	5
Proportion with eggs	0.042	0.610	0.784	0.965	
Mean No. per group with eggs	1.00	1.14	1.92	2.30	
Mean No. eggs per group	0.042	0.684	1.550	2.220	

translocation by any means suppressed the operation of these resistance mechanisms and hence their effects.

Larvae hatched from eggs in the absence of resistance when the lumen of the drills and adjacent xylem vessels were invaded by fungal hyphae and first and some second stage larvae fed exclusively on the fungus before entering the wood. Later stage larvae fed through fungus-invaded wood and the high nutritive status of the fungus was reflected in the incremental changes in larval gallery diameter relative to the volumes of wood frass excreted within early and late stage instars. The conversion ratio fell by a factor of 20 from the second to the fifth instar (Madden, unpublished data).

The number of instars varied from seven to 12 depending on conditions of aeration and moisture content within the infested log. King (1966) reported that optimal temperature for *A. areolatum* growth was between 20–25°C in cultural studies and Coutts (1965) found that the optimal moisture content of pine sapwood was between 60–70% (oven dry weight). Different-sized adult insects emerged at the same time from infested billets which had different initial moisture contents, and which had been incubated at the same temperature. It was also observed that larvae arising from eggs in the same oviposition drill but, encountering different conditions of wood moisture, and thus fungal growth and activity, were significantly different in size.

Microorganisms other than the fungus recovered from the midgut of *S. noctilio* larvae and larval frass included *Saccharomyces, Flavobacterium, Azotobacter* and *Acetobacter* (Madden 1975). The nitrogen content of larval frass in the galleries of large, actively feeding larvae was six times greater than that of the surrounding generally fungus-infested wood (0.03% Kjeldahl)and its moisture content was greater by 1.5–2.0-fold. Although nitrogen fixation could occur, the results suggested that the grazing activity of the larva within the gallery resulted in a net migration of both water and nitrogen via the fungus to the larval space for the ninhydrin positive materials contained in alcohol extracts of frass were comparable to those in extracts of the fungus. The differentials in moisture and nitrogen contents between frass and wood in wet or dry logs containing slowly developing larvae were either poorly represented or absent (Madden, unpublished data). Adults emerge through the bark in the summer following pupation in late spring.

Madden (1977, and Fig. 7.1) has proposed that damage to trees resulted in an initial loss of water through the stomata and increased water tension of the tree system. As a result of these changes, translocation was inhibited and a respiratory stress within the phloem was created with a reduction of both the osmotic and turgor status of the tree. Rupture of the xylem vessels by the ovipositor resulted in the invasion of these elements by air and the creation of optimal conditions for fungal growth and invasiveness and subsequent development of the siricid larvae.

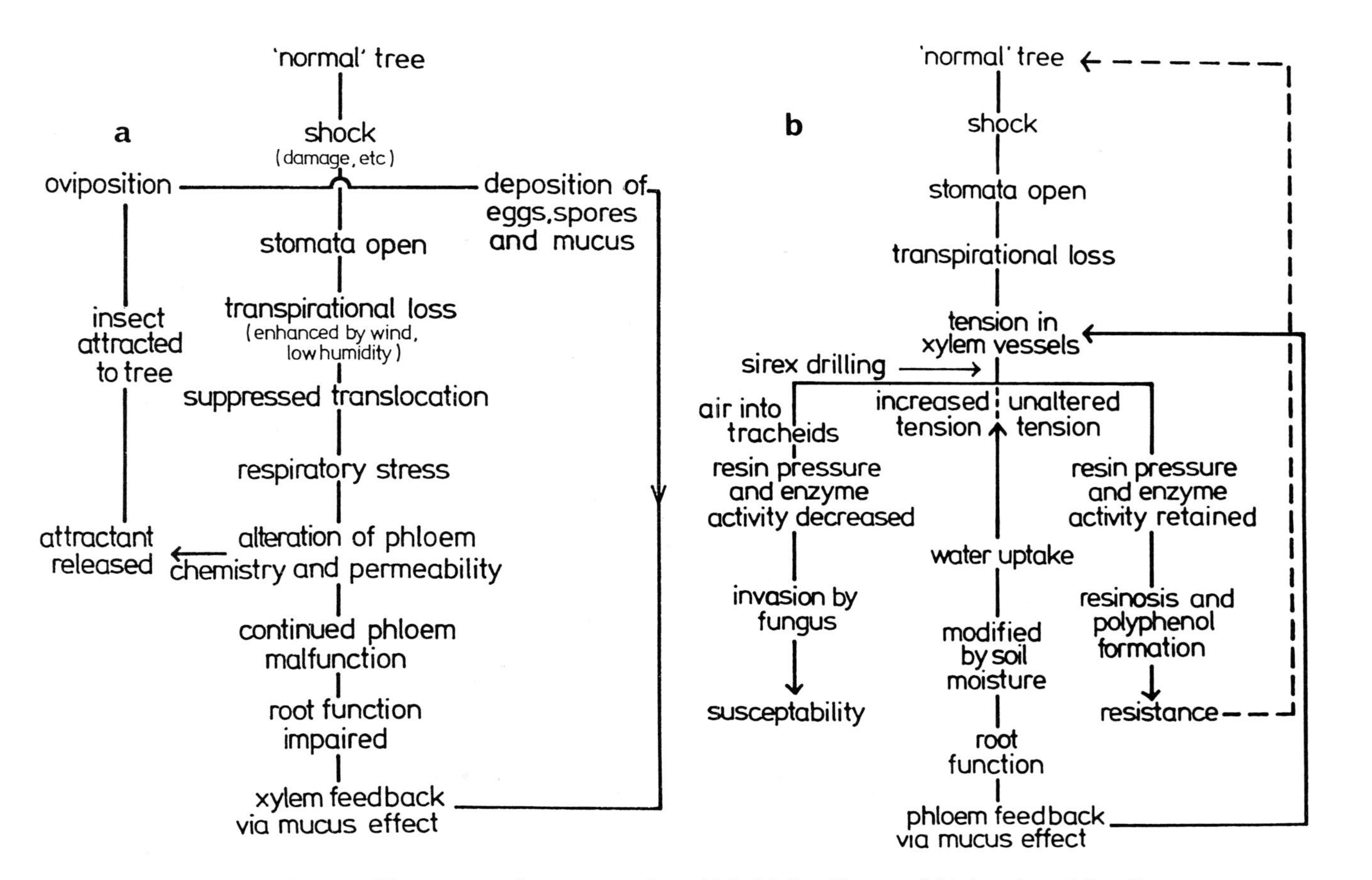

Figure 7.1. Possible sequence of events occurring within (a) the phloem and (b) the xylem of *P. radiata* which may influence the establishment, or otherwise, of *A. areolatum* and its symbiont partner, *S. noctilio*.

The interdependence of the insect-fungus relationship was also reflected in the behavior and biology of certain natural enemies of *S. noctilio* and other siricids. Madden (1968b) and Spradbery (1970a and b) found that the cynipid, *Ibalia leucospoides* Hochenw. and the ichneumonids, *Rhyssa persuasoria* (L.) and *Megarhyssa nortoni nortoni* Cressou, important parasitoids of siricids, exploit materials of fungal origin in their location of hosts. In the former instance, volatiles produced during fungal establishment resulted in the detection of drills containing mature eggs or early instar larvae while materials associated with mature fungal growth, and particularly concentrated in fresh, wet frass, resulted in the exploitation of larval hosts deep in the wood by the latter group of parasitoids during the spring prior to pupation.

Bedding (1967, 1968, 1972) reported on a unique instance of female dimorphism in the entomophagous nematode, *Deladenus siricidicola* Bedding. The free-living form was mycetophagous and fed specifically on *Amylostereum areolatum* whereas in the presence of the siricid host and its frass a non-mycetophagous, infective female was formed which invaded the larva and resulted in sterilization of the adult female. The mycetophagous form possessed the diagnostic features of the Neotylenchidae while the entomophagous form compared to those of the Allotonematidae.

CONCLUSIONS

The association of *A. areolatum* with *S. noctilio* has been found to be not only essential to a continued existence of the insect, through its debilitating effects on the host tree and the provision of larval food, and the fungus, through its active transport to new host trees, but also to the existence and success of siricid parasitoids and nematodes. In the former case the fungus provided olfactory uses which resulted in the detection of hosts and in the latter, the *S. noctilio* female was the vehicle of dispersal for the nematode and the fungus upon which the nematode fed.

Research on *S. noctilio* has provided much information on the distribution of siricids, their host trees and natural enemies and the basis for definitive studies on the association of different fungi with the different siricids and their role in larval nutrition.

LITERATURE CITED

Bedding, R.A. 1967. Parasitic and free-living cycles in entomogenous nematodes of the genus *Deladenus*. Nature 214 (5084): 174–175.

——. 1968. *Deladenus wilsoni* n.sp. and *D. siricidicola* n.sp. (Neotylenchidae), entomophagous-mycetaphagous nematodes parasitic in siricid woodwasps. Nematologica 14: 515–525.

——. 1972. Biology of *Deladenus siricidicola* (Neotylenchidae) an entomophagous-mycetophagous nematode parasitic in siricid woodwasps. Nematologica 18: 482–493.

Boros, C. B. 1968. The relationship between the woodwasp *Sirex noctilio* F. and the woodrot fungus *Amylostereum* sp. M. Agr. Sci. Thesis, Univ. of Adelaide. 65 pp.
Buchner, P. 1928. Holznahrung und symbiose. Berlin, J. Springer. 64 pp.
———. 1965. Endosymbiosis of animals with plant microorganisms. New York Interscience Publishers, John Wiley and Sons. 909 pp.
Cameron, E. A. 1965. The Siricinae (Hymenoptera: Siricidae) and their parasites. Commonwealth Inst. Biol. Cont. Tech. Bull. 5: 1-31.
Cartwright, K. St. G. 1929. Notes on the fungus associated with *Sirex cyaneus*. Ann. Appl. Biol. 16: 182-187.
———. 1938. A further note on fungus association in the Siricidae. Ann. Appl. Biol. 25: 430-432.
Coutts, M. P. 1965. *Sirex noctilio* and the physiology of *Pinus radiata*. Bull. Commonwealth Forest Timber Bur. Canberra 41: 1-79.
———. 1968. Rapid physiological change in *Pinus radiata* following attack by *Sirex noctilio* and its associated fungus, *Amylostereum* sp. Aust. J. Sci. 30: 275-277.
———. 1969a. The mechanism of pathogenicity of *Sirex noctilio* on *Pinus radiata* I. Effects of the symbiotic fungus *Amylostereum* sp. (Thelephoraceae). Aust. J. Biol. Sci. 22: 915-924.
———. 1969b. The mechanism of pathogenicity of *Sirex noctilio* on *Pinus radiata* II. Effects of *S. noctilio* mucus. Aust. J. Biol. Sci. 22: 1153-1161.
Coutts, M. P. and J. E. Dolezal. 1966. Polyphenols and resin in the resistance mechanism of *Pinus radiata* attacked by the woodwasp, *Sirex noctilio*, and its associated fungus. Leaflet 101. Commonwealth Forest Timber Bur. Canberra.
———. 1969. Emplacement of fungal spores by the woodwasp, *Sirex noctilio*, during oviposition. For. Sci. 15: 412-416.
Fong, L. K. and R. K. Crowden. 1973. Physiological effects of mucus from the woodwasp, *Sirex noctilio* F., on the foliage of *Pinus radiata* D. Don. Aust. J. Biol. Sci. 26: 365-378.
Francke-Grosmann, H. 1939. Über das Zusammenleben von Holzwespen (Siricinae) mit Pilzen. Z. angew. Entomol. 25: 647-680.
———. 1957. Über das Schiksal der Siricidenpilze Wahrend der Metamorphose. Wanderversamml. Deutsch. Entomol. 8: 37-43.
Gaut, I. P. C. 1969. Identity of the fungal symbiont of *Sirex noctilio*. Aust. J. Biol. Sci. 22: 905-914.
Gilbert, J. M., and L. W. Miller. 1952. An outbreak of *Sirex noctilio* in Tasmania. Aust. J. For. 16:63-69.
Graham, K. 1967. Fungal-insect mutualism in trees and timber. Ann. Rev. Entomol. 12: 105-106.
Henderson, I. F., and W. D. Henderson. 1963. A dictionary of biological terms. 8th ed. Longman, London. 640 pp.
Hillis, W. E., and T. Inoue. 1968. The formation of polyphenols in trees. IV. The polyphenols found in *Pinus radiata* after *Sirex* attack. Phytochem. 7: 13-22.
Kile, G. A. and C. R. A. Turnbull. 1974a. Drying in the sapwood of radiata pine after inoculation with *Amylostereum areolatum* and *Sirex* mucus. For. Res. Aust. 6: 35-40.
———. 1974b. The effect of radiata pine resin and resin components on the growth of the *Sirex* symbiont, *Amylostereum areolatum*. For. Res. Aust. 6: 27-34.
King, J. M. 1966. Some aspects of the biology of the fungal symbiont of *Sirex noctilio*. Aust. J. Bot. 14: 25-30.
Madden, J. L. 1968a. Physiological aspects of host tree favourability for the woodwasp, *Sirex noctilio* F. Proc. Ecol. Soc. Aust. 3: 147-149.
———. 1968b. Behavioral responses of parasites to the symbiotic fungus associated with *Sirex noctilio* F. Nature 218: 189-190.
———. 1971. Some treatments which render Monterey pine (*Pinus radiata*) attractive to the woodwasp, *Sirex noctilio* F. Bull. Entomol. Res. 60: 467-472.
———. 1974. Oviposition behaviour of the woodwasp, *Sirex noctilio* F. Aust. J. Zool. 22: 341-351.
———. 1975. Bacteria and yeasts associated with *Sirex noctilio*. J. Invert. Pathol. 25: 283-287.
———. 1977. Physiological reactions of *Pinus radiata* to attack by woodwasp, *Sirex noctilio* F. (Hymenoptera: Siricidae). Bull. Entomol. Res. 67: 405-426.
Parkin, E. A. 1942. Symbiosis and siricid woodwasps. Ann. Appl. Biol. 29: 268-279.

Rawlings, G. S. 1949. Recent observations on the *Sirex noctilio* population in *Pinus radiata* forests in New Zealand. N.Z. J. For. 5: 411–421.

———. 1951. The establishment of *Ibalia leucospoides* in New Zealand. For. Res. Notes N.Z. 1: 1–14.

———. 1953. Rearing of *Sirex noctilio* and its parasite *Ibalia leucospoides*. For. Res. Notes N.Z. 1: 20–34.

Spradbery, J. P. 1970a. The biology of *Ibalia drewseni* Borries (Hymenoptera: Ibaliidae), a parasite of siricid woodwasps. Proc. Roy. Entomol. Soc. Lond. 45: 104–113.

———. 1970b. Host finding by *Rhyssa persuasoria* (L.) on ichneumonid parasite of siricid woodwasps. Anim. Behav. 18: 103–114.

———. 1973. A comparative study of the phytotoxic effects of siricid woodwasps on conifers. Ann. Appl. Biol. 75: 309–320.

Stillwell, M. A. 1966. Woodwasps (Siricidae) in conifers and the associated fungus, *Stereum chailletii*, in Eastern Canada. For. Sci. 12: 121–128.

Talbot, P. H. B. 1964. Taxonomy of the fungus associated with *Sirex noctilio*. Aust. J. Bot. 12: 46–52.

Wolf, F. 1967. Les Siricidae en Belgique et les problemes quils soulevent. Ann. Acad. Fac. Sci. Agronom. Gembloux 1966–67.

chapter 8

Commensalism of the Trichomycetes

by STEPHEN T. MOSS*

ABSTRACT

The Trichomycetes is an ecologically defined group of lower fungi which exhibits wide diversity in its vegetative and reproductive organization. The four recognized orders—Amoebidiales, Eccrinales, Asellariales, Harpellales—are united by their production of sporangiospores, possession of a holdfast and an obligate association with the cuticular surfaces of marine, freshwater and terrestrial, mandibulate arthropods. The phylogenetic affinities of the Harpellales and Asellariales appear to be with the Kickxellales (Zygomycotina) but those of the Amoebidiales and Eccrinales are less certain. Except for the ectozoic species *Amoebidium parasiticum,* trichomycete thalli are attached to, but do not penetrate, the cuticle lining the fore, mid or hind guts of their hosts. The relationship between host and fungus is considered as commensalistic; commensalism in this context is defined as an interspecific relationship in which the host provides nutrients and a protected habitat for the commensal but neither benefits nor suffers from the association. Endocommensalistic Trichomycetes utilize food material within the digestive tract but the absence of any detectable competition for nutrients precludes a parasitic relationship. Similarly, *A. parasiticum* does not obtain nutrients directly from the host tissues or body fluids but is normally attached to regions on the external surface where it may benefit from feeding and/or respiratory currents of the host. Whereas *A. parasiticum* has a low degree of host specificity the endocommensalistic forms are generally restricted to certain genera or families of hosts and several species may cohabit a single host gut. A high degree of host specificity combined with an obligate relationship with the host requires specializations of the commensal to ensure successful transmittance of propagules in order to maintain the level of infestation within a host population. Furthermore, the loss of thalli from

*Department of Biological Sciences, Portsmouth Polytechnic, Portsmouth, U.K.

INSECT-FUNGUS SYMBIOSIS /Batra (ed.) / Allanheld, Osmun, Montclair, NJ

the host at ecdysis and non-transmittance of the commensal with host eggs imposes a temporal restriction on the trichomycete life cycle and necessitates close coordination of host and commensal life cycles. Structural and developmental attributes of the commensal relationship of Trichomycetes include: appendaged spores of the Harpellales and some aquatic Eccrinales which aid retention of released spores in the vicinity of the host population; rapid holdfast production followed by attachment of the ingested spore to the host cuticle; coordination of the vegetative growth and reproductive development of the commensal with metabolism, ecdysis and death of the host.

INTRODUCTION

The trichomycete-arthropod relationship is probably the least studied and understood of all fungal-arthropod associations. This is, perhaps, partially attributable to the apparent non-economic importance of the association, the absence of any anatomical modification of the arthropod host which may have stimulated their greater study by zoologists, the inability to axenically culture all but two genera (*Amoebidium,* Amoebidiales and *Smittium,* Harpellales) and the reluctance of mycologists to acquaint themselves with arthropod anatomy. Nevertheless, in this chapter those features which augur for a commensalistic relationship between the Trichomycetes and their arthropod hosts are presented.

The form class Trichomycetes was established in 1948 by the three French zoologists Duboscq, Léger and Tuzet (1948) for an assemblage of thallial, lower fungi obligately associated with the cuticular surfaces of certain Arthropoda. With the exception of *Amoebidium parasiticum* Cienkowski (1861), which is restricted in its attachment to the external surfaces of freshwater arthropods, the Trichomycetes are attached to, but do not generally penetrate, the cuticle lining the digestive tracts of their hosts. It was, however, nearly a century previous to the publication by Duboscq et al. (1948) that Leidy (1849a) described the first trichomycete, *Enterobryus elegans,* attached to the cuticle lining the hindgut of the milliped *Julus marginatus* Say (Diplopoda, Myriapoda) and considered it a colourless alga belonging to the Confervaceae. Since 1849 over 150 species have been validly described from marine, freshwater and terrestrial Arthropoda. The unusual morphology of these fungi, ranging from the amoeboid sporangiospores produced within the unbranched thalli of the ectozoite *A. parasiticum* to the often highly branched members of the endozoite Harpellales with their exogenous, deciduous, appendaged sporangia and biconical zygospores, has resulted in species being classified with the Algae, Sporozoa (Protozoa) and Mycota. There has been similar conjecture as to the relationship of the Trichomycetes with their hosts, having been variously described as parasitic, symbiotic and commensalistic. The assignment of these different terms to describe the relationship results from both the limited information available on the trichomycete-arthropod association and also the confusion which presently exists in the definitions of these terms.

Symbiosis, as originally defined by de Bary (1879), is a heterospecific association in which the symbiotes exist in a close spatial and physiological relationship, with no suggestion of mutual benefit, a connotation frequently intended by present workers. When the organisms in a heterospecific relationship are reciprocally dependent the association is mutualistic, and when one organism benefits at the expense of the other it is termed parasitic. Commensalism, as used in this chapter, is defined as an interspecific relationship in which the host, the arthropod, provides nutrients for the commensal, the trichomycete, but in which the host neither benefits nor suffers from the association. This definition of commensalism approximates that originally intended by van Beneden (1876), although more recent usage of the term has included non-nutritional benefits including protection and transport (Cloudsley-Thompson 1965). It follows that commensalism, mutualism and parasitism should be considered as special cases of symbiosis, although it is often difficult and may be not desirable to separate one type of relationship from another.

In nature the Trichomycetes are obligately dependent on their hosts; they grow, reproduce and die either within the digestive tract or upon the external surface of the host; the infestive, dispersive spore is normally the only stage of the life cycle which is not dependent upon the host. The host, however, is not dependent upon the commensal—the trichomycete—as not all potential host populations are infested and those which are infested do not appear to benefit or suffer from the infestation. When applied to the trichomycete-arthropod relationship the terms "benefits" and "suffers" require qualification as both are dependent on the degree to which minor changes in the arthropod's biology can be detected. Criteria such as blood pressure, heartbeat, body temperature, etc., which are routinely used in the detection of disease and body stress in higher organisms, cannot be used for the Arthropoda. In studies of the Trichomycetes, workers have used criteria such as changes in feeding, growth patterns, moulting, maturation, reproduction and most commonly death of the host as indicators of beneficial or detrimental effects. Correspondingly, the assignment of the term commensalism to this association results from the possible failure to detect minor deviations from normality of the infested host. In this chapter those features of trichomycete biology which may indicate a commensalistic relationship are presented.

THE TRICHOMYCETES

The Trichomycetes contains four orders of lower, thallial fungi—Amoebidiales (Léger and Duboscq 1929a), Eccrinales (Léger and Duboscq 1929a), Harpellales (Lichtwardt and Manier 1978), Asellariales (Lichtwardt and Manier 1978)—characterized by an association with the cuticular surfaces of arthropods and the presence of a holdfast. The orders are distinguished by their asexual reproductive apparatus: the Amoebidiales are unbranched and produce amoeboid sporangiospores; the Eccrinales are unbranched and form endogenous, single-spored sporangia in basipetal succession from the thallus apex; the Harpellales are branched or unbranched and form exogenous, unisporous, dehiscent, appendaged sporangia (trichospores); the branched thalli of the Asellariales fragment to form arthrospores. Sexual reproduction has only been confirmed in the Harpellales and owing to the production of zygospores in this order the Trichomycetes are considered a class of the Zygomycotina (Lichtwardt 1973b).

The morphology and life cycles of the Trichomycetes are not typical of other fungal taxa as they possess modifications to both morphology and life cycle which adapt them to their unique ecological niche and obligate dependence on arthropod hosts. Correspondingly, it seems pertinent to present a brief description of each order.

Amoebidiales. Amoebidiales are distinguished from other Trichomycetes by the production of amoeboid sporangiospores. Their thalli are unbranched, coenocytic and attached to the host cuticle by a basal, secreted holdfast (Fig. 8.2; Whisler and Fuller 1968). The order contains one family with two genera, both of which are associated with freshwater Crustacea

Table 8.1 Habitat and Host Range of the Amoebidiales

AMOEBIDIALES Léger and Duboscq (1929a)		ARTHROPOD HOST		
Family	Genus	Class	Order	Family
AMOEBIDIACEAE Lichtenstein (1917)	*Amoebidium*[a] Cienkowski (1861)	CRUSTACEA	CLADOCERA CYCLOPOIDA AMPHIPODA ISOPODA	Many
		INSECTA	DIPTERA EPHEMEROPTERA ODONATA PLECOPTERA TRICHOPTERA	Many
	Paramoebidium Léger & Duboscq (1929c)	INSECTA	DIPTERA	SIMULIIDAE
			EPHEMEROPTERA	BAETIDAE ECDYONURIDAE LEPTOHLEBIIDAE
			PLECOPTERA	CHLOROPERLIDAE NEMURIDAE

[a]Monotypic

and Insecta (Table 8.1). *Amoebidium* contains the single species *A. parasiticum* and is the only trichomycete found attached to the external surface of its host (Figs. 8.1–8.5). The second genus, *Paramoebidium,* grows attached to the cuticle lining the hindguts of insect larvae and nymphs (Fig. 8.6). Since Léger and Duboscq (1929a) described *P. inflexum* from the hindgut of *Nemura variegata* Oliv. (Plecoptera, Insecta) several species have been described although their validity is uncertain as many of the characters used as taxonomic criteria, e.g. thallus size, degree of thallus curvature, size of the holdfast, may be unstable and influenced by the environment. Indeed Whisler (1965) has cast doubt on the validity of a generic distinction in this order.

Reproduction of *A. parasiticum* is by cleavage of the entire thallus cytoplasm to form either rigid, walled, uninucleate sporangiospores or amoeboid sporangiospores whereas *Paramoebidium* spp. are known to produce only amoeboid spores (Figs. 8.6–8.8). Between host moults the thallus of *A. parasiticum* cleaves into rigid, ellipsoidal sporangiospores which are released through 'breaks' in the thallus wall (Figs. 8.2–8.4). These spores secrete holdfast material through "polar pit-fields" (Whisler and Fuller 1968) and upon contact with the host become attached (Fig. 8.5),

Stage Infested	Region of Thallus Attachment	Habitat	Principal References
Adult Larva Nymph	External surface	Freshwater	Chatton 1906a, b; Cienkowski 1861; Kuno 1973; Lichtenstein 1917; Lichtwardt 1973a, 1976; Manier & Raibaut 1970; Poisson 1931b; Trotter & Whisler 1965; Whisler 1960, 1962, 1965, 1968; Whisler & Fuller 1968.
Larva Nymph	Procto-daeum	Freshwater	Chatton & Roubaud 1909; Duboscq, Léger & Tuzet 1948; Léger & Duboscq 1929c; Lichtwardt 1973a, 1976; Manier 1950, 1955, 1962a; Tuzet & Manier 1955a; Whisler 1965.

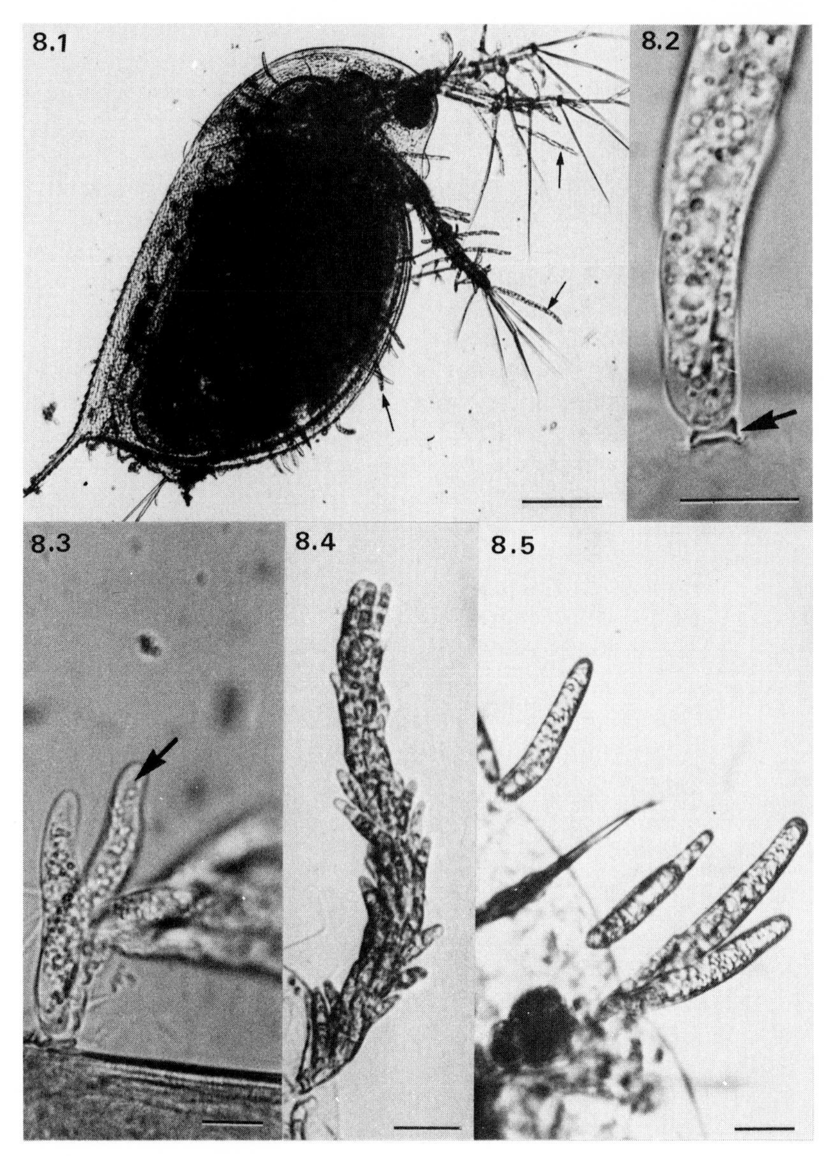

Figures 8.1–8.5. Amoebidium parasiticum (Amoebidiales). *8.1.* Vegetative thalli (arrows) attached to antennae and lower edge of the carapace of *Daphnia* sp., Cladocera. (Bright-field illumination, bar=500μm.) *8.2.* Base of thallus with secreted holdfast (arrow). Note the partial cleavage of the cytoplasm to form the rigid, elongate sporangiospores. (Bright-field illumination, bar=10μm.) *8.3.* Base of same thallus as Fig. 8.2 but later in sporangiospore development. The formed sporangiospores (arrow) project through the thallus wall. (Bright-field illumination, bar= 10 μm.) *8.4.* Entire thallus cytoplasm cleaved into rigid sporangiospores. (Bright-field illumination, stained lactophenol / cotton-blue, bar=40μm.) *8.5.* Young thalli developing from recently attached, rigid sporangiospores. (Bright-field illumination, bar=10μm.)

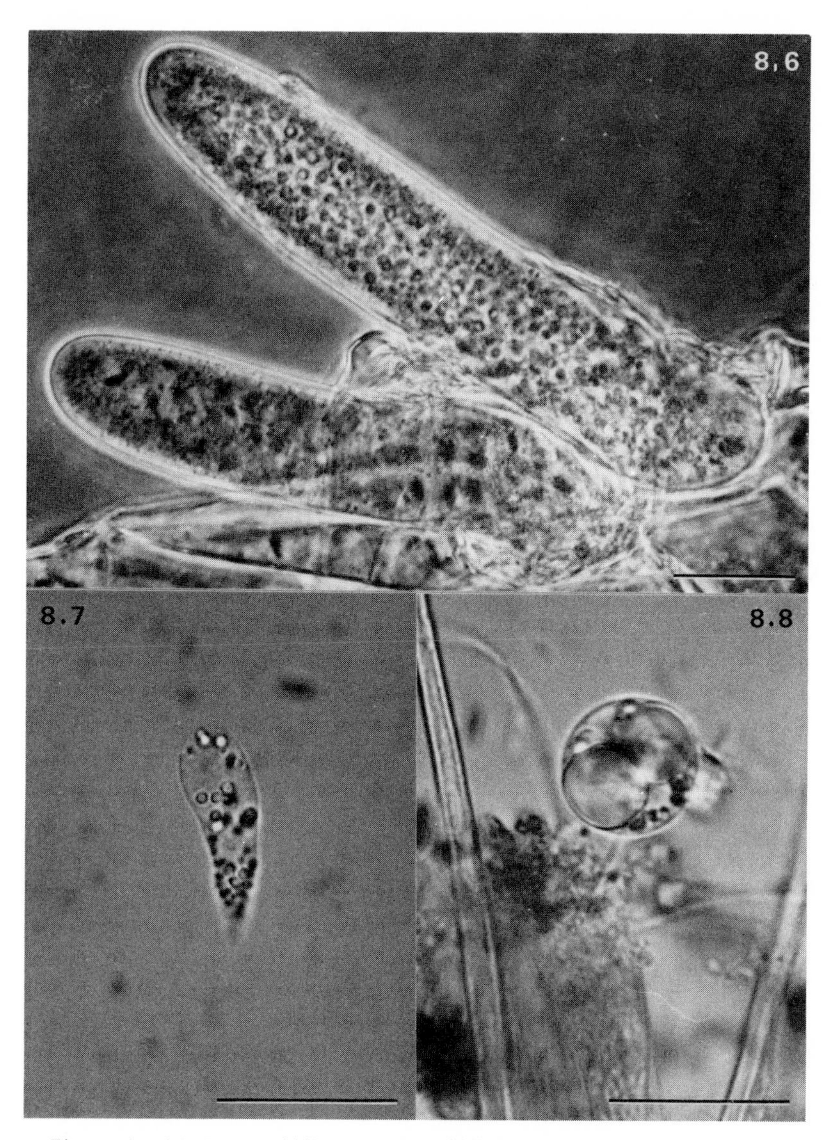

Figures 8.6–8.8. Paramoebidium sp. (Amoebidiales). *8.6 (top).* Thallus from the hindgut of a mayfly, Ephemeroptera, nymph with cytoplasm cleaved into amoeboid sporangiospores. (Phase-contrast illumination, bar = 40 μm.) *8.7 (bottom left)* Released and *8.8 (bottom right)* encysted amoebae. Note the characteristic hemispherical and tapered posterior ends of the migrating amoeba in *8.7.* (Bright-field illumination, bars = 20 μm.)

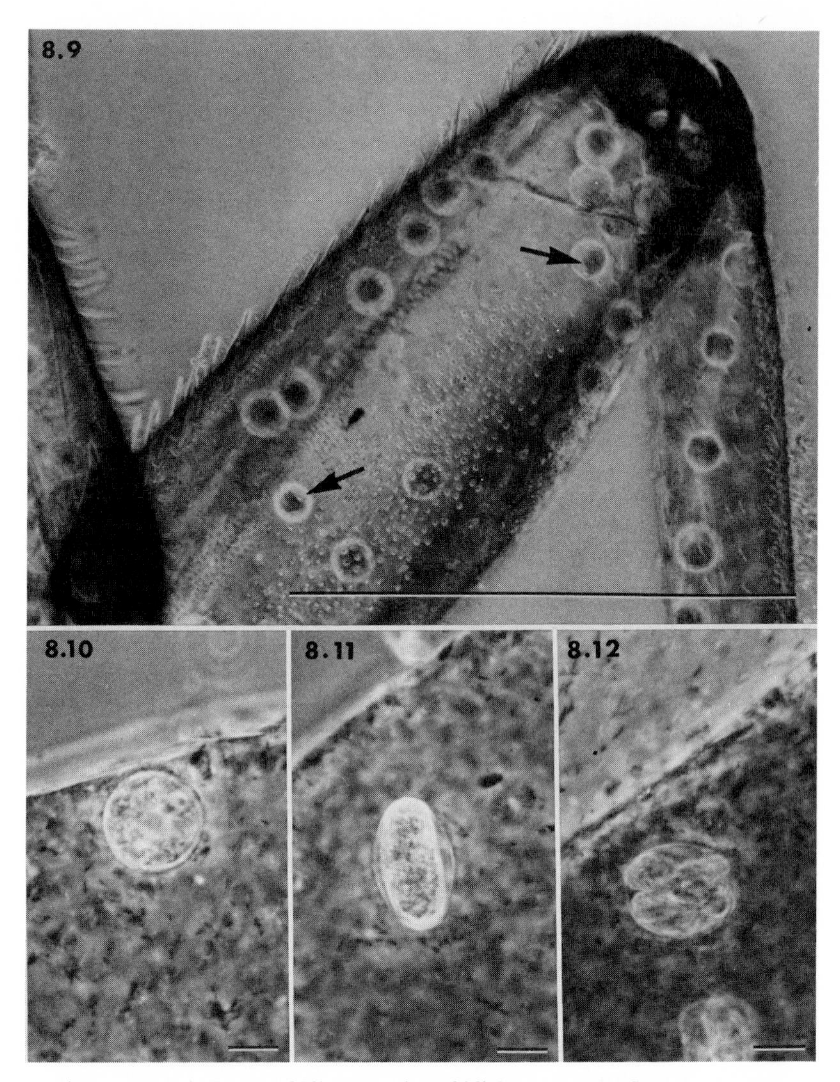

Figures 8.9–8.12. Paramoebidium sp. (Amoebidiales). *8.9.* Leg of mayfly exuvium containing cysts (arrows). (Phase-contrast illumination, bar = 1mm.) *8.10.* Cyst prior to formation of cytospores. (Phase-contrast illumination, bar = 20μm.) *8.11–8.12.* Cysts containing one (*8.11*) and several (*8.12*) cytospores. (Phase-contrast illumination, bars = 20μm.)

thus increasing the density of infestation on the same host. Upon moulting, injury or death of the host there is a morphogenetic switch from rigid to amoeboid sporangiospore production in *Amoebidium*. Similarly, the endozoic *Paramoebidium* spp., not known to produce the rigid sporangiospores characteristic of *Amoebidium,* also produce amoeboid sporangiospores, either at the time of moulting or upon death of the host. The released amoebae of *Amoebidium* spp. and *Paramoebidium* spp. are indistinguishable and their subsequent development identical. The normally uninucleate amoebae swarm from the thallus and migrate over the exuvium or cadaver of the host. These amoebae do not possess pseudopodia but are rounded anteriorly, attenuated posteriorly (Fig. 8.7) and their movement is associated with rotation of the cytoplasm. Following a swarm period of up to several hours the amoebae encyst (Figs. 8.8–8.10) and remain dormant for days or even weeks. Subsequent cleavage of the cyst cytoplasm produces a number, usually one to 16, of ellipsoidal cystospores (Figs. 8.11, 8.12) which upon release and contact, with *Amoebidium,* or ingestion, by *Paramoebidium,* a host, develop into new thalli. The stimulus for cystospore formation and subsequent liberation is not known. Swarming amoebae have not been observed to engulf food material nor has their reported fusion to form a zygote (Poisson 1931b; Tuzet and Manier 1951b) been confirmed. The life cycle of *A. parasiticum* is represented in Figure 8.13.

Eccrinales. The Eccrinales is the most morphologically diverse order of Trichomycetes; a character which may be partially attributable to their wide host range (Crustacea, Insecta, Myriapoda) and the varied habitats of these hosts (marine, freshwater, terrestrial; Table 8.2). The order is characterized by unbranched thalli and the production of unisporous sporangia formed by basipetal septation from the thallus apex and contains three families (Eccrinaceae, Palavasciaceae, Parataeniellaceae) with 15 genera, of which seven are monotypic (Table 8.2).

No species has been axenically cultured although limited growth has been followed in hanging drop and water slide mounts (Manier 1954; Tuzet and Manier 1954a) and on agar (Lichtwardt 1964b). Vegetative thalli of the Eccrinales are characteristically unbranched (Figs. 8.14–8.19), nonseptate, multinucleate and attached to the cuticle lining the digestive tract of the host by a secreted, basal holdfast (Figs. 8.18, 8.19). In *Alacrinella limnoriae* Manier and Ormières (1961a) ex Manier (1968), the holdfast is associated with a highly lobed basal region of the thallus (Manier and Ormières 1961a), whilst in *Enteromyces callianassae* Lichtwardt (1961a), found in the stomachs of the anomuran (Crustacea), marine, mud shrimps *Upogebia affinus* Say (McCloskey and Caldwell 1965), *Callianassa uncinata* Edwards and *C. brachyophthalma* Edwards (Lichtwardt 1961a) and the brachyuran *Uca pugilator* Latreille (Tuzet and Manier 1962), many individual thalli are attached to a common point on the host cuticle by a

Table 8.2 Habitat and Host Range of the Eccrinales

ECCRINALES Léger & Duboscq (1929a)		ARTHROPOD HOST						
Family	Genus	Class	Order	Family	Stage Infested	Region of Thallus Attachment	Habitat	Principal References
	Alacrinella[a] Manier & Ormières (1961a) ex Manier (1968)	CRUSTACEA	ISOPODA	LIMNORIIDAE	Adult	Proctodaeum	Marine	Galt 1971; Manier 1968; Manier & Ormières 1961a.
	Arundinula Léger & Duboscq (1906)	CRUSTACEA	DECAPODA	GALATHEIDAE LITHODIDAE PAGURIDAE PORCELLANIDAE POTAMOBIIDAE	Adult	Stomodaeum & Proctodaeum	Marine Freshwater	Duboscq, Léger & Tuzet 1948; Galt 1971; Léger & Duboscq 1905, 1906, 1911; Lichtwardt 1962; Manier 1950, 1968; Manier & Ormières 1962.
	Astreptonema Hauptfleisch (1895)	CRUSTACEA	AMPHIPODA	COROPHIIDAE GAMMARIDAE	Adult	Proctodaeum	Marine Marine & Freshwater	Duboscq, Léger & Tuzet 1948; Galt 1971; Hauptfleisch 1895; Léger & Duboscq 1933; Manier 1950, 1961, 1964a, 1968; Moss 1975; Poisson 1929; Scheer 1976a.
	Eccrinidus[a] Léger & Duboscq (1906) Manier (1969a)	MYRIAPODA	DIPLOPODA	GLOMERIDAE	Adult	Proctodaeum	Terrestrial	Duboscq, Léger & Tuzet 1948; Léger & Duboscq 1906; Leidy 1850; Maessen 1955; Manier 1950, 1969a; Tuzet & Manier 1954b.
	Eccrinoides Leger & Duboscq (1929a)	CRUSTACEA	ISOPODA	PORCELLIONIDAE TYLIDAE	Adult	Proctodaeum	Marine Terrestrial	Duboscq, Léger & Tuzet 1948, Léger & Duboscq 1906, 1929a; Maessen 1955; Manier 1969b; Poisson 1931a.
		MYRIAPODA	DIPLOPODA	GLOMERIDAE	Adult	Proctodaeum	Terrestrial	
ECCRINACEAE Léger & Duboscq (1929a) emend. Manier & Lichtwardt (1968)	*Enterobryus* Leidy (1849a)	INSECTA	COLEOPTERA	HYDROPHILIDAE PASSALIDAE	Adult	Proctodaeum	Freshwater Terrestrial	Cronin & Johnson 1958; Duboscq, Léger & Tuzet 1948; Galt 1971; Granata 1908; Léger & Duboscq 1905, 1916, 1929a; Leidy 1849a,b,c, 1850; Lichtwardt 1954a,b, 1957a,b, 1958, 1960a,b,c; Maessen 1955; Manier 1950, 1969a,b; Manier, Gasc & Bouix 1972, 1974; Manier & Lichtwardt 1968; Manier & Théodoridès 1965; Poisson 1931a; Rajagopalan 1967; Robin 1853; Thaxter 1920; Tuzet & Manier 1948a, 1951a, 1957a,b; Tuzet & Manier 1967; Whisler 1963.
		MYRIAPODA	DIPLOPODA	IULIDAE ODONTOPYGIDAE POLYDESMIDAE SPIROSTREPTIDAE	Adult	Proctodaeum	Terrestrial	
		CRUSTACEA	DECAPODA	OCYPODIDAE	Adult	Proctodaeum	Marine	

	Enteromyces[a] Lichtwardt (1961a)	CRUSTACEA	DECAPODA	CALLIANASSIDAE OCYPODIDAE	Adult	Stomodaeum	Marine	Galt 1971; Lichtwardt 1961a; McCloskey & Caldwell 1965; Tuzet & Manier 1962.
	Nodocrinella[a] Scheer (1977)	CRUSTACEA	ISOPODA	TRICHONISCIDAE	Adult	Proctodaeum	Terrestrial	Scheer 1977.
	Paramacrinella[a] Manier & Grizel (1971)	CRUSTACEA	AMPHIPODA	AORIDAE	Adult	Proctodaeum	Marine	Manier & Grizel 1971.
	Ramacrinella[a] Manier & Ormières (1961b) ex Manier (1968)	CRUSTACEA	AMPHIPODA	AORIDAE	Adult	Proctodaeum	Marine	Manier 1968; Manier & Ormières 1961b.
	Taeniella Léger & Duboscq (1911)	CRUSTACEA	DECAPODA	CALLIANASSIDAE CANCERIDAE GALATHEIDAE GRAPSIDAE LITHODIDAE PAGURIDAE PORTUNIDAE	Adult	Proctodaeum	Marine	Duboscq, Léger & Tuzet 1948; Galt 1971; Johnson 1966, Léger & Duboscq 1911; Manier 1961, 1968; Manier & Ormières 1962.
	Taeniellopsis Poisson (1927)	CRUSTACEA	AMPHIPODA	TALITRIDAE	Adult	Proctodaeum	Marine	Manier 1950, 1969b; Poisson 1927, 1929.
PALAVASCIACEAE Duboscq, Léger & Manier (1948) ex Manier & Lichtwardt (1968)	*Palavascia* Tuzet & Manier (1947a) ex Lichtwardt (1964a)	CRUSTACEA	ISOPODA	ONISCIDAE SPHAEROMIDAE	Adult	Proctodaeum	Marine	Duboscq, Léger & Tuzet 1948; Johnson 1966; Lichtwardt 1961b, 1964a, 1973a; Manier 1950, 1961, 1963a, 1968, 1969b; Manier & Lichtwardt 1968; Tuzet & Manier 1947a, 1948b.
PARATAENIELLACEAE Manier & Lichtwardt (1968)	*Lajassiella*[a] Tuzet & Manier (1950b) ex Manier (1968)	INSECTA	COLEOPTERA	SCARABAEIDAE	Larva	Proctodaeum	Terrestrial	Manier 1968; Manier & Théodoridès 1965; Tuzet & Manier 1950b.
	Parataeniella Poisson (1929)	CRUSTACEA	ISOPODA	ARMADILLIDAE ONISCIDAE TRICHONISCIDAE	Adult	Proctodaeum	Terrestrial	Lichtwardt & Chen 1964; Manier 1950, 1964c, 1969b; Poisson 1928, 1929; Scheer 1976b.

[a]Monotypic

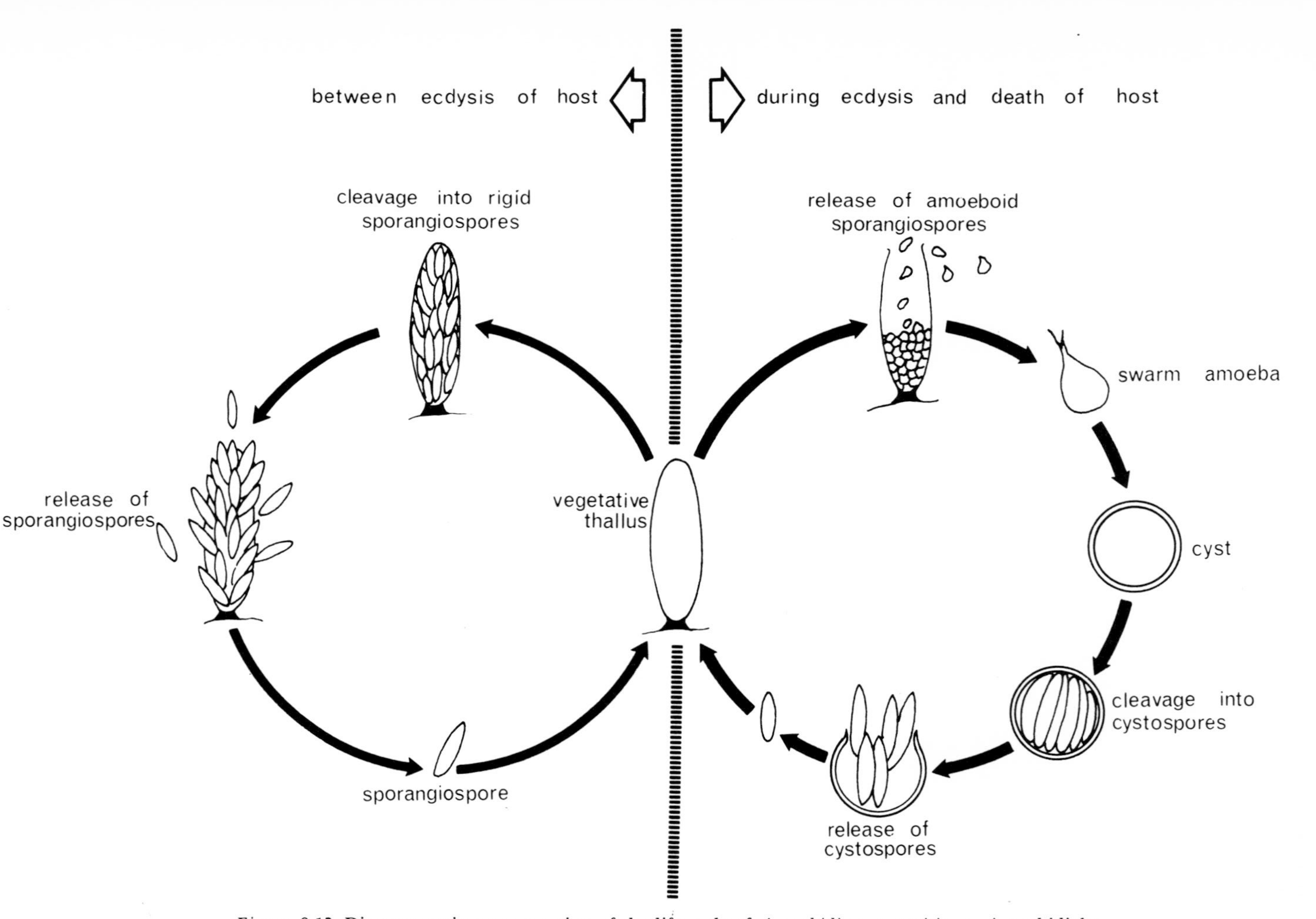

Figure 8.13. Diagrammatic representation of the life-cycle of *Amoebidium parasiticum*, Amoebidiales.

secreted, multiple holdfast and may superficially appear branched. True branching is only found in *Ramacrinella raibauti* Manier and Ormières (1961b) ex Manier (1968), where branches originate close to the holdfast (Manier and Ormières 1961b) and in *Enterobryus tuzetae* Manier, Gasc and Bouix (1972) and *E. bifurcatus* Whisler (1963), where thalli attached to the cuticle lining the posterior region of the proctodaeum of the diplopod hosts *Pachybolus ligulatus* Voges and *Californiobolus uncigerus* Wood respectively have two branches, one on each side of the holdfast (Manier, Gasc and Bouix 1972; Whisler 1963). In addition to the normal vegetative thalli, species of *Alacrinella, Astreptonema, Enteromyces* and *Ramacrinella* possess unbranched, coenocytic microthalli which are attached to the same region of the cuticle as the normal, macrothalli. However in certain species of *Enterobryus* and in *Nodocrinella hylonisci* Scheer (1977) morphologically dissimilar thalli, which occur in different regions of the host gut, have also been attributed to the same species (Lichtwardt 1957a; Manier, Gasc and Bouix 1972; Scheer 1977). Until these fungi can be axenically cultured it will not be possible to determine whether these temporally and spatially associated but morphologically dissimilar thalli are different forms of the same species.

Asexual reproduction is normally limited to the distal region of the thallus and is by successive septation from the apex to form a series of unisporous, basipetally maturing sporangia (Figs. 8.15, 8.16). Mature sporangiospores are usually released by rupture or dissolution of the lateral wall of the sporangium (Fig. 8.22). Except for the Parataeniellaceae the Eccrinales produce sporangiospores of two basic types. One spore type is multinucleate (normally four or eight nuclei), thin-walled, elongate, nonseptate (Fig. 8.15) and upon release from the sporangium germinates within the gut to increase the infestation within the same host (Fig. 8.17). Lichtwardt (1954a) and Cronin and Johnson (1958) have noted rudimentary holdfasts on these multinucleate sporangiospores which probably facilitate their immediate attachment to the host cuticle following release from the parent thallus. Some genera (i.e. *Alacrinella, Arundinula, Astreptonema, Enterobryus, Enteromyces, Lajassiella, Nodocrinella, Palavascia, Taeniella*) retain the wall of the germinated, thin-walled sporangiospore at the apex of the thallus as a slightly reflexed region (Fig. 8.17) which may abort, be sterile or function as an apical sporangium. In *Ramacrinella raibauti* and *Paramacrinella microdeutopi* Manier and Grizel (1971) the original spore wall is not retained at the apex of the thallus but close to the holdfast at the base of the thallus (Manier and Ormières 1961b; Manier and Grizel 1971). The second spore type in the Eccrinales is usually uninucleate, oval to ellipsoidal and frequently thick-walled (Fig. 8.16). Following their release through the lateral wall of the sporangium these thick-walled sporangiospores are passed from the digestive tract of the host with faecal matter, or with the cuticle at the time of moulting, and function to infest susceptible hosts which may ingest them whilst feeding. It is usual for each

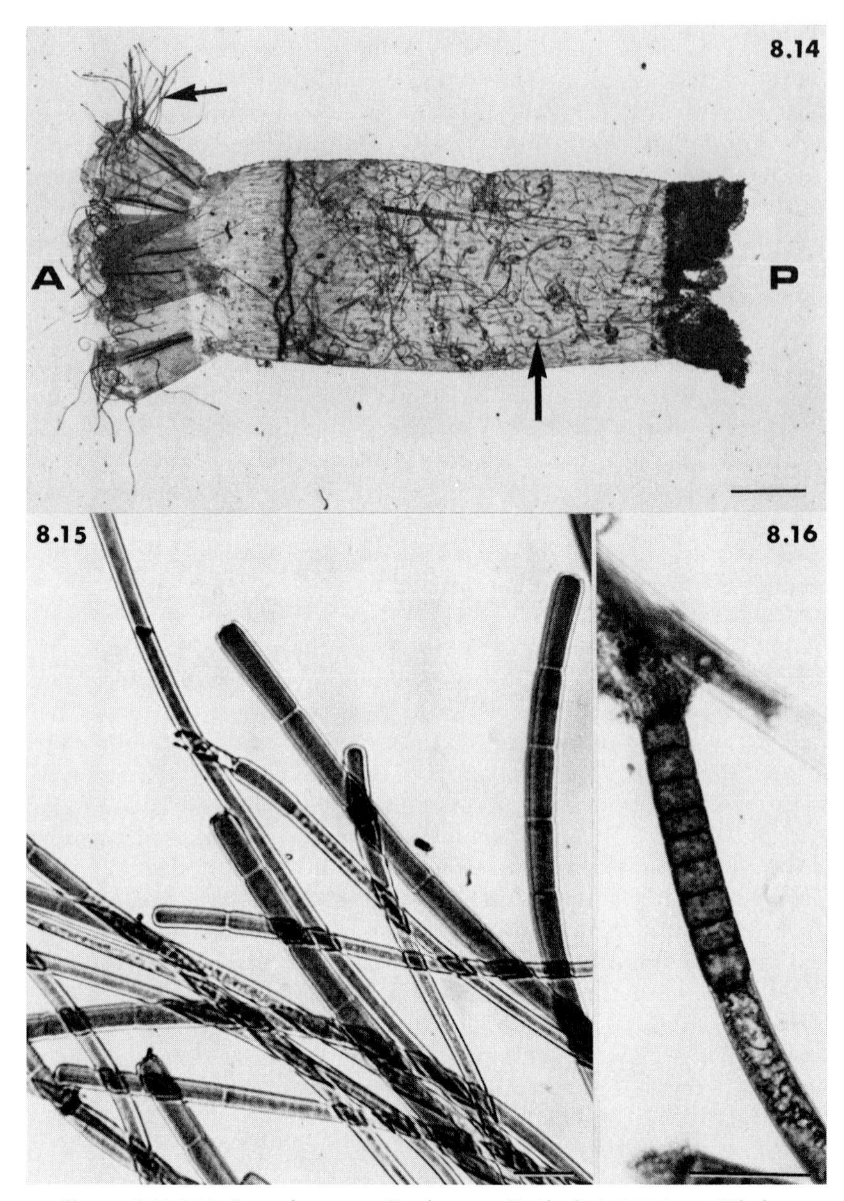

Figures 8.14–8.16. Enterobryus sp. (Eccrinaceae, Eccrinales). *8.14 (top).* Hindgut lining from *Thyropygus* sp., Diplopoda, with thalli producing thin-walled, multinucleate sporangiospores (arrow) attached within the anterior region (A) and thalli producing uninucleate, dispersive sporangiospores (dart) attached within the posterior region (P). (Bar = 1mm.) *8.15 (bottom left).* Distal region of thalli producing multinucleate sporangiospores. (Bar = 50μm.) *8.16 (bottom right).* Distal region of thallus with uninucleate, dispersive sporangiospores. (Bar = 50μm.) (*8.14–8.16.* Bright-field illumination, stained lactophenol / cotton-blue.)

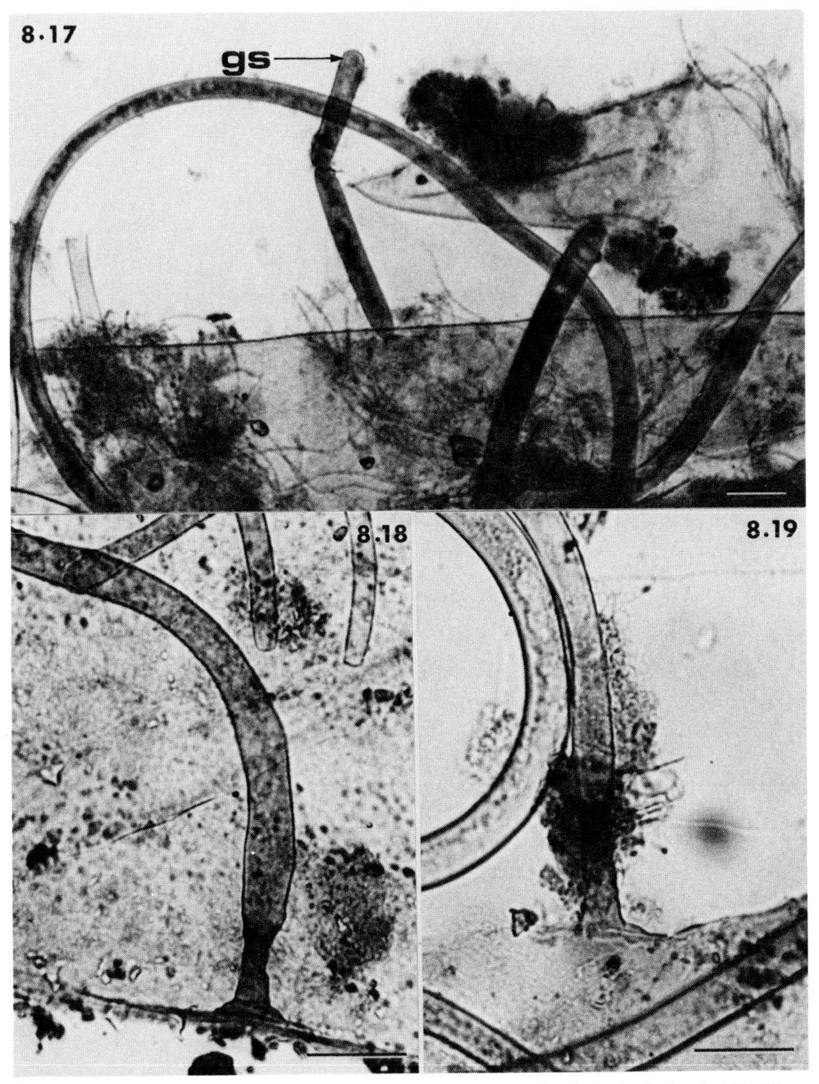

Figures 8.17–8.19. Enterobryus sp. (Eccrinaceae, Eccrinales). *8.17 (top).* Thalli attached to hindgut cuticle from *Thyropygus* sp., Diplopoda. Note the wall of the germinated, elongate sporangiospore (gs) borne distally and at an angle upon the developing thallus. (Bar=40μm.) *8.18 (bottom left), 8.19 (bottom right).* Holdfasts of mature thalli. The holdfast does not penetrate the host cuticle (*8.18*) and is frequently surrounded by bacteria (*8.19*). Bars=40μm.) (*8.17–8.19.* Bright-field illumination, stained lactophenol / cotton-blue.)

thallus to produce only one type of sporangiospore, however in *Astreptonema* spp. (Hauptfleisch 1895; Léger and Duboscq 1933; Manier 1964a), *Parataeniella* spp. (Lichtwardt and Chen 1964; Poisson 1929), *Enterobryus borariae* Lichtwardt (Lichtwardt 1958) and *E. dixidesmi* Lichtwardt (Lichtwardt 1960c) the thalli may produce both types of spore. The thin-walled, multinucleate sporangiospores are normally produced first and at the apex of the thallus and the thick-walled, uninucleate sporangiospores later and in the more proximal region. In the Parataeniellaceae, thalli may either produce binucleate, probably equivalent to the multinucleate, sporangiospores singly within terminal sporangia, convert to a single sporangium and produce only uninucleate sporangiospores or form binucleate spores singly in terminal sporangia and many uninucleate spores in a single, proximal sporangium (Lichtwardt and Chen 1964; Poisson 1929). Lichtwardt (1954b, 1958) recognized nine spore types in the Eccrinales but suggested that these may be no more than morphological variations of the two basic types. *Palavascia* spp. are atypical in that only one type of sporangiospore is produced. The sporulating regions of *Palavascia* spp. thalli frequently project from the hindguts of their marine, isopod hosts (Lichtwardt 1961b) and form thick-walled sporangiospores, each with two, short, lateral appendages (Lichtwardt 1973a) which, following release, infest other hosts. Certain sporangia in *Palavascia* spp. do not form sporangiospores but germinate *in situ* to produce lateral hyphae (microthalli?) which segment distally (Johnson 1966; Lichtwardt 1961b; Tuzet and Manier 1947a, 1948b). Subsequent development of these terminal cells (spores?), as with those produced by the microthalli of other genera, has not been determined. The presence of appendaged spores in the Eccrinales is restricted to the thick-walled, dispersive sporangiospores of a few genera associated with freshwater and marine Crustacea. The appendages are formed within the sporangium and only become apparent following spore release. The spores of *Palavascia sphaeromae* Tuzet and Manier (1948b) ex Manier (1968) possess two, short, lateral appendages (Lichtwardt 1973a) and those of *Arundinula capitata* Léger and Duboscq (1906), associated with the hermit crab *Pagurus maculatus* Hell., have a single, spine-like appendage at each pole (Duboscq, Léger and Tuzet 1948). Appendages have not been found associated with the spores of other species of *Arundinula* (Lichtwardt 1962; Manier and Ormières 1962). Longer, often several hundred micrometers, flexuose appendages occur in *Taeniella carcini* Léger and Duboscq (1911) found within the hindguts of marine crabs (Decapoda, Crustacea, Figs. 8.20, 8.21) and *Astreptonema* spp. from the hindguts of freshwater and marine Amphipoda, Crustacea (Figs. 8.22, 8.23). Frequently adjacent, released spores of *T. carcini* are held together by what Johnson (1966) described as a "granular-gelatinous material", however if such spores are left in seawater the gelatinous material is seen to comprise the highly folded appendages, two at each pole. Similar appendages are attached to the spores of *A. gammari* (Léger and Duboscq 1933;

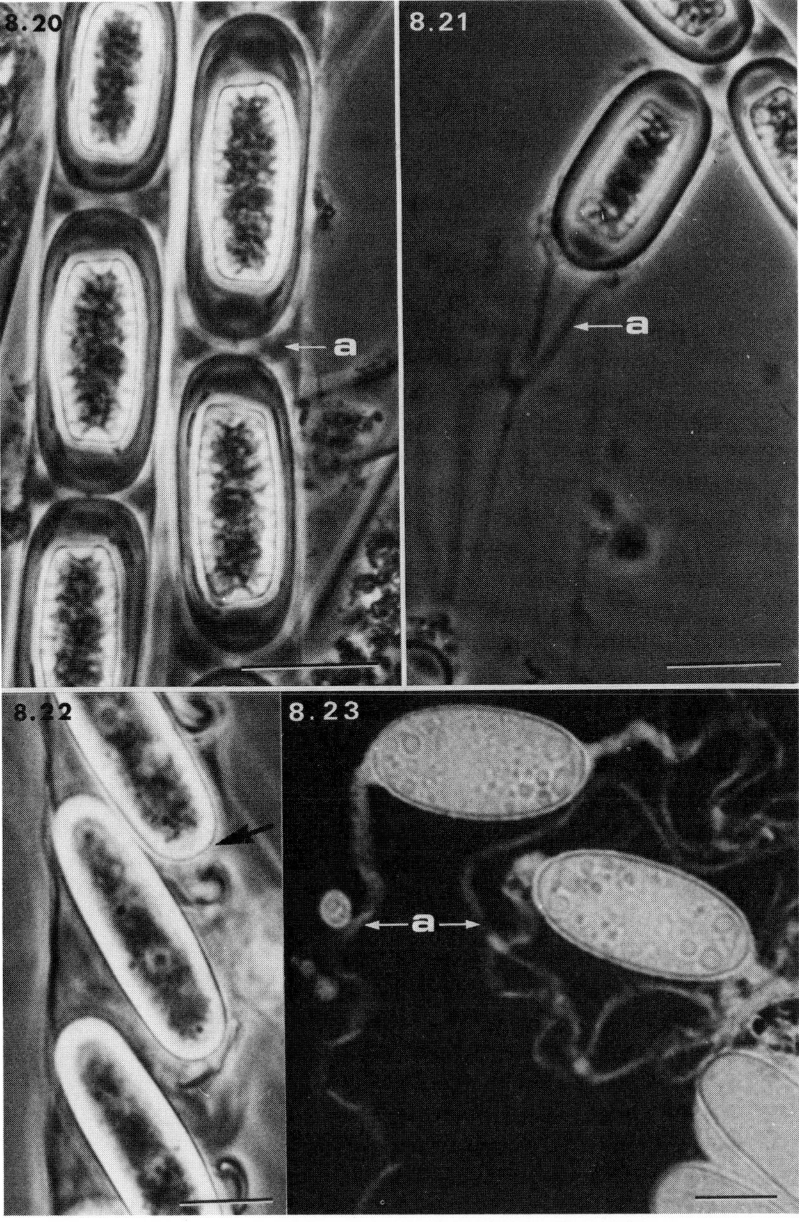

Figures 8.20–8.23. 8.20–8.21. Taeniella carcini (Eccrinaceae, Eccrinales). *8.20.* Mature, thick-walled sporangiospores contained within the distal region of thallus. The dense regions (a) between spores are the folded appendages. (Phase-contrast illumination, bar = 10 μm.) *8.21.* Released sporangiospore showing the two appendages (a) at each pole. (Phase-contrast illumination, bar = μm). *8.22–8.23. Astreptonema gammari* (Eccrinaceae, Eccrinales). *8.22.* Mature, unreleased sporangiospores. Release is by rupture, or dissolution (?), of the lateral wall of the thallus. (Phase-contrast illumination, bar = μm.) *8.23.* Released sporangiospores showing the single, mucilaginous appendage (a) at each pole. (India ink mount, bar = 10μm.) each pole. (Indian ink mount, bar = 10μm.)

Manier 1964a) except that only a single appendage occurs at each pole (Fig. 8.23). Preliminary observations (Moss 1972) have shown that the appendages of *A. gammari* are extensions of an outer, mucilaginous, sporangiospore wall which is formed early in spore differentiation by the extracellular deposition of golgi-derived material and thus differ in both ontogeny and structure from the sporangial appendages of the Harpellales.

Sexual reproduction has not been confirmed in any species of Eccrinales. The early reports of karyogamy followed by reduction division involving adjacent protoplasts in the sporulating regions of *Palavascia philosciae* Tuzet and Manier (1947a) ex Manier and Lichtwardt (1968), *P. sphaeromae* (Manier 1950; Tuzet and Manier 1948b), *Eccrinidus flexilis* Léger and Duboscq (1906) Manier (1969a) and *Enterobryus duboscqui* Tuzet and Manier (1948a) ex Manier (1968) have not been confirmed. However Galt (1971) has reported scalariform fusions between thalli of *Taeniella carcini* in the exuviae of *Pagurus kennerlyi* Stimpson (Decapoda) and speculated that these fusions may be involved in the formation of the thick-walled, resistant spores. Further work is necessary before these stages can be interpreted as sexual. A generalized life cycle of the Eccrinales is represented in Figure 8.24.

Harpellales. The Harpellales is the most thoroughly investigated order of Trichomycetes. Attachment of their thalli is restricted to the digestive tracts of freshwater larval and nymphal stages of Insecta (Table 8.3). The order is characterized by the formation of exogenous, dehiscent, unisporous sporangia (Moss and Lichtwardt 1976) termed trichospores (Manier and Lichtwardt 1968), which upon release are seen to possess one to several basally attached, non-motile appendages, and, in those species in which sexual reproduction has been observed, biconical zygospores (Moss and Lichtwardt 1977; Moss, Lichtwardt and Manier 1975). The order is subdivided into two families distinguished by their regions of attachment and vegetative morphology. The Harpellaceae contains three genera (*Carouxella, Harpella, Stachylina*) characterized by thalli which are attached to the peritrophic membrane lining the midguts of dipteran larvae (Table 8.3, Fig. 8.25). Thalli of the Genistellaceae with thirteen genera (*Genistella, Genistellospora, Glotzia, Graminella, Orphella, Pennella, Pteromaktron, Simuliomyces, Smittium, Spartiella, Stipella, Trichozygospora, Zygopolaris*) are branched and attached to the intucking of the external cuticle lining the proctodaea of dipteran larvae and ephemeropteran and plecopteran nymphs (Table 8.3, Fig. 8.26). Reproductively the two families are indistinguishable and it is probable that this subdivision of the order is unnatural (Moss and Young 1978).

Vegetative thalli are nonseptate (Harpellaceae) or irregularly septate (Genistellaceae) and attached to, but, except for *Stachylina minuta* Gauthier (1961) (Fig. 8.37), do not penetrate, the host cuticle by a holdfast (Fig. 8.35). This may be a well defined, localized secretion at the base of the

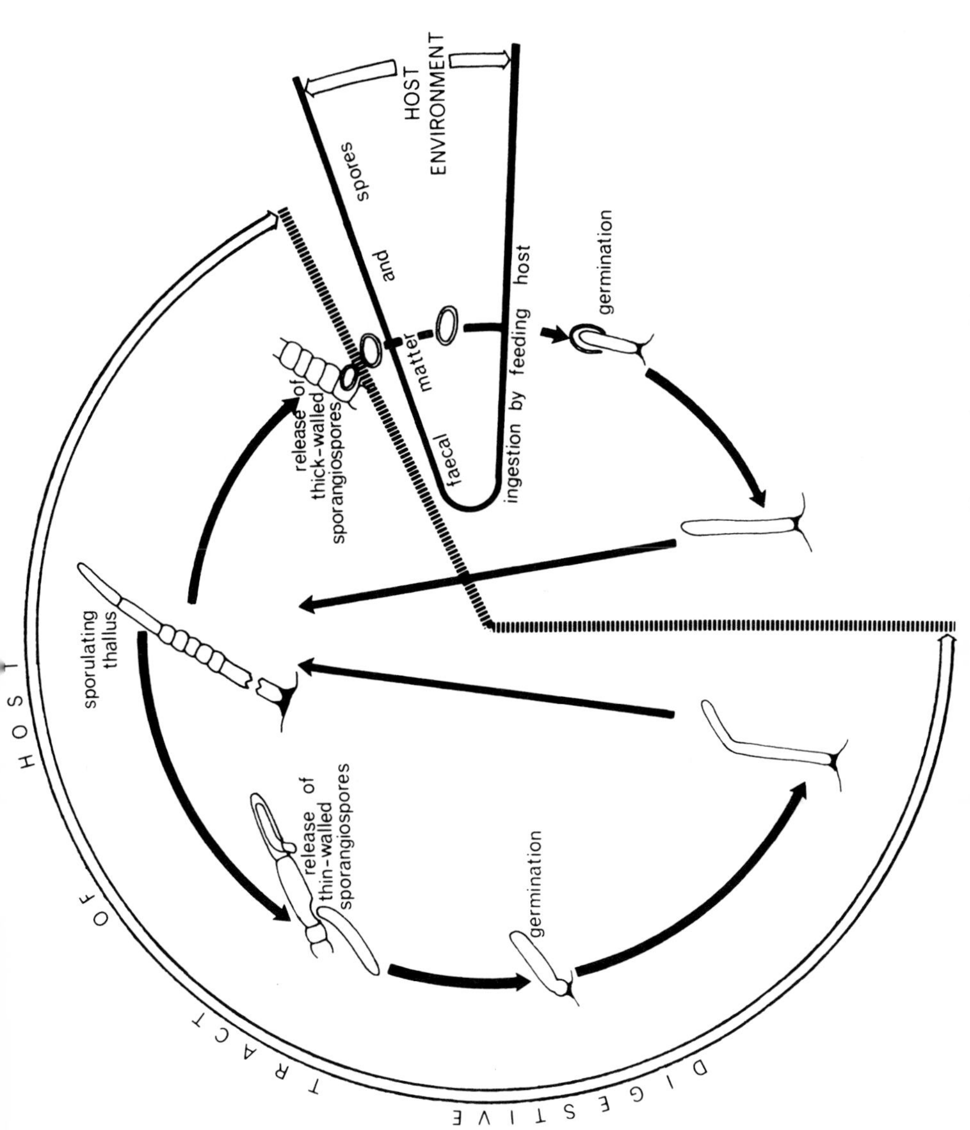

Figure 8.24. Diagrammatic representation of the life-cycle of the Eccrinales.

Table 8.3 Habitat and Host Range of the Harpellales

HARPELLALES Manier (Lichtwardt and Manier 1978)		ARTHROPOD HOST						
Family	Genus	Class	Order	Family	Stage Infested	Region of Thallus Attachment	Habitat	Principal References
HARPELLACEAE Léger & Duboscq (1929b)	*Carouxella*[a] Manier, Rioux & Manier (1961) ex Manier & Lichtwardt (1968)	INSECTA	DIPTERA	CERATOPOGONIDAE	Larva	Peritrophic membrane Lining Mesenteron	Freshwater	Manier 1969b; Manier & Lichtwardt 1968; Rioux & Whisler 1961, 1965.
	Harpella[a] Léger & Duboscq (1929b)	INSECTA	DIPTERA	SIMULIIDAE	Larva			Crosby 1974; Frost & Manier 1971; Léger & Duboscq 1929b; Lichtwardt 1967; Moss 1970; Reichle & Lichtwardt 1972.
	Stachylina Léger & Gauthier (1932)	INSECTA	DIPTERA	CHIRONOMIDAE	Larva			Gauthier 1961; Kobayasi et al. 1969; Leger & Gauthier 1932; Lichtwardt 1972; Manier & Coste-Mathiez 1971; Moss 1974, 1976.
	Genistella[a] Léger & Gauthier (1932)	INSECTA	EPHEMEROPTERA	BAETIDAE	Nymph			Gauthier 1960; Léger & Gauthier 1932, 1935; Manier 1962a, 1973a, Tuzet & Manier 1953, 1955b; Whisler 1963.
	Genistellospora[a] Lichtwardt (1972)	INSECTA	DIPTERA	SIMULIIDAE	Larva			Lichtwardt 1972; Moss & Lichtwardt 1976, 1977.
	Glotzia Gauthier (1936) ex Manier & Lichtwardt (1968)	INSECTA	EPHEMEROPTERA	BAETIDAE	Nymph			Gauthier 1936; Lichtwardt 1972; Manier 1968; Manier & Lichtwardt 1968.
	Graminella[a] Léger & Gauthier (1937) ex Manier (1962a)	INSECTA	EPHEMEROPTERA	BAETIDAE	Nymph			Léger & Gauthier 1937; Manier 1962a.

GENISTELLACEAE Léger & Gauthier (1932)	*Orphella*[a] Léger & Gauthier (1931)	INSECTA	PLECOPTERA	NEMURIDAE	Nymph			Léger & Gauthier 1931, 1932.
	Pennella Manier (1963b) ex Manier (1968)	INSECTA	DIPTERA	SIMULIIDAE	Larva			Frost & Manier 1971; Lichtwardt 1972; Manier 1963b, 1968; Williams & Lichtwardt 1971.
	Pteromaktron[a] Whisler (1963)	INSECTA	EPHEMEROPTERA	BAETIDAE	Nymph	Proctodaeum	Freshwater	Whisler 1963.
	Simuliomyces[a] Lichtwardt (1972)	INSECTA	DIPTERA	SIMULIIDAE	Larva			Lichtwardt 1972.
	Smittium Poisson (1936)	INSECTA	DIPTERA	CERATOPOGONIDAE CHIRONOMIDAE CULICIDAE SIMULIIDAE	Larva			Clark, Kellen & Lindegren 1963, El-Buni & Lichtwardt 1976a,b; Farr & Lichtwardt 1967; Kobayasi et al. 1969; Lichtwardt 1964b, 1972; Manier 1962a, 1969b; Manier & Coste-Mathiez 1971; Manier & Mathiez 1965; Manier, Rioux & Juminer 1964; Poisson 1936; Tuzet & Manier 1947b, 1953; Williams & Lichtwardt 1972a,b.
	Spartiella[a] Tuzet & Manier (1950a) ex Manier (1968)	INSECTA	EPHEMEROPTERA	BAETIDAE	Nymph			Manier 1962b, 1968; Tuzet & Manier 1950a, 1953.
	Stipella[a] Léger & Gauthier (1932)	INSECTA	DIPTERA	SIMULIIDAE	Larva			Léger & Gauthier 1932; Manier 1950, 1963b; Moss 1970; Tuzet & Manier 1950a.
	Trichozygospora[a] Lichtwardt (1972)	INSECTA	DIPTERA	CHIRONOMIDAE	Larva			Lichtwardt 1972, 1976; Moss & Lichtwardt 1976, 1977.
	Zygopolaris[a] Moss, Lichtwardt & Manier (1975)	INSECTA	EPHEMEROPTERA	BAETIDAE	Nymph			Moss, Lichtwardt & Manier 1975; Moss & Lichtwardt 1977.

[a]Monotypic

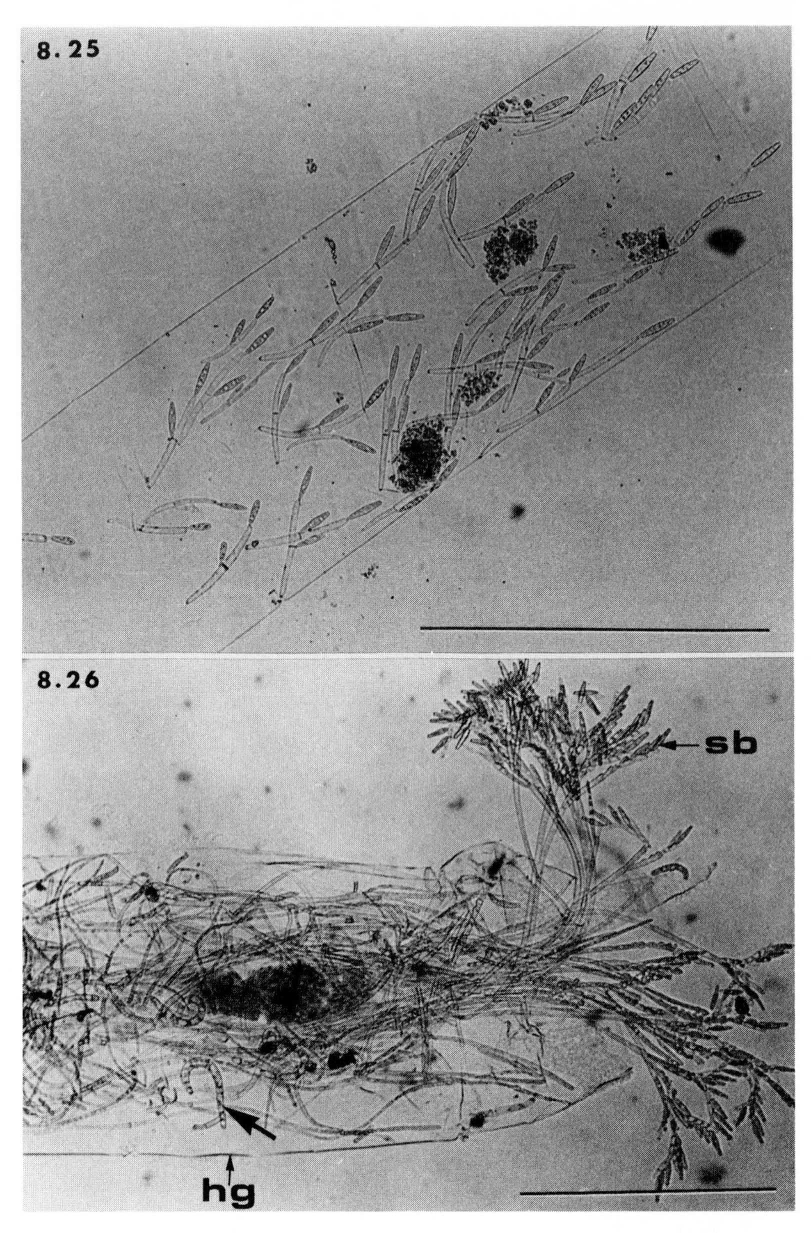

Figures 8.25–8.26. 8.25 (top). Stachylina grandispora (Harpellaceae, Harpellales). Thalli attached to the peritrophic membrane removed from the midgut of a *Tanytarsus* sp., Diptera, larva. (Bright-field illumination, bar=500μm.) *8.26 (bottom). Zygopolaris ephemeridarum* (Genistellaceae, Harpellales). Thalli attached within, and sporulating branches (sb) projecting from, the cuticular lining (hg) removed from the hindgut of a *Baetis* sp., Ephemeroptera, nymph. Thalli of *Paramoebidium* (Amoebidiales) (arrow) are also present. (Bright-field illumination, bar=500 μm.)

thallus (e.g. *Harpella, Stachylina, Genistellospora, Graminella, Simuliomyces, Trichozygospora, Zygopolaris)* or a more diffuse structure in the form of a mucilaginous secretion which enrobes the base of the thallus and causes it to lie adpressed to the cuticle (e.g. *Stipella, Pennella)* (Fig. 8.34). Other types of attachment mechanisms are also recognized. In *Pteromaktron protrudens* Whisler (1963) the holdfast is described as "much branched and corraloid" (Whisler 1963) with no mention of a mucilaginous sheath or an adhesive pad, whereas in *Glotzia ephemeridarum* Lichtwardt (1972) the basal region of the axial hypha of the thallus has a unilateral series of small holdfasts (Fig. 8.36). The limited fine-structural information available has shown that certain morphologically similar holdfasts (e.g. *Harpella melusinae* Léger and Duboscq (1929b), *Stachylina grandispora* Lichtwardt (1972), *Genistellospora homothallica* Lichtwardt (1972) possess different substructures and modes of secretion of the adhesive material. In *H. melusinae* the basally secreted holdfast is associated with pores in the wall of a delimited basal region (Reichle and Lichtwardt 1972), whilst in the mature vegetative thallus of *S. grandispora* the holdfast material forms a thin layer around the structurally unmodified base of the thallus (Moss 1972). A more conspicuous holdfast occurs in *G. homothallica* (Moss, unpublished) and although in the mature thallus there is no associated modification of the cell wall, the mature sporangiospore does possess several pores in an apically thickened region of the wall. It has been suggested (Moss and Lichtwardt 1976) that holdfast material is secreted through these pores in order that the sporangiospore may attach to the host gut immediately after trichospore germination. In some species, such as *G. homothallica,* holdfast formation may occur for only a limited period following germination whereas in others, e.g. *H. melusinae,* it may be a continuous process. *Stachylina minuta* is the only species of Harpellales, and indeed Trichomycetes, in which the holdfast is known to penetrate the host cuticle (Fig. 8.37) although not the underlying epithelial tissue.

Trichospore formation is preceded by regular septation of the entire thallus, Harpellaceae (Fig. 8.27), or distal regions of lateral branches, Genistellaceae (Figs. 8.30, 8.31), into specialized, uninucleate generative cells. Trichospores are produced unilaterally and in basipetal succession, one from the distal region of each generative cell (Figs. 8.27, 8.30, 8.31). Early in trichospore formation the single generative cell nucleus divides, and one daughter nucleus migrates into the spore initial whilst the other remains within the generative cell (Moss 1974). In some genera (e.g. *Harpella)* trichospores are borne directly on their generative cells whereas in others (e.g. *Genistella,* Fig. 8.31) they develop terminally on short, nonseptate, lateral branches. Except for the specialized subsidiary cells of *Pteromaktron protrudens* (Whisler 1963) these spore-bearing branches are not delimited from their generative cells by septa. Mature trichospores are characterized by one to several basally attached appendages. These are formed either early (*Genistellospora homothallica,* Moss and Lichtwardt

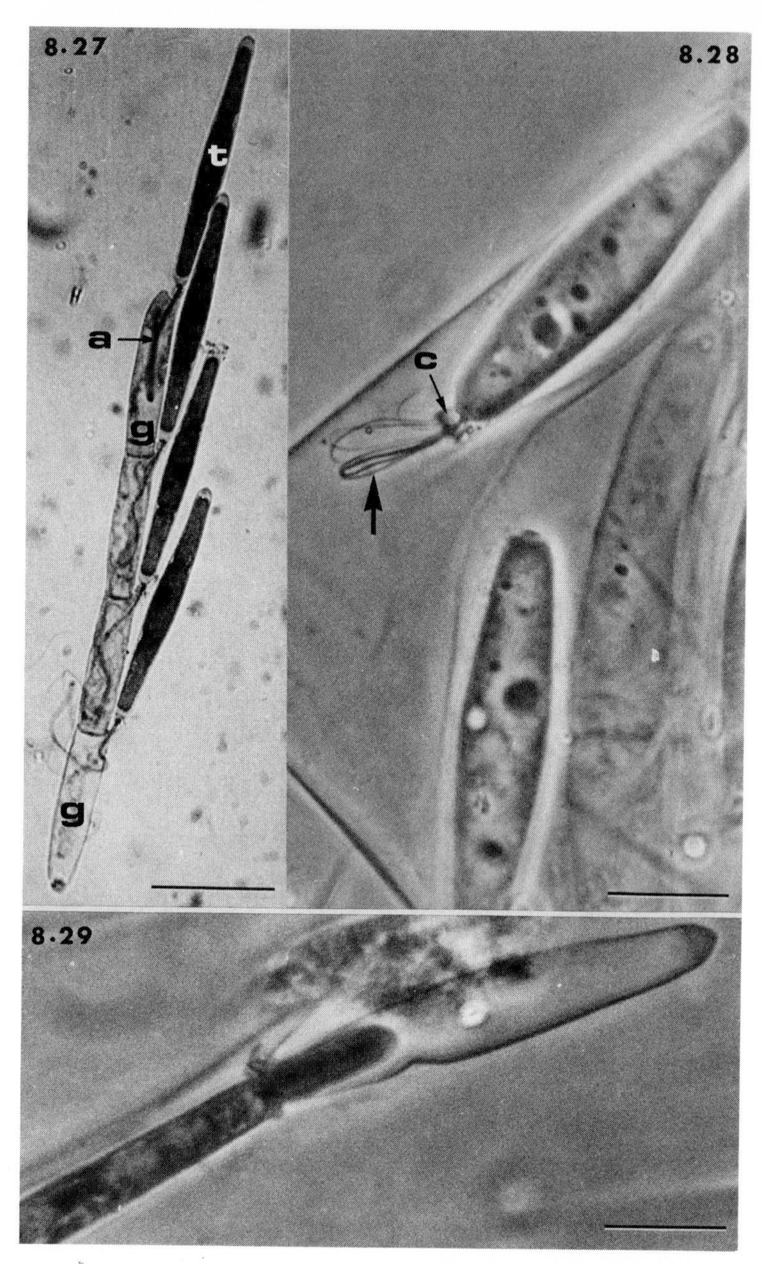

Figures 8.27–8.29. Stachylina grandispora (Harpellaceae, Harpellales). *8.27 (top left).* Mature thallus with a trichospore (t) produced termino-laterally from each generative cell (g). Note the single appendage (a) within the generative cell. (Bright-field illumination, stained lactophenol / cotton-blue, bar = 20μm.) *8.28 (top right).* Released trichospore with the highly folded appendage (arrow) extending from within the collar region (c) of the trichospore. Phase-contrast illumination, bar = 20μm.) *8.29 (bottom).* Distal region of young thallus with the original, trichospore wall (= sporangium wall) still attached. (Phase-contrast illumination, bar = 20μm.)

1976) or late (*Stachylina grandispora,* Moss 1976) in development of trichospore and external to the plasmalemma of the generative cell by the deposition of intracellularly synthesized material (Moss 1976; Moss and Lichtwardt 1976). Mature appendages are contiguous with the wall at the base of the trichospore and are considered extensions of the sporangial wall. The number of appendages and their arrangement within the generative cell are characteristic for each genus. In *S. grandispora* the appendage is folded within a single invagination of the plasmalemma (Moss 1976) which extends into the lumen of the generative cell (Figs. 8.27, 8.28). Each of the 5–7 appendages of *G. homothallica* is formed adpressed to the longitudinal wall of the generative cell and within a separate invagination of the plasmalemma (Moss and Lichtwardt 1976), whereas the two and four appendages found in *Genistella* spp. (Figs. 8.31–8.33) and *Harpella melusinae* respectively are spiralled around the distal region of the generative cell wall (Manier 1973a; Reichle and Lichtwardt 1972). Similarly, the substructure of appendages has been found to differ. Appendages of *H. melusinae* have a regular banded structure (Reichle and Lichtwardt 1972), those of *S. grandispora* (Moss, 1976), *Genistella ramosa* Léger and Gauthier (1932) (Manier 1973a) and *G. homothallica* (Moss and Lichtwardt 1976) have no distinctive substructure and those of *Smittium mucronatum* Manier and Mathiez (1965) ex Manier (1969b) (Manier and Coste-Mathiez 1968) and *S. culicis* (Tuzet and Manier 1947b) Manier (1969b) comprise in transverse section several electron-opaque concentric layers (Moss and Lichtwardt 1976). Trichospores are released from their generative cells by dissolution, or rupture, of the wall at either the base of the trichospore (e.g. *Genistella)* or at the base of the supporting branch *(Stachylina, Smittium, Trichozygospora).* In the latter case a region of the supporting branch remains attached to the released trichospore as a collar from within which the appendage(s) emerges (Fig. 8.28). Upon spore release the appendages, which may be several hundred micrometers long, are withdrawn from the generative cell and trail behind the spore as fine, non-motile filaments (Figs. 8.32, 8.33). Detached trichospores are passed from the gut into the host environment and only germinate after ingestion by a host. Germination is normally polar (Fig. 8.29). It has been shown (El-Buni and Lichtwardt 1976b) that in axenic cultures of *Smittium* spp. trichospore germination is either by a process resembling germ tube formation or by the partial or complete emergence of the sporangiospore from the sporangium (Fig. 8.29). Although the dispersive and infestive propagule is the sporangium containing a single sporangiospore (trichospore), it is the sporangiospore which, subsequent to germination, attaches to the host cuticle. Following spore release the thallus degenerates.

The Harpellales is the only order of Trichomycetes in which sexual reproduction has been confirmed. Zygospores are known in many genera and with the exception of *G. homothallica* (Lichtwardt, 1972) their formation is preceded by hyphal conjugation. In the unbranched Harpel-

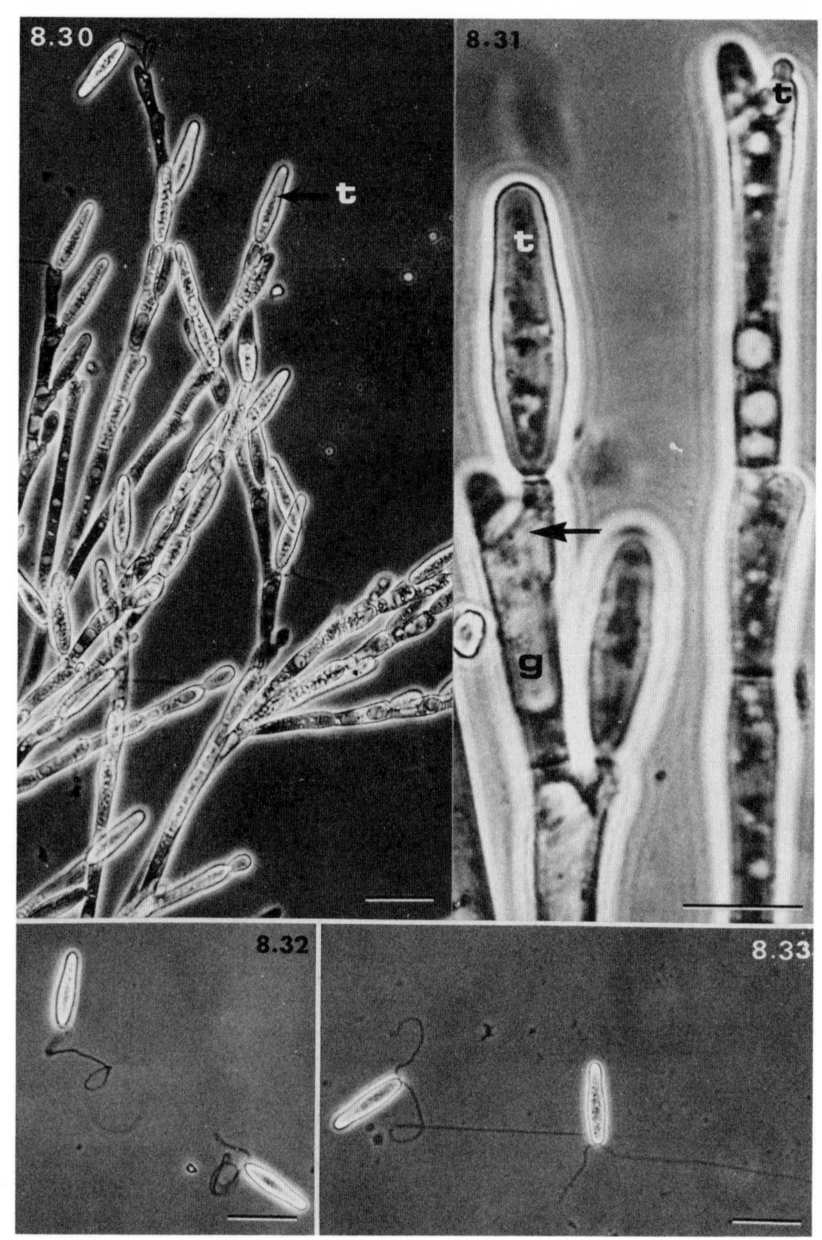

Figures 8.30–8.33. Genistella sp. (Genistellaceae, Harpellales). *8.30.* Sporulating branches showing the basipetal development of trichospores (t) from their generative cells. (Bar=40μm.) *8.31.* Mature and immature trichospores (t) showing the appendages (arrow) spirally arranged within the generative cell (g). (Bar=20μm.) Figs. *8.32–8.33.* Released trichospores with partially (*8.32*) and fully extended (*8.33*) appendages; one long and one short appendage. (Bars=40μm.) (*8.30–8.33.* Phase-contrast illumination.)

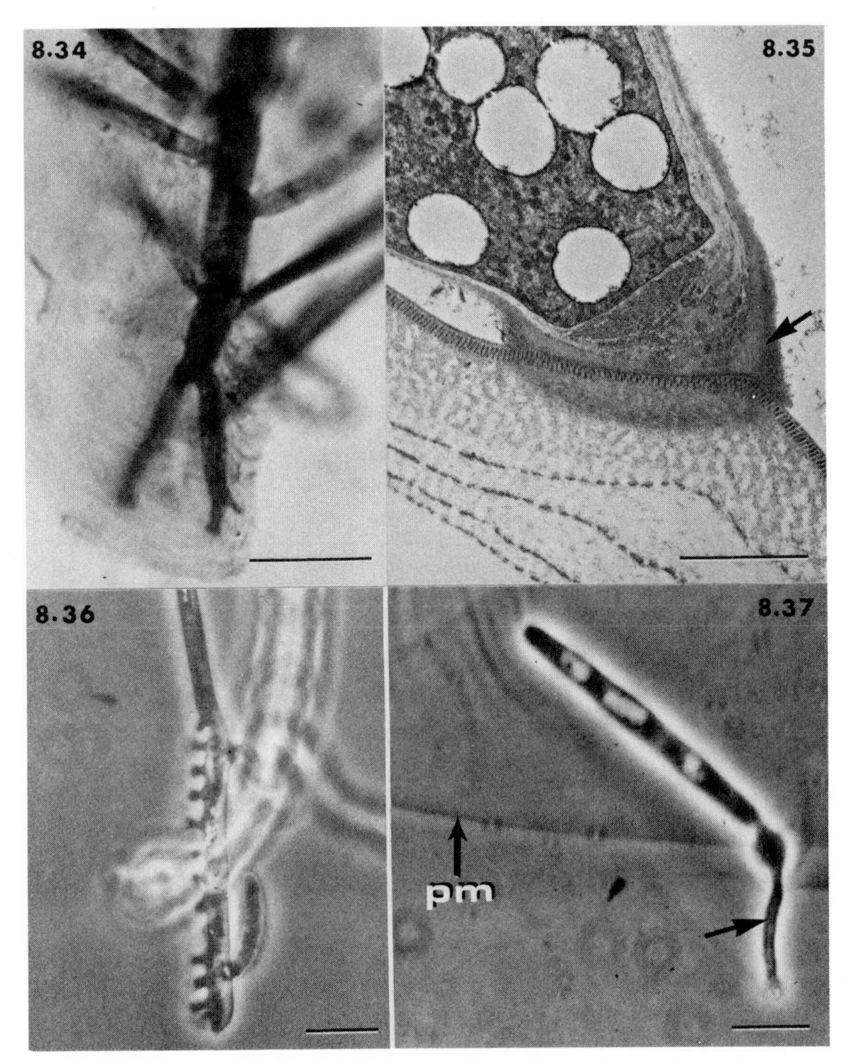

Figures 8.34–8.37. Holdfasts of the Harpellales. *8.34 Stipella vigilans* (Genistellaceae). Basal, branched region of the axial hypha of thallus. Mucilage associated with this region attaches the thallus to the host cuticle. (Bright-field illumination, stained lactophenol / cotton-blue, bar = 20μm.) *8.35. **Harpella** sp.* (Harpellaceae). Transmission electron micrograph; L.S. base of thallus, secreted holdfast material (arrow) and the associated peritrophic membrane. Note that the holdfast material does not penetrate all layers of the peritrophic membrane. (Fixative - potassium permanganate, bar = 1μm.) *8.36. Glotzia ephemeridarum* (Genistellaceae). Base of thallus showing the unilateral series of peglike holdfasts. (Phase-contrast illumination, bar = 20μm.) *8.37. Stachylina minuta* (Harpellaceae). Immature thallus with a dilated base and hyphal extension (arrow) through the peritrophic membrane (pm). (Phase-contrast illumination, bar=20μm.)

laceae conjugation can only occur between cells of adjacent thalli, frequently several conjugations producing a scalariform arrangement. In the Genistellaceae conjugation may occur between cells of the same or different thalli (Figs. 8.38, 8.39). There is, however, no evidence of physiological or morphological heterothallism. Zygospores are produced upon outgrowths, termed zygosporophores, from either one of the two conjugants or in *Glotzia* spp. from the conjugation tube (Lichtwardt 1972). Mature zygospores are biconical and polarly thickened. Moss et al. (1975) recognize four types of zygospore distinguished by their position and angle of attachment to the zygosporophore. Attachment may be median with the zygospore perpendicular to the zygosporophore (*Harpella, Simuliomyces, Spartiella, Stipella* (Fig. 8.39), submedian and oblique to the zygosporophore (*Genistella, Glotzia, Smittium, Trichozygospora),* parallel with attachment median (*Genistellospora, Pennella)* or coaxial to the zygosporophore with polar attachment (*Carouxella, Zygopolaris* (Fig. 8.38). All zygospores examined have been found to contain a single nucleus but its genetic state has not been established. Based on fragmentary ultrastructural evidence, Moss and Lichtwardt (1977) hypothesized that in *H. melusinae* karyogamy followed by meiosis occurs within the conjugation tube and that a single haploid nucleus migrates into the zygospore. However, based on the formation of these spores subsequent to conjugation, their thickened walls, and possession of storage material, they are considered zygospores (Moss and Lichtwardt 1977).

Zygospores of most genera detach from the apex of the zygosporophore and do not possess appendages. However in *Genistellospora homothallica* the membrane-bound, glycogen-rich contents of the zygosporophore remain attached to the released zygospore (Lichtwardt 1972) and may serve as a food reserve for subsequent development of the zygospore (Moss and Lichtwardt 1977) whilst in *Zygopolaris ephemeridarum* Moss, Lichtwardt and Manier (1975) fibrous material withdrawn from the zygosporophore remains attached to the released zygospore (Moss and Lichtwardt 1977). Only in *Genistella, Smittium* and *Trichozygospora* does the released zygospore possess an appendage(s) and a collar. The collar is formed from the distal region of the zygosporophore wall and in *Trichozygospora ephemeridarum* Lichtwardt (1972) appendage ontogeny and structure is similar to that of the trichospore appendages (Moss and Lichtwardt 1977). The protoplast of the mature zygospore comprises, in at least some species, a folded filament with thickened apices. Whisler (1963), during attempts to culture *Genistella ramosa,* and Lichtwardt (1972), by observing zygospores of *Glotzia ephemeridarum* in water mounts, showed that germination occurs by "forceful ejection" of the protoplast through the apical walls, aided by the conically-thickened protoplast wall (Fig. 8.40). It is presumed that these zygospores germinate only when ingested by a host.

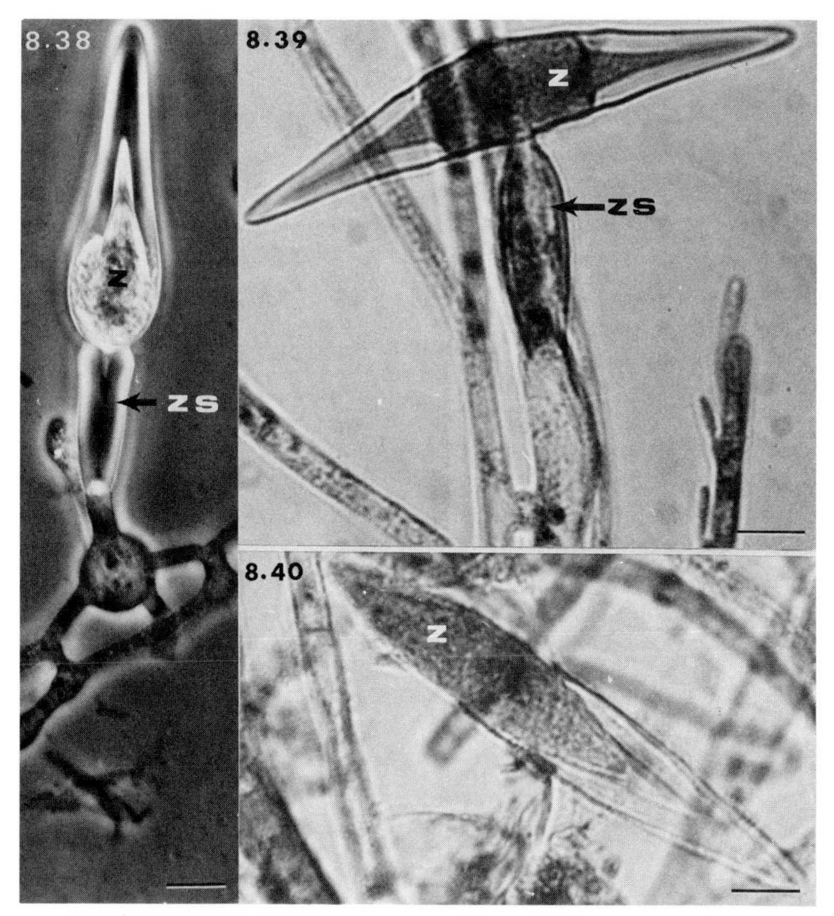

Figures 8.38–8.40. Zygospores of the Harpellales. *8.38 (left). Zygopolaris ephemeridarum* (Genistellaceae). Mature, apically thickened zygospore (z) attached polarly upon the zygosporophore (zs) originating from one of the conjugated cells. (Phase-contrast illumination, bar=10μm.) *8.39–8.40. Stipella vigilans* (Genistellaceae). *8.39 (top right).* Mature, polarly thickened zygospore (z) attached medianly upon the zygosporophore (zs). *8.40 (bottom right).* Germinated zygospore (z) from hindgut of a *Simulium* sp., Diptera, larva. (*8.39, 8.40.* Bright-field illumination, stained lactophenol / cotton-blue, bars=10μm.)

The life-cycle of the Harpellales is presented in Figure 8.41. Trichospores released from the digestive tract of an infested host during defecation, ecdysis or death will only germinate following their ingestion by a host. In *Pteromaktron protrudens* and *Zygopolaris ephemeridarum* (Fig. 8.26) the sporulating branches project from the hindgut and their trichospores are released directly into the host environment (Moss et al. 1975; Whisler 1963). Only under exceptional conditions (Moss 1972) do trichospores germinate within the same individual without first leaving the host. Correspondingly, maintenance of the infestation in any population is dependent on trichospores being retained within the host environment and their subsequent ingestion by feeding hosts.

Although many genera produce zygospores their role in the life cycle is uncertain. The presence of thickened walls and storage material does indicate that they may serve as resting spores and enable an inoculum to survive conditions lethal or at least detrimental to either the host population or to the growth of vegetative thalli and production of trichospores. Perhaps the reason that zygospores are only found infrequently is that investigators have tended to examine healthy host populations. Examination of larval stages subjected to adverse conditions (e.g. low oxygen tension, extremes of temperature, poor food supply) may yield more information on this phase of the life cycle. Indeed it has been noted that zygospores are more frequent in larvae and nymphs immediately prior to ecdysis, within the exuviae and at low temperatures (Moss et al. 1975).

Asellariales. The Asellariales is an order of branched Trichomycetes in which the arthrospore is the only known reproductive propagule. Thalli occur within the hindguts of terrestrial Collembola and dipteran larvae, Insecta, and marine, freshwater and terrestrial Isopoda, Crustacea (Table 8.4). The order contains a single family, the Asellariaceae, with three genera—*Asellaria, Orchesellaria* and *Trichoceridium.*

Vegetative thalli are branched, irregularly septate and attached to the

Table 8.4 Habitat and Host Range of the Asellariales

ASELLARIALES[b]		ARTHROPOD HOST		
Family	Genus	Class	Order	Family
ASELLARIACEAE Manier (1950) ex Manier & Lichtwardt (1968)	*Asellaria* Poisson (1932a)	CRUSTACEA	ISOPODA	ARMADILLIDIIDAE ASELLIDAE LIGIIDAE
	Orchesellaria Manier (1958) ex Manier & Lichtwardt (1968)	INSECTA	COLLEMBOLA	ENTOMOBRYIDAE ISOTOMIDAE
	Trichoceridium[a] Poisson (1932b)	INSECTA	DIPTERA	TIPULIDAE

[a]Monotypic
[b]Manier and Lichtwardt (Lichtwardt and Manier 1978)

cuticle of their host by either a definite, secreted holdfast or a mucilaginous sheath associated with a modified basal cell (Fig. 8.42; Figs. 8.45, 8.47). In *Asellaria* spp. the basal cell may be bulbous (*A. ligiae* Tuzet and Manier (1950a) ex Manier (1968), digitate (*A. armadillidi* Tuzet and Manier (1953) ex Manier (1968) and *A. aselli* Scheer, (1944) Scheer (1972b) (Figs. 8.45, 8.47) or spoon-shaped with marginal protuberances (*A. caulleryi* Poisson (1932a). In *Orchesellaria* spp. the basal cell is normally bilobed with a restricted region of holdfast secretion (*O. mauguioi* Manier (1964b) ex Manier (1969b), Fig. 8.42) or, as in *O. lattesi* Manier (1958) ex Manier (1969b), elongate with a holdfast extending the length of the basal cell. *Trichoceridium,* with the single species *T. ramosum* Poisson (1932b), has a well defined secreted holdfast. In *Asellaria* and *Trichoceridium* the thallus comprises a single, axial hypha from which the lateral branches originate (Fig. 8.45) whereas in *Orchesellaria* the several principal branches originate from the holdfast region (Fig. 8.42).

Reproduction involves regular septation of the lateral branches into a series of uninucleate cells (Fig. 8.46), followed by their disarticulation to form one-celled arthrospores (Figs. 8.42, 8.43). Arthrospores may germinate either within the same host gut, be passed from the gut with fecal matter or be rejected with the cuticle at the time of moulting. Infestation of hosts occurs subsequent to the ingestion followed by germination and attachment of the arthrospores (Fig. 8.44) to the host gut.

Sexual reproduction has not been found although Lichtwardt (1973a) has observed conjugations between cells of *A. ligiae* which resemble those which precede zygospore formation in the Harpellales. The life cycle of the Asellariales is presented in Figure 8.48.

Stage Infested	Region of Thallus Attachment	Habitat	Principal References
Adult Adult Adult	Proctodaeum	Terrestrial Freshwater Marine and Freshwater	Lichtwardt 1973a, 1976; Manier 1950, 1963a, 1968a, 1969b, 1973b; Poisson 1932a; Scheer 1972a,b.
Imago	Proctodaeum	Terrestrial	Manier 1958, 1964b, 1969b, Manier & Lichtwardt 1968; Moss 1975.
Larva	Procotodaeum	Terrestrial	Manier 1950, Poisson 1932b.

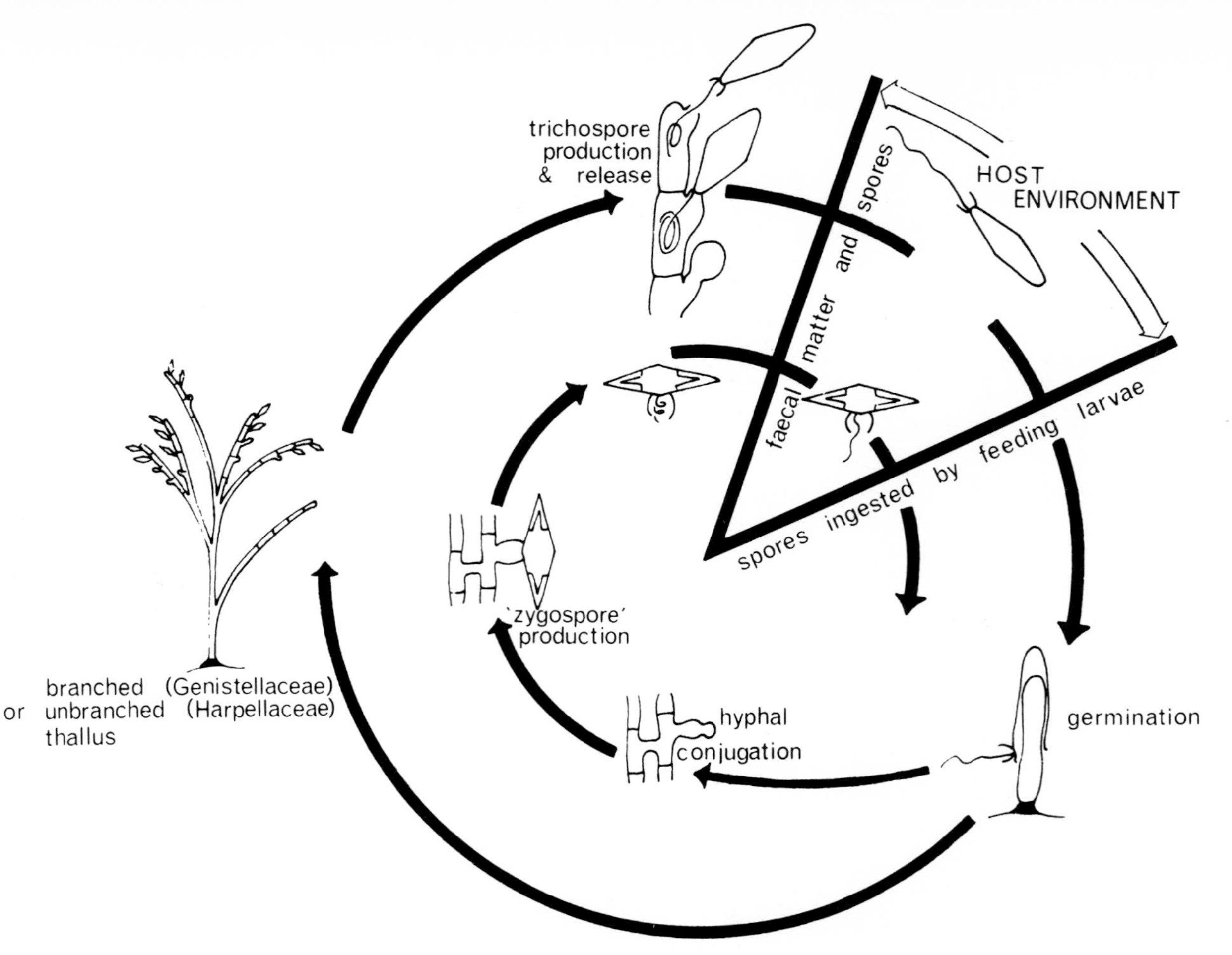

Figure 8.41. Diagrammatic representation of the life-cycle of the Harpellales.

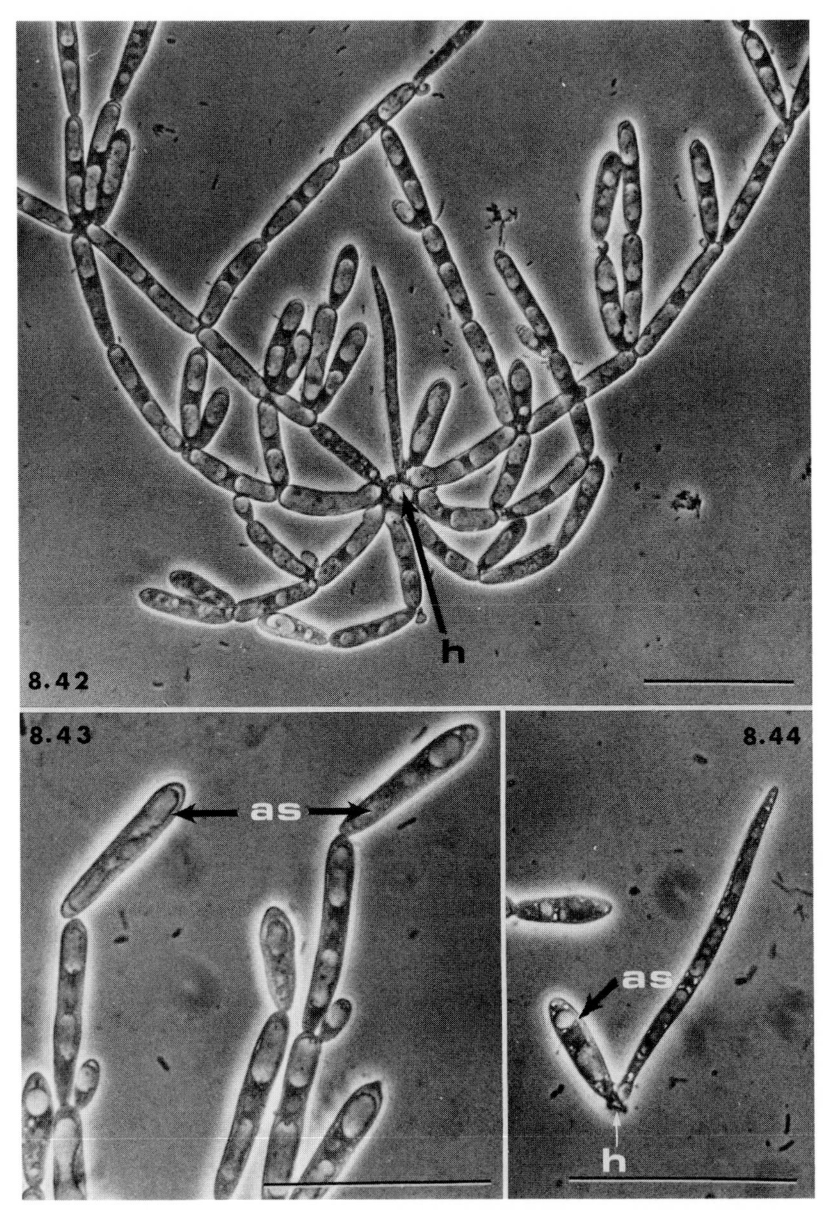

Figure 8.42–8.44. Orchesellaria mauguioi (Asellariales). *8.42 (top).* Vegetative thallus with branches originating from the holdfast region (h). *8.43 (bottom left).* Fragmentation of branches to form arthrospores (as). *8.44 (bottom right).* Germinated arthrospore (as) with holdfast (h) and first vegetative branch. (*8.42–8.44.* Phase-contrast illumination, bars=40μm.)

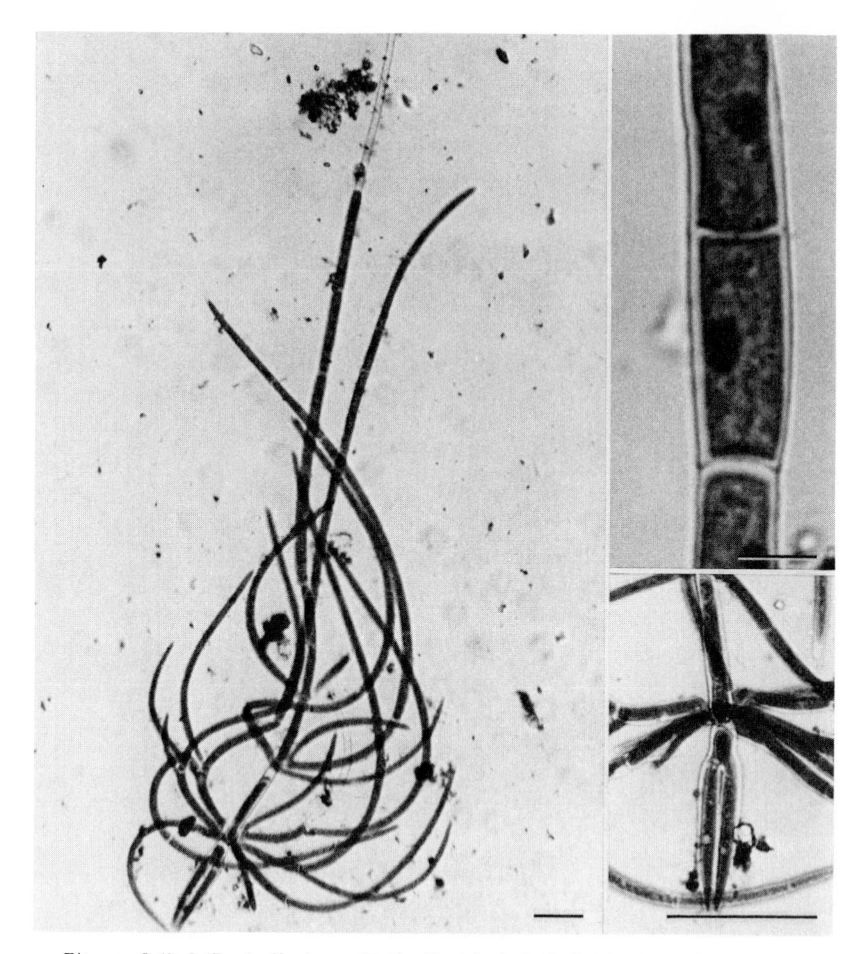

Figures 8.45–8.47. Asellaria aselli (Asellariales). *8.45 (left).* Vegetative thallus. (Bar = 25μm.) *8.46 (top right).* Septation of lateral branch into uninucleate arthrospores. (Bar=5μm.) *8.47 (bottom right).* Bifurcate, basal, attachment cell of thallus. (Bar=25μm.) (*8.45–8.47.* Bright-field illumination, stained lactophenol cotton-blue.)

TRICHOMYCETE HOSTS

Trichomycetes are restricted in their host range to the mandibulate classes of the Arthropoda; these are the Insecta, Crustacea and Myriapoda but exclude the sucking Arachnida and the geographically restricted Onychophora (Table 8.5). However, within the mandibulate forms only herbivores and omnivores have been found infested with the endocommensal Trichomycetes. The reason for this apparent exclusion of carnivores is not fully understood although it may be speculated that obligate carnivores lack enzymes or other exogenous substrates necessary for germination

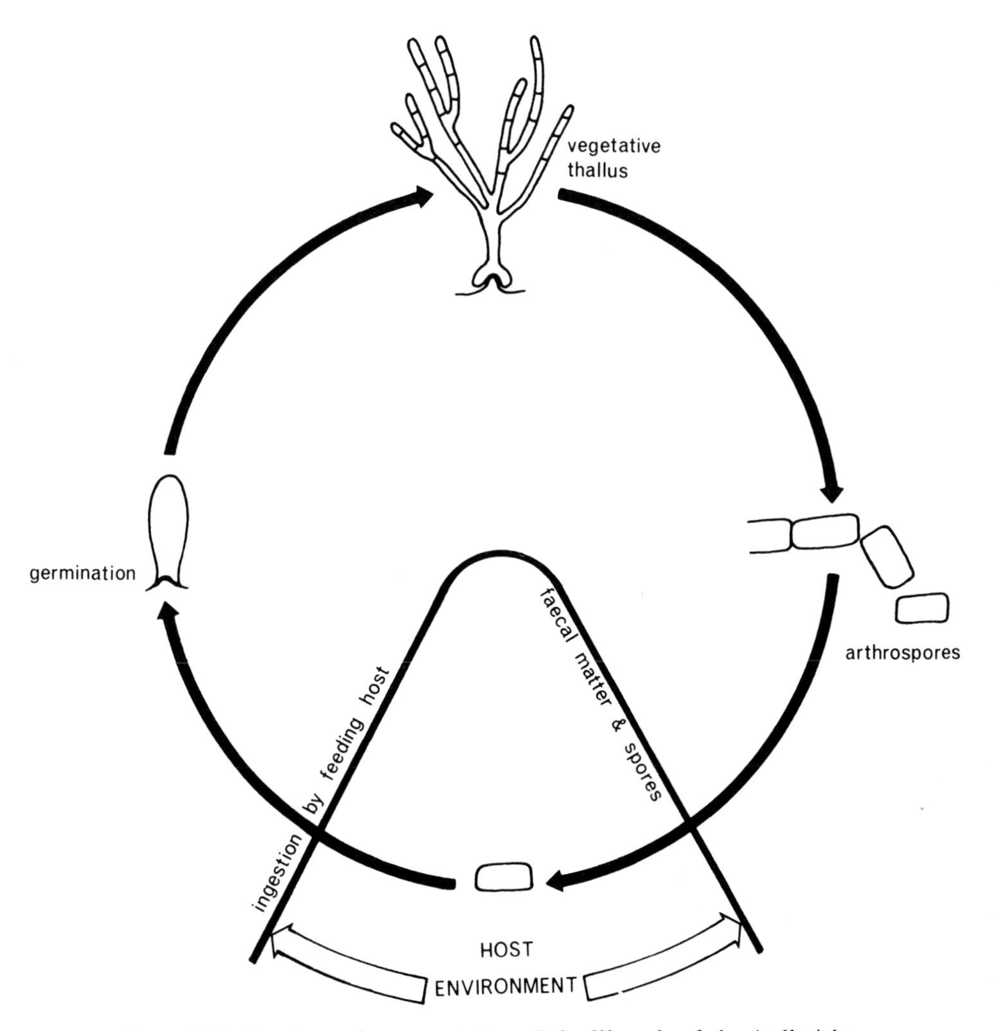

Figure 8.48. Diagrammatic representation of the life-cycle of the Asellariales.

and/or growth of ingested trichomycete spores, the only means by which hosts become infested. Trichomycetes are commensals of terrestrial, freshwater and marine hosts but have not been found associated with the aerial, adult stages of Insecta (Table 8.5). Species of Eccrinales and Asellariales occur associated with hosts from all habitats whereas the Amoebidiales and Harpellales are restricted to freshwater hosts (refer to Tables 8.1–8.4 for a complete distribution of Trichomycetes). Trichomycetes have been collected from all the major continents and their geographical distribution appears to be limited only by that of their hosts.

Table 8.5 Arthropod Orders Infested with Trichomycetes.

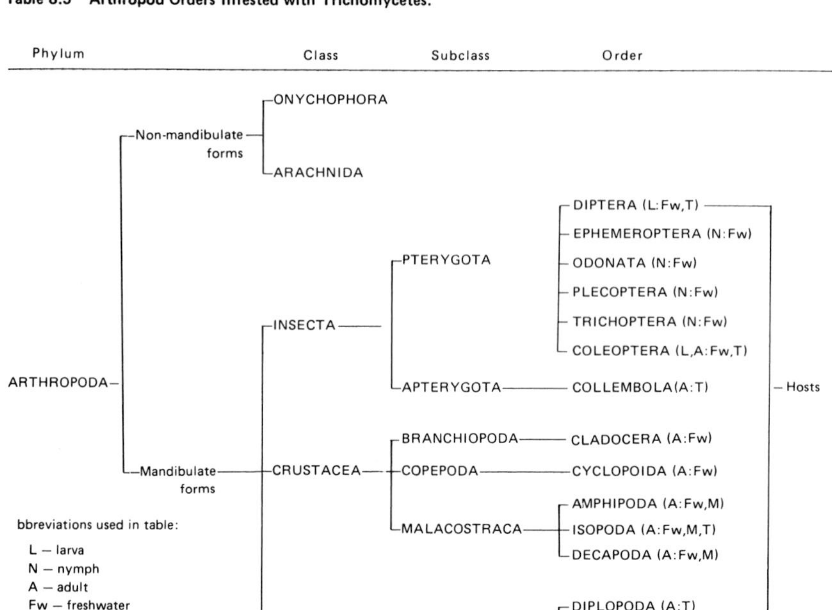

Phylum		Class	Subclass	Order	
ARTHROPODA	Non-mandibulate forms	ONYCHOPHORA			
		ARACHNIDA			
	Mandibulate forms	INSECTA	PTERYGOTA	DIPTERA (L:Fw,T)	Hosts
				EPHEMEROPTERA (N:Fw)	
				ODONATA (N:Fw)	
				PLECOPTERA (N:Fw)	
				TRICHOPTERA (N:Fw)	
				COLEOPTERA (L,A:Fw,T)	
			APTERYGOTA	COLLEMBOLA(A:T)	
		CRUSTACEA	BRANCHIOPODA	CLADOCERA (A:Fw)	
			COPEPODA	CYCLOPOIDA (A:Fw)	
			MALACOSTRACA	AMPHIPODA (A:Fw,M)	
				ISOPODA (A:Fw,M,T)	
				DECAPODA (A:Fw,M)	
		MYRIAPODA		DIPLOPODA (A:T)	
				CHILOPODA	

bbreviations used in table:

L – larva
N – nymph
A – adult
Fw – freshwater
M – marine
T – terrestrial

Trichomycete thalli are attached to four basic regions of the host cuticle: (a) the external surface; (b) the intucking of the external cuticle which lines the foregut (stomodaeum); the intucking of the external cuticle lining the hindgut (proctodaeum); (d) in freshwater dipteran larvae the chitinous peritrophic membrane of the midgut (mesenteron). In the Arthropoda two types of peritrophic membrane are recognized. The type present in most arthropods is produced by the entire midgut epithelium and is known as the delamination type. The production of the second type, the secretory type, is restricted to a belt of cells at the anterior extremity of the mesenteron and the resultant peritrophic membrane forms a continuous tube which: is separated from the midgut epithelium by an ectoperitrophic space; continuously moves posteriorly from its region of formation to ultimately enter and disintegrate in the hindgut; serves to protect the epithelial cells from the food bolus; and acts as a selective filter (Zhuzhikov 1970) allowing digestive enzymes to enter the lumen of the digestive tract and products of extracellular digestion to pass in the opposite direction. The secretory type of membrane is characteristic of dipteran larvae and is the only type of peritrophic membrane to which Trichomycetes, i.e. the Harpellaceae, are attached.

Amoebidium, Amoebidiales, is the only genus of Trichomycetes found attached to the external surface and is considered an ectocommensal. All other Trichomycetes are associated with the cuticle lining the digestive tract and are termed endocommensals. Whether this terminology to describe the spatial distribution of Trichomycetes is universally acceptable is uncertain as traditionally the digestive tract is considered external. To overcome confusion Starr (1975) suggested that "modifers [sic] to the terms exhabitant and inhabitant" could be used to indicate the precise location of symbiotic organisms within or upon their hosts. Adoption of this system is of limited use with respect to the Trichomycetes and at best it could only indicate the region of thallus attachment, which is not always that of vegetative growth and sporulation. For example, although attachment in several species of aquatic Eccrinales (e.g. *Astreptonema gammari, Taeniella carcini*) and Harpellales (e.g. *Pteromaktron protrudens, Zygopolaris ephemeridarum*) is to the cuticle lining the hindgut, the sporulating regions of these thalli characteristically project from the anus into the host environment. Similar difficulty arises in describing the exact position of Harpellaceae within dipteran larvae. The peritrophic membrane to which Harpellaceae are attached moves continuously towards the hindgut from the region of its formation and thus the position of thalli within the host is not constant. For simplicity only the terms ectocommensal and endocommensal will be used for Trichomycetes attached to the external surface (i.e. *Amoebidium*) and the chitinous gut linings (all other genera) respectively, although it is recognized that the digestive tract and external surface are regions of a continuum.

With few exceptions (e.g. *Amoebidium, Paramoebidium, Eccrinoides, Enterobryus*) trichomycete genera are restricted to a single order and frequently a single family of Arthropoda (Tables 8.1–8.4) and the recorded specificity of some Trichomycetes beyond family rank may be due to insufficient field collections of closely related hosts. Host specificity is difficult to test experimentally since only *Smittium* spp. (Harpellales) (Clark, Kellen and Lindegren 1963; Lichtwardt 1964) and *Amoebidium* (Amoebidiales) (Whisler 1962) have been axenically cultured. However Williams and Lichtwardt (1972a) have indicated that the trichospores of *Smittium* sp. isolated from mosquito larvae more readily infest mosquito larvae than either Chironomidae or Simuliidae, although all have a similar feeding habit and diet. Moss (1972) has also shown that whereas in the laboratory infestation of Chironomidae larvae by *Stachylina grandispora* (Harpellaceae, Harpellales) is limited only by the feeding habits of the host, omnivores and herbivores being infested and obligate carnivores not being infested, in nature there is a more restricted host range, possibly related to the diverse but constant ecological niches of larval species.

ADAPTATIONS TO COMMENSALISM

The adoption of a commensalistic mode of life by an organism normally necessitates modifications to its morphology, ecology, physiology and biochemistry. The present ability to axenically culture only *Amoebidium parasiticum* (Amoebidiales) and *Smittium* spp. (Harpellales) of the Trichomycetes has made it virtually impossible to characterize the metabolic pathways within the host and hence physiological and biochemical modifications. Furthermore, both *A. parasiticum* and *Smittium* spp. are less host-specific than are other Trichomycetes. *A. parasiticum* occurs on the external surfaces of a wide range of freshwater Insecta and Crustacea (Table 8.1) and *Smittium* spp. occur in the hindguts of several families of dipteran larvae (Table 8.3). This less restricted host range may indicate that their growth requirements are less critical than those of other more specific Trichomycetes and in this respect may be atypical. However, considerable information has accumulated on the morphological and ecological modifications of Trichomycetes which contribute towards their successful colonization of the cuticular surfaces of Arthropoda.

Symbiotes attached to the cuticular surfaces of arthropods have a temporal restriction imposed on their relationship determined by the duration of the intermoult period. Trichomycetes attached to the external surface or to the cuticle lining the fore and hind guts are shed with the host cuticle at each moult, whilst those attached to the peritrophic membrane are passed from the host with fecal matter following fragmentation of the membrane within the hindgut. Consequently, success of the trichomycete-host relationship depends upon coordination of their life cycles in order that growth and sporulation of the commensal occurs prior to ecdysis, followed by the transmission of infestive propagules to other hosts. Since those Trichomycetes (Amoebidiales, Harpellales) in holometabolous and hemimetabolous Insecta are usually associated only with the larval or nymphal stages and those (Amoebidiales, Eccrinales, Asellariales) infesting adult Crustacea, Diplopoda and Collembola are not transmitted from parent to progeny with the fertilized egg, it is necessary for the development of mechanisms which facilitate transmission of infestive propagules to potential hosts in order to maintain the level of infestation within host populations.

Morphological. Certain morphological adaptations common to each of the four trichomycete orders may have developed independently and represent convergence to adapt the commensals to similar ecological niches and may not necessarily be indicative of close phylogeny. These include possession of a holdfast, limited vegetative growth, conversion of the entire or majority of the thallus cytoplasm into infestive propagules, sequential maturation of asexual spores in the Eccrinales and Harpellales and spore appendages in genera of the Eccrinales and Harpellales infesting aquatic hosts.

The presence of an efficient holdfast mechanism is essential for trichomycete thalli since their growth and sporulation require permanent attachment to the host cuticle. If detached there is no mechanism to allow reattachment. Additionally, endocommensal Trichomycetes have a restricted region of the gut in which they can successfully grow. Correspondingly attachment of ingested spores contained within the food bolus, which in some hosts may pass through the entire digestive tract in less than an hour, must be rapid. Evidence for the rapid attachment of ingested trichomycete spores has been provided by Williams and Lichtwardt (1972a). These workers showed that trichospores of *Smittium culisetae* Lichtwardt (1964b) become attached to the hindgut linings of mosquito larvae within 30 minutes of ingestion. The rigid sporangiospores of *A. parasiticum* (Whisler and Fuller 1968), the sporangiospore contained within the sporangium (trichospore) of many and possibly all Harpellales (Moss and Lichtwardt 1976), and the multinucleate, thin-walled sporangiospores of the Eccrinales (Cronin and Johnson 1958; Lichtwardt 1954a) all possess holdfast initials or mechanisms for the rapid secretion of holdfast material immediately upon contact with a specific location within the host gut. However, the stimulus which results in the consistent rather than random distribution of thalli within the host has not been established. The higher density of *A. parasiticum* thalli on the head and anal regions of the external surfaces of freshwater Insecta and Crustacea may not be due to selectivity in the region of spore attachment but selectivity against subsequent growth in environmentally hostile regions. A similar hypothesis cannot support the selectivity of hosts and position of endocommensal Trichomycetes.

Vegetative thalli of all Trichomycetes are limited and culminate in sporulation by conversion of the entire or majority of the vegetative thallus into sporangiospores (Amoebidiales, Eccrinales, Harpellales) or arthrospores (Asellariales) prior to moulting of the host. The Amoebidiales are holocarpic, in Eccrinales the distal region of the unbranched thallus into which most of the cytoplasm migrates segments to form unisporous sporangia, in Harpellales the entire (Harpellaceae) or lateral branches (Genistellaceae) form generative cells each with a single trichospore, and in the Asellariales lateral branches and frequently the entire thallus fragments into arthrospores. The sequential development of infestive spores in the Eccrinales, Harpellales, and to a lesser extent the Asellariales may be a mechanism to ensure production and release of at least some propagules in the event of a premature interruption in the host-fungal relationship. Ability to axenically culture only two trichomycete taxa has made it impossible to determine whether limited vegetative growth is an expression of the genome or controlled by external factors within the host gut. However, maximum utilization of the reproductive potential of the cytoplasmic complement is characteristic of many obligate symbiotes of the digestive tract which rely on infestive units, released with fecal matter, for propagation.

As there is no direct transfer of the infestation from adult to progeny, no motility of the released, infestive propagules and the frequently hostile environment into which these spores are released, mechanisms are required to ensure that spores remain in the vicinity of the host population where they may be ingested by potential hosts. This is particularly important for species of Harpellaceae (*Carouxella, Harpella, Stachylina*) which normally produce between four and eight trichospores per thallus and in which no endogenous means of reinfestation has been recorded. Although in a heterogeneous population of closely related arthropods the relatively low host specificity characteristic of many Trichomycetes (Chapman 1966; Tuzet, Rioux and Manier 1961; Williams and Lichtwardt 1972a) increases the chances of maintaining an infestation in the locality.

The unisporous sporangia (trichospores) of the Harpellales, the sporangiospores of the Amoebidiales and Eccrinales and the arthrospores of the Asellariales are the only dispersive, asexual propagules of the Trichomycetes and it is these which must be ingested, or in the ectocommensal *Amoebidium* contacted, by the host. The infestive propagules of aquatic hosts released into the host environment are immediately subjected to dispersal by water currents. However, dispersal of trichomycete spores is not necessarily advantageous, as many hosts have a restricted ecological distribution and removal of spores from the immediate vicinity of the host population may result in total loss of the infestation. For example, *Simulium* spp. larvae infested with Harpellales are found, often close to the source, in fast-flowing, well-aerated streams and released spores are immediately subjected to dispersal by the fast-flowing water. Correspondingly retention of infestive spores close to the host population is of paramount importance. The non-motile appendages attached to the infestive spores of many Trichomycetes associated with aquatic hosts may aid in retention of these spores. The trichospores of the Harpellales and the sporangiospores of many genera of Eccrinales (*Arundinula, Astreptonema, Palavascia, Taeniella*) which are endocommensals of aquatic Crustacea possess appendages. These appendages are usually flexuous, several hundred micrometers long and immediately following release from the host gut become entangled with plants, debris, etc., so attaching the spore close to the host population and in a position where it is likely to be ingested by feeding hosts. Appendages also increase the effective dimensions of the spore which aids their removal from the water by filter feeders (e.g. *Simulium* larvae). It is interesting to note that whereas the thick-walled, dispersive sporangiospores of many Eccrinales associated with aquatic hosts are appendaged, those of their terrestrial counterparts are not.

Ecological. As arthropods do not apparently benefit or suffer from their association with Trichomycetes it seems improbable that modifications to their physiology, ecology or morphology have evolved either to encourage or restrict their association with the fungus. It is more probable that successful associations have developed only with those arthropods with life cycles, physiology, feeding habits and ecology conducive to growth, reproduction and transmission of the commensal Trichomycetes.

In both Trichomycetes infesting aquatic arthropods where the released, appendaged spores are retained in the vicinity of the hosts, and in Trichomycetes infesting terrestrial arthropods where the released, infestive propagule is contained within the fecal pellets, successful transmission of the infestation is more likely to occur if there is a high host density and/or an intimate association between individuals. Although none of the recognized gregarious or socially gregarious arthropod groups (e.g. termites) has been found to harbour Trichomycetes, many hosts do exhibit a restricted habitat imposed by environmental pressures. The resultant superficially gregarious habit which is characteristic of many infested Arthropoda normally results from either the specialized environmental conditions required for oviposition by the adult, survival of the larval, pupal or adult stages and/or emergence of the adult from the eggs, pupae or nymphs. *Simulium* spp. larvae, for example, are found attached to substrates in well-aerated, fast-flowing, silt-free water, often restricted to single stones, leaves, etc. This results from the turbulent water necessary for deposition of the eggs by the adult, for the subsequent emergence of the adult from the pupal case, and respiration of larvae and pupae. The required low content of suspended organic matter is essential in order to prevent damage and occlusion of the larval head-rakes used for the filtering of food material, e.g. diatoms, from the water. Similarly many littoral Crustacea (e.g. Gammaridae, Asellidae) infested with Asellariales and Eccrinales aggregate in rock crevices, under seaweed, in the splash zone, etc. at low tide to avoid desiccation. The high population densities caused by such environmental pressures on some hosts results in an increase of fecal pellets in the vicinity of the host population and a correspondingly higher inoculum potential of associated Trichomycetes. Transmission of infestive spores of the Trichomycetes by a form of indirect proctodaeal feeding also occurs in those arthropod hosts which exhibit parental care of the young. In the millipedes (Myriapoda) and passalid beetles (Insecta), both of which are frequently infested by Eccrinales, the parents protect their eggs and the progeny develop in an environment rich with parental fecal matter and infestive spores. On the other hand, the young of many Gammaridae are protected beneath the abdomen of the adult, a position where fecal matter containing released trichomycete spores may be readily ingested.

Although sporulation of Trichomycetes normally occurs prior to ecdysis, it is not uncommon for spores to be retained within the cast exuvium. In those arthropod hosts which intentionally ingest their exuviae to conserve calcium (Myriapoda, Crustacea) or unintentionally ingest exuviae because of their omnivorous diet and/or common tubicolous habit, spores within the cast moult may serve as a means of infestation.

Present evidence suggests that the Trichomycetes should be considered as opportunistic commensals. Successful associations having developed only with arthropods which possess, rather than have developed, characteristics which enhance reinfestation and aid transmission of propagules between adult hosts and from generation to generation. With the exception of their absence from the non-mandibulate Arthropoda, there is no clear phylogenetic relationship between infested hosts and it is perhaps the modes of life, habitats and diets of arthropods which are the selective factors for successful trichomycete-arthropod associations.

Coordination of host and trichomycete life cycles. The periodic moulting of arthropod hosts and concomitant loss of trichomycete thalli imposes a temporal limitation on the growth period available to these commensals. It is also essential that sporulation and release of infestive spores occurs prior to moulting, an event which may occur every 24 hours under optimum conditions for host development but as infrequently as every few months under suboptimum (e.g. winter) conditions. To accommodate this wide range of intermoult period; to prevent excessive vegetative growth which may occlude the digestive tract; to ensure sporulation prior to moulting and, in those orders which exhibit dimorphism of asexual spores (Amoebidiales, Eccrinales), the production of non-dispersive spores during the intermoult period and dispersive spores immediately prior to moulting, a coordination of host and trichomycete life cycle is advantageous.

The morphogenetic switch from rigid to amoeboid sporangiospore production of the ectocommensal *Amoebidium parasiticum* is the best documented evidence for a coordination of trichomycete and arthropod life cycles (Cienkowski 1861; Chatton 1906a and b; Duboscq, Léger and Tuzet 1948; Kuno 1973; Lichtenstein 1917; Lichtwardt 1973a, 1976; Manier 1950; Tuzet and Manier 1951b; Whisler 1965). During the intermoult period thalli of *A. parasiticum* produce rigid, elongate sporangiospores which may infest other hosts or, more commonly, attach to the same host to increase the density of thalli. Immediately prior to moulting, or upon injury or death of the host there is a switch from rigid to amoeboid sporangiospore production. The amoebae swarm from the thalli, frequently over the exuvium, encyst and produce cytospores (Figure 8.13). Thus, immediately before ecdysis there is a morphogenetic switch of spore-type to produce a host-independent, reproductive and possibly resistant phase of the life cycle. The amoebagenic factor(s) which initiates this phase of the reproductive cycle has not been characterized. It appears to be none of the insect hormones tested—ecdysterone, diosgenin and a juvenile hormone (Kuno 1973)—but is heat stable, water-soluble and dialyzable from homogenates of mosquito larvae (Whisler 1965). Ecdysis also stimulates amoebae release of the ectocommensal *Paramoebidium* (Léger and Duboscq 1929c; Lichtwardt 1976; Manier 1950; Whisler 1963, 1965).

Field collections of insect larvae and nymphs infested with Harpellales have shown zygospores to be most common immediately before moulting, in cadavers and in exuviae (Moss, Lichtwardt and Manier 1975; Whisler 1963). In many insects there is synchrony of metamorphosis from the infested larvae to the uninfested pupae or, in hemimetabolous insects, emergence of the uninfested imago from the infested nymph. This results in the absence of hosts from the environment during the pupal and imago instars. In such a situation it is essential for the production of propagules which are capable of surviving independently in the host environment and which can serve as an inoculum for subsequent generations. Formation of zygospores by the Harpellales immediately before moulting may provide such a resistant spore. Attempts to induce zygospore formation in axenic cultures of *Smittium* have proved unsuccessful (El-Buni 1975) and the stimulus for their formation has not been identified. Their production within the pre-moult host, however, may indicate a host-integrated response.

There is evidence of a similar host coordinated production of spore types in the Eccrinales (Galt 1971; Manier, Gasc and Bouix 1972, 1974). The order is characterized by the formation of multinucleate sporangiospores which germinate within the same host and so increase the infestation endogenously and by uninucleate, frequently thick-walled sporangiospores which pass from the host to infest other individuals (Figure 8.24). The multinucleate spores are generally produced during the intermoult period whereas the uninucleate sporangiospores are most abundant immediately prior to moulting. Since none of the Eccrinales can be axenically cultured, this apparent coordination of dispersive spore production with moulting cannot be experimentally tested. Galt (1971), however, found that in non-moulting, winter Crustacea only multinucleate spores were produced, an observation which may indicate that the production of dispersive spores is stimulated by a factor associated with moulting. Members of the Asellariales show no indication of a host coordinated life cycle.

Intimate relationships between invertebrate hormones and the morphogenesis of symbiotic microorganisms have been well documented (Cleveland 1947, 1959; Gordon 1970) but until species of Eccrinales, Asellariales and many more Harpellales can be axenically cultured it may not be possible to provide indisputable evidence of similar relationships in the endocommensal Trichomycetes.

TAXONOMIC AFFINITIES

Previous literature has invariably considered the Trichomycetes as a well defined, natural class of lower fungi. This assumption has been based on their obligate association with the cuticle lining the digestive tracts or external surfaces of their hosts, simple organization, possession of a holdfast and their morphological dissimilarity to other fungal taxa. Although this chapter, by virtue of its title, may appear to perpetuate this concept, it does so only for convenience and correspondingly it seems necessary to present evidence contrary to this opinion and to indicate possible taxonomic relationships.

The diversity in vegetative and reproductive organization of the Trichomycetes, ranging from the unbranched, holocarpic Amoebidiales which reproduce by either rigid or amoeboid sporangiospores, to the frequently highly branched Genistellaceae (Harpellales) with their deciduous, appendaged sporangia and biconical zygospores may indicate polyphylogeny. The ubiquity of holdfasts in the Trichomycetes does not necessarily support close taxonomic affinity as adhesive organs have developed independently in many groups of lower plants and their presence in the Trichomycetes may be no more than an adaptation by unrelated organisms to a habitat where attachment to the substrate is of paramount importance for survival.

Recent taxonomic treatments (Lichtwardt 1973a and b, 1976) have placed the Trichomycetes in the Zygomycotina on account of their coenocytic thalli, asexual reproduction by sporangiospores and the presence of zygospores in many genera of the Harpellales. Since 1954 data obtained from cell wall analysis (Lichtwardt 1954a; Sangar and Dugan 1973; Whisler 1963), immunology (Sangar et al. 1972), sterol production (Starr 1977) and fine-structural studies (Farr and Lichtwardt 1967; Lichtwardt 1973a; Manier 1973a and b; Manier and Coste-Mathiez 1968; Manier and Grizel 1972; Moss 1975, 1976; Moss and Lichtwardt 1976, 1977; Moss and Young 1978; Reichle and Lichtwardt 1972; Tuzet and Manier 1967; Whisler and Fuller 1968) have provided information which may be of taxonomic relevance.

The micromorphology of the septum is considered to be of particular significance in fungal taxonomy. In phycomycetes the typical septum is of the complete or abscissional type whereas those characteristic of ascomycetes and basidiomycetes are simple-perforate, with or without associated Woronin bodies, and dolipore respectively. The septal apparatus of all the Harpellales examined, whether vegetative (Farr and Lichtwardt 1967; Lichtwardt 1973a; Moss 1975) at the base of the trichospores (Manier 1973a; Manier and Coste-Mathiez 1968; Moss 1975, 1976; Moss and Lichtwardt 1976; Reichle and Lichtwardt 1972) or delimiting zygospores from their zygosporophores (Moss and Lichtwardt 1977) is not of the typical phycomycetes type but comprises a crosswall flared around a single central pore. Occluding the pore and retained within it by the flared crosswall layer is a rigid, electron-opaque, biumbonate, non-membrane bound plug. Intercellular continuity is maintained by the plasmalemma which is continuous around the rim of the pore. A morphologically identical septal apparatus has also been found in the only two genera of the Asellariales examined at the ultrastructural level (Manier 1973b; Moss 1975) and it may not be too speculative to predict that this type of septum is characteristic of the order. In the Eccrinales septa develop only between adjacent sporangia and thus constitute part of the sporangial wall. In *Astreptonema gammari* (Eccrinaceae), the only species of Eccrinales in which septal structure has been examined (Moss 1975), the septa are initially perforate, but, following migration of cytoplasm into the developing sporangium, become complete by the deposition of golgi-derived wall material onto the crosswall. There is no reason to believe that in such a morphologically uniform group as the Eccrinales there should be gross variation of septal structure. On this premise there appears to be at least two distinct septal types in the Trichomycetes: 1) the plugged pore of the Harpellales and Asellariales; 2) the complete septum of the Eccrinales; neither type resembles in ontogeny nor mature form the septal types characteristic of the phycomycetes, ascomycetes or basidiomycetes. The Amoebidiales are non-septate. Young (1969) and more recently Benny and Aldrich (1975) found a septal apparatus morphologically identical to that of the Harpellales and Asellariales in members of the Kickxellales (Zygomycotina), a character which may indicate close phylogeny of the Harpellales, Asellariales and Kickxellales. Further evidence to support this relationship is provided by the results of an immunological study by Sangar et al., (1972) in which a serological relationship was found between axenic isolates of *Smittium* spp. (Genistellaceae, Harpellales) and *Linderina pennispora* Raper and Fennel and *Dipsacomyces acuminosporous* Benjamin of the Kickxellales. Similar studies using isolates of *Amoebidium parasiticum* (Amoebidiales) showed no affinity with the isolates of *Smittium* spp. or with the two species of the Kickxellales. Unfortunately no other genus of Trichomycetes has been axenically cultured in order to allow wider application of this technique.

Bartnicki-Garcia (1970) considers fungal cell wall chemistry as phylogenetically significant. However, insufficient information is available on cell wall composition in the Trichomycetes to enable critical appraisal of its importance in this group. Nevertheless, initial studies on two species of Harpellales (*Pteromaktron protudens* and *Genistella* sp.) by Whisler (1963) and more recently on isolates of *Smittium culisetae* by Sangar and Dugan (1973), showed them to contain chitin in their walls. Conversely, the Eccrinales examined have given positive reactions for cellulose and negative for chitin (Whisler 1963) whereas Amoebidiales seem to possess neither chitin nor cellulose as major constituents but do contain a high percentage of protein and some galactose (Trotter and Whisler 1965), constituents found also in the thraustochytriaceous fungi (Darley, Porter and Fuller 1973). No information is available on the wall composition of the Asellariales.

In addition to septal structure, serological evidence and wall chemistry to support a relationship between the Harpellales and Kickxellales, they share certain morphological characters. Both possess monosporous, elongate, dehiscent sporangia (termed merosporangia in the Kickxellales and trichospores in the Harpellales) and the merosporangium-pseudophialide-sporocladium complex of the Kickxellales may be the homologue of the trichospore-collar-generative cell complex of the Harpellales (Moss and Young 1978). Perhaps the most significant similarity of the kickxellaceous merosporangium and the harpellaceous trichospore is the presence of an extracellular structure contiguous with the septum delimiting the sporangium and extending into the subtending cell (i.e. sporocladium in the

Kickxellales, generative cell and/or collar region of the generative cell in the Harpellales). In *Kickxella alabastrina* Coemans and other unnamed Kickxellales (Young 1974) and *Linderina pennispora* (Benny and Aldrich 1975) an extraplasmalemmal structure contiguous with the crosswall at the base of the merosporangium extends into the pseudophialide. This resembles in position and structure the immature trichospore appendages in the generative cells of the Harpellales. It has been suggested (Moss and Young 1978) that these unique extraplasmalemmal structures in the Kickxellales and Harpellales are homologous.

Available evidence strongly supports a phylogenetic relationship between the Harpellales, Asellariales and Kickxellales. The position of the Eccrinales and Amoebidiales is less certain but does not appear to be with each other or with the Harpellales and Asellariales and it is, perhaps, better at the present time to consider the Trichomycetes as an ecologically defined rather than phylogenetically related class of arthropodophilic lower fungi.

CONCLUSION

It would be naive to believe that more than a very superficial knowledge of the trichomycete-arthropod relationship has been elucidated. For a complete understanding of this heterospecific association it will be necessary to investigate fully all aspects of host and commensal biology, namely, chemotherapy, ecology, immunology, morphology, nutrition, physiology and serology. At present many fundamental aspects of trichomycete biology are uncertain, a situation which is not improved by the ability to culture only two genera.

The assignment of the term commensalism for the trichomycete-arthropod relationship is, perhaps, due to the lack of evidence to support either a mutualistic or parasitic association. Owing to the position of most Trichomycetes within the hindguts of their hosts where the food bolus is fecal material and of no further nutrient value to the host, the extracellular habit and the absence of any detectable detrimental effect on the host, the association is considered non-parasitic. It is also unlikely that the presence of some Eccrinales (e.g. *Enteromyces*) in the foreguts and Harpellaceae in the midguts of their hosts presents a situation where the fungus is in direct competition with the host for nutrients, or at least never to the extent of causing stress. Similarly, the attachment of *Amoebidium parasiticum* thalli to the external surfaces of the head and anal regions of aquatic arthropods does not appear to harm the host but may enable the fungus to obtain nutrients from the respiratory and feeding currents created by the host. There is no evidence that the Trichomycetes provide growth factors for their arthropod hosts or that they function as catabolic agents supplying energy sources derived from ingested food material which cannot be directly digested by the host. However Starr (1977) has identified desmosterol from axenic cultures of *Smittium* (Harpellales) and hypothesizes that, for hosts developing under suboptimal conditions, infesting Trichomycetes may provide the exogenous sterols required by arthropods (Svoboda and Robbins 1975) for the synthesis of moulting and sex hormones. This suggested basis for a mutualistic relationship deserves further investigation, although in nature infested hosts show no advantage over non-infested hosts with respect to moulting and maturation. Conversely, the obligate association of Trichomycetes with their hosts may result from the inability of the fungus to synthesize compounds necessary for either metabolism, germination, growth and/or reproduction, and the failure to culture them due to the inability to simulate the gut environment. If metabolic dependence is taken as a criterion for parasitism and if the Trichomycetes require essential compounds produced by the host, then maybe they should be considered parasites. However, in the absence of any detectable deviation from normality by the infested hosts, immunological response and penetration of host tissues, the association between the Trichomycetes and the Arthropoda should be presently categorized as commensalistic.

It is hypothesized that the present distribution of Trichomycetes amongst terrestrial and aquatic arthropods and their position within the digestive tract may have evolved from free living, aquatic fungi growing in an oligotrophic environment. Association of these fungi with the external surfaces of aquatic arthropods, especially in regions close to feeding currents around the head and the fecal-rich anal segments, may have provided them with an environment rich in nutrients. Subsequent development by these fungi of a holdfast, their migration into the digestive tract for protection and a continuous supply of nutrients and the loss of an ability to synthesize certain compounds necessary for development could well have resulted in the present obligate endocommensalism characteristic of the Trichomycetes.

LITERATURE CITED

Bartnicki-Garcia, S. 1970. Cell wall composition and other biochemical markers in fungal phylogeny. Pp. 81–103 *in* J. B. Harborne, ed., Phytochemical phylogeny. Academic Press, London and New York.

Benny, G. L., and H. C. Aldrich. 1975. Ultrastructural observations on septal and merosporangial ontogeny in *Linderina pennispora* (Kickxellales; Zygomycetes). Canad. J. Bot. 52: 2325–2335.

Chapman, M. E. 1966. Isolation and experimental studies on some Trichomycetes. M. A. Thesis, University of Kansas, 42 pp.

Chatton, E. 1906a. Sur la biologie, la spécification et la position systématique des *Amoebidium*. Arch. Zool. Exp. Gén. 5: 17–31.

———. 1906b. Sur la morphologie et l'évolution de l'*Amoebidium recticola,* nouvelle espèce commensale des Daphnies. Arch. Zool. Exp. Gén. 5: 33–38.

Chatton, E. and Roubaud. 1909. Sur un *Amoebidium* du rectum des larves de simulies (*Simulium argyreatum* Meig. et *S. fasciatum* Meig.). C. R. Séanc. Soc. Biol. 66: 701–703.

Cienkowski, L. 1861. Ueber parasitische schläuche auf Crustaceen und einigen insektenlarven (*Amoebidium parasiticum* m.) Bot. Ztg. 19: 169–174.

Clark, T. B., W. R. Kellen, and J. E. Lindegren. 1963. Axenic culture of two Trichomycetes from California mosquitoes. Nature (London). 197: 208–209.
Cleveland, L. R. 1947. Sex produced in the protozoa of *Cryptocercus* by molting. Science 105: 16–17.
———. 1959. Sex induced with ecdysone. Proc. Natl. Acad. Sci. [USA] 45: 747–753.
Cloudsley-Thompson, J. L. 1965. Animal conflict and adaptation. G. T. Foulis & Co., London. 160 pp.
Cronin, E. T., and T. W. Johnson. 1958. A halophilic *Enterobryus* in the mole crab *Emerita talpoida* Say. J. Elisha Mitchell Sci. Soc. 74: 167–172.
Crosby, T. K. 1974. Trichomycetes (Harpellales) of New Zealand *Austrosimulium* larvae (Diptera: Simuliidae). J. Nat. Hist. 8: 187–192.
Darley, W. M., D. Porter, and M. S. Fuller. 1973. Cell wall composition and synthesis via Golgi-directed scale formation in the marine Eucaryote, *Schizochytrium aggregatum*, with a note on *Thraustochytrium* sp. Arch. Mikrobiol. 90: 89–106.
DeBary, A. 1879. Die erscheinung der symbiose. Verlag von Karl, J., Trubner, Strassburg.
Duboscq, O., L. Léger, and O. Tuzet. 1948. Contribution à la connaissance des Eccrinides. Les Trichomycètes. Arch. Zool. Exp. Gén. 86: 29–144.
El-Buni, A. M. 1975. Factors affecting sporulation, growth and spore germination in species of *Smittium* (Trichomycetes) Ph.D. Thesis, University of Kansas. 136 pp.
El-Buni, A. M., and R. W. Lichtwardt. 1976a. Asexual sporulation and mycelial growth in axenic cultures of *Smittium* spp. (Trichomycetes). Mycologia 68: 559–572.
———. 1976b. Spore germination in axenic cultures of *Smittium* spp. (Trichomycetes). Mycologia 68: 573–582.
Farr, D. F., and R. W. Lichtwardt. 1967. Some culturable and ultrastructural aspects of *Smittium culisetae* (Trichomycetes) from mosquito larvae. Mycologia 59: 172–182.
Frost, S., and J.-F. Manier. 1971. Notes on Trichomycetes (Harpellales: Harpellaceae and Genistellaceae) in larval blackflies (Diptera: Simuliidae) from Newfoundland. Canad. J. Zool. 49: 776–778.
Galt, J. H. 1971. Studies on some protists associated with Crustacea: the Ellobiopsidae and the Trichomycetes. M.S. Thesis, Univ. of Washington. 150 pp.
Gauthier, M. 1936. Sur un nouvel Entophyte du groupe des Harpellacées Lég. et Dub., parasites des larves d'Ephémérides. C. R. Hebd. Séanc. Acad. Sci. Paris 202: 1096–1098.
———. 1960. Un nouveau Trichomycète rameux, parasites des larves de *Baetis pumilus* (Burm.). Trav. Lab. Hydrobiol. Piscic. Univ. Grenoble, years 50, 51: 225–227.
———. 1961. Une nouvelle espèce de *Stachylina: St. minuta* n. sp. parasite des larves de Chironomides Tanytarsiens. Trav. Lab. Hydrobiol. Piscic. Univ. Grenoble, years 52, 53: 1–4.
Gordon, R. 1970. A neuroendocrine relationship between the nematode *Hammerschmidtiella diesingi* and its insect host, *Blatta orientalis*. J. Parasitol. 61: 101–110.
Granata, L. 1908. Di un nuovo parassita dei millipiedi (*Capillus* n. g. *intestinalis* n. sp.). Biologica, Torino 2: 3–16.
Hauptfleisch, P. 1895. *Astreptonema longispora* n.g., n. sp., eine neus Saprolegniacee. Ber. Deut. Bot. Ges. 13: 83–88.
Johnson, T. W. 1966. Trichomycetes in species of *Hemigrapsus*. J. Elisha Mitchell Sci. Soc. 82: 1–6.
Kobayasi, Y., N. Hiratsuka, Y. Otani, K. Tubaki, S. Udagawa, and M. Soneda. 1969. The second report on the mycological flora of the Alaskan Arctic. Bull. Natl. Sci. Mus. Tokyo 12: 311–430.
Kuno, G. 1973. Biological notes of *Amoebidium parasiticum* found in Puerto Rico. J. Invert. Pathol. 21: 1–8.
Léger, L., and O. Duboscq. 1905. Les Eccrinides, nouveau groupe de Protophytes parasites. C. R. Hebd. Séanc. Acad. Sci., Paris 141: 425–427.
———. 1906. L'évolution des *Eccrina* des *Glomeris*. C. R. Hebd. Séanc. Acad. Sci., Paris 142: 590–592.
———. 1911. Sur les Eccrinides des Crustacées Décapodes. Ann. Univ. Grenoble, Fr. 23: 139–141.
———. 1916. Sur les Eccrinides des Hydrophilides. Arch. Zool. Exp. Gén. 56: 21–31.
———. 1929a. *Eccrinoides henneguyi* n. g., n. sp. et la systématique des Eccrinides. Arch. Anat. Micr., Fr. 25: 309–324.

———. 1929b. *Harpella melusinae* n. g., n. sp. Entophyte eccriniforme, parasite des larves de Simulie. C. R. Hebd. Séanc. Acad. Sci., Paris 188: 951–954.

———. 1929c, L'évolution des *Paramoebidium,* nouveau genre d'Eccrinides, parasite des larves aquatiques d'Insectes. C. R. Hebd. Séanc. Acad. Sci., Paris 189: 75–77.

———. 1933. *Eccrinella (Astreptonema?) gammari* Lég. et. Dub., Eccrinide des Gammares d'eau douce. Arch. Zool. Exp. Gén. 75: 283–292.

Léger, L., and M. Gauthier. 1931. *Orphella coronata* n. g., n. sp., Entophyte parasite des larves de Némurides. Trav. Lab. Hydrobiol. Piscic. Univ. Grenoble, year 23: 67–72.

———. 1932. Encomycetes nouveaux des larves aquatiques d'Insectes. C. R. Hebd. Séanc. Acad. Sci., Paris 194: 2262–2265.

———. 1935. La spore des Harpellacées (Léger et Duboscq), champignons parasites des Insectes. C. R. Hebd. Séanc. Acad. Sci., Paris 200: 1458–1460.

———. 1937. *Graminella bulbosa* nouveau genre d'Entophyte parasite des larves d'Ephémérides du genre *Baetis.* C. R. Hebd. Séanc. Acad. Sci., Paris 202: 27–29.

Leidy, J. 1849a. *Enterobrus,* a new genus of Confervaceae. Proc. Acad. Nat. Sci. Philadelphia 4: 225–227.

———. 1849b. Remarks upon several new species of Entophyta, *Enterobrus spiralis* and *Enterobrus attenuatus,* and a new species of Gregarina. Proc. Acad. Nat. Sci. Philadelphia 4: 245.

———. 1849c. Descriptions of a new genera and species of Entophyta. Proc. Acad. Nat. Sci. Philadelphia 4: 249–250.

———. 1850. Descriptions of a new Entophyta growing within animals. Proc. Acad. Nat. Sci. Philadelphia 5: 8–9.

———. 1853. A flora and fauna within living animals. Smithsonian Contributions to Knowledge 5: 1–67.

Lichtenstein, J. L. 1917. Sur un *Amoebidium* à commensalisme interne du rectum des larves d'*Anex imperator* Leach: *Amoebidium fasciculatum* n. sp. Arch. Zool. Exp. Gén. 56: 49–62.

Lichtwardt, R. W. 1954a. Morphological, cytological, and taxonomic observations on species of *Enterobryus* from the hindgut of certain millipeds and beetles. Ph.D. Thesis, Univ. of Illinois. 241 pp.

———. 1954b. Three species of Eccrinales inhabiting the hindguts of millipeds, with comments on the Eccrinids as a group. Mycologia 46: 564–585.

———. 1957a. *Enterobryus attenuatus* from the passalid beetle. Mycologia 49: 463–474.

———. 1957b. An *Enterobryus* occurring in the milliped *Scytonotus granulatus* (Say). Mycologia 49: 734–739.

———. 1958. An *Enterobryus* from the milliped *Boraria carolina* (Chamberlin). Mycologia 50: 550–561.

———. 1960a. An *Enterobryus* (Eccrinales) in a common greenhouse milliped. Mycologia 52: 248–254.

———. 1960b. Taxonomic position of the Eccrinales and related fungi. Mycologia 52: 410–428.

———. 1960c. New species of *Enterobryus* from southeastern United States. Mycologia 52: 743–752.

———. 1961a. A stomach fungus in *Callianassa* spp. (Decapoda) from Chile. Lunds Univ. Arsskrift 57: 3–10.

———. 1961b. A *Palavascia* (Eccrinales) from the marine isopod *Sphaeroma quadridentatum* Say. J. Elisha Mitchell Sci. Soc. 77: 242–249.

———. 1962. An *Arundinula* (Trichomycetes, Eccrinales) in a crayfish. Mycologia 54: 440–447.

———. 1964a. Validation of the genus *Palavascia* (Trichomycetes). Mycologia 56: 318–319.

———. 1964b. Axenic culture of two new species of branched Trichomycetes. Amer. J. Bot. 51: 836–842.

———. 1967. Zygospores and spore appendages of *Harpella* (Trichomycetes) from larvae of Simuliidae. Mycologia 59: 482–491.

———. 1972. Undescribed genera and species of Harpellales (Trichomycetes) from the guts of aquatic insects. Mycologia 64: 167–197.

———. 1973a. The Trichomycetes: what are their relationships? Mycologia 65: 1–20.

———. 1973b. Trichomycetes. Pages 237-243 *in* G. C. Ainsworth, F. K. Sparrow and A. S. Sussman, eds. The Fungi, IV B. Academic Press, London and New York.

———.1976. Trichomycetes. Pages 671-751 in E. B. Gareth Jones, ed. Recent advances in aquatic mycology. Elek Science, London.

Lichtwardt, R. W., and A. W. Chen. 1964. A *Parataeniella* (Trichomycetes, Eccrinales) in an isopod. Mycologia 56: 163-169.

Lichtwardt, R. W., and J. -F. Manier. 1978. Validation of the Harpellales and Asellariales. Mycotaxon 7: 441–442.

McCloskey, L. R., and S. P. Caldwell. 1965. *Enteromyces callianassae* Lichtwardt (Trichomycetes, Eccrinales) in the mud shrimp *Upogebia affinus* (Say). J. Elisha Mitchell Sci. Soc. 81: 114–117.

Maessen, K. 1955. Die zooparasitären Eccrinidales. Parasit. Schrreihe 2: 1–129.

Manier, J.-F. 1947. *Paratrichella pentodoni* n. g., n. sp., Entophyte parasite des larves de *Pentodon punctatus* de Vill. Ann. Sci. Nat. Zool. 9: 275–279.

———. 1950. Recherches sur les Trichomycètes. Ann. Sci. Nat. Bot. 11: 53–162.

———. 1954. Essais de culture des *Eccrina flexilis* Léger et Duboscq, Trichomycètes endocommensaux des *Glomeris marginata* Villers. Ann. Parisitol. Hum. Comp. 29: 265–270.

———. 1955. Nouvelles observations sur *Stipella vigilans* Léger et Gauthier et sur *Paramoebidium chattoni* Duboscq, Léger et Tuzet. Leurs cultures. Ann. Sci. Nat. Zool. 17: 63–66.

———. 1958. *Orchesellaria lattesi* n. g., n. sp., Trichomycete rameux Asellariidae commensal d'un Apterygote collembole *Orchesella villosa* L. Ann. Sci. Nat. Zool. 20: 131–139.

———. 1961. Eccrinides de Crustacés récoltés sur les côtes du Finistere (*Eccrinella corophii* n. sp., *Palavascia sphaeromae* Tuz. et Man., *Taeniella carcini* Lég. et Dub., *Arundinula* sp.). Cah. Biol. Mar. 2: 313–326.

———. 1962a. Présence de Trichomycètes dans le rectum des larves d'Ephémeres des torrents du Massif de Néouvieille (Hautes-Pyrénées). Bull. Soc. Hist. Nat., Toulouse 97: 241–254.

———. 1962b. Revision du genre *Spartiella* Tuzet et Manier, 1950 (sa place dans la classe des Trichomycètes). Ann. Sci. Nat. Zool. 4: 517–525.

———. 1963a. Trichomycètes parasites d'Isopodes Oniscoidea Ann. Sci. Nat. Bot. 4: 557–577.

———. 1963b. Trichomycètes de larves de Simulies (Harpellales du proctodeum). Ann. Sci. Nat. Bot. 4: 737–750.

———. 1964a. Nouvelle contribution à l'etude des Trichomycètes (Eccrinales parasites d'Amphipodes). Ann. Sci. Nat. Bot. 5: 767–772.

———. 1964b. *Orchesellaria mauguioi* n. sp., Trichomycète Asellariale parasite du rectum de *Isotomurus palustris* (Müller) 1776, (Insecte Apterygote Collembole). Rev. Ecol. Biol. Sol. 1: 443–449.

———. 1964c. Endophytes parasites d'Arthropodes cavernicoles récoltés dans des grottes de l'Ariège et de la Haute-Garonne. Ann. Spéléol. 19: 803–812.

———. 1968. Validation de Trichomycètes par leur diagnose Latine. Ann. Sci. Nat. Bot. 9: 93–108.

———. 1969a. Changement de nom pour *Eccrina flexilis* Léger et Duboscq, 1906. Ann. Sci. Nat. Bot. 10: 469–471.

———. 1969b. Trichomycètes de France. Ann. Sci. Nat. Bot. 10: 565–672.

———. 1973a. L'ultrastructure de la trichospore de *Genistella ramosa* Léger et Gauthier, Trichomycète Harpellale parasite du rectum des larves *Baetis rhodani* Pict. C. R. Hebd. Séanc. Acad. Sci., Paris 276: 2159-2162.

———. 1973b. Quelques aspects ultrastructuraux du Trichomycète Asellariale, *Asellaria ligiae* Tuzet et Manier, 1950 ex Manier, 1968. C. R. Hebd. Séanc. Acad. Sci., Paris 276: 3429-3431.

Manier, J.-F., and F. Coste-Mathiez. 1968. L'ultrastructure du filament de la spore de *Smittium mucronatum* Manier et Mathiez, 1965 (Trichomycète, Harpellale). C. R. Hebd. Séanc. Acad. Sci., Paris 266: 341-342.

———. 1971. Trichomycètes Harpellales de larves de Diptères Chironomidae; création de cinq nouvelles espèces. Bull. Soc. Mycol. Fr. 87: 91–99.

Manier, J.-F., C. Gasc, and G. Bouix. 1972. *Enterobryus tuzetae* n. sp. (Trichomycètes-Eccrinales) de l'intestin posterieur de *Pachybolus ligulatus* (Voges) (Diplopodes-Spirobolidae) récoltés au Dahomey (Afrique). Biologica Gabonica 3-4: 305–322.

———. 1974. Sur quelques *Enterobryus* (Trichomycètes Eccrinales) parasites de Myriapodes Diplopodes du Sud-Dahomey. Bull. Inst. Fr. Afr. Noire 36: 614-641.
Manier, J.-F., and H. Grizel. 1971. *Paramacrinella microdeutopi* n. g., n. sp., Trichomycète parasite de *Microdeutopus anomalus* H. Rathke (Amphipode). Ann. Sci. Nat. Bot. 12: 1-8.
———. 1972. L'ultrastructure de l'enveloppe et du "pavillon" des Trichomycètes Eccrinales. C. R. Hebd. Séanc. Acad. Sci., Paris 274: 1159-1160.
Manier, J.-F., and R. W. Lichtwardt. 1968. Révision de la systématique des Trichomycètes. Ann. Sci. Nat. Bot. 9: 519-532.
Manier, J.-F., and F. Mathiez. 1965. Deux Trichomycètes Harpellales Genistellacées, parasites de larves de Chironomides. Ann. Sci. Nat. Bot. 6: 183-196.
Manier, J.-F., and R. Ormières. 1961a. *Alacrinella limnoriae* n. g., n. sp., Trichomycète Eccrinidae, parasite du rectum de *Limnoria tripunctata* Menzies (Isopode). Vie Milieu 12: 285-295.
———. 1961b. *Ramacrinella raibauti* n. g., n. sp., Eccrinide ramifié commensal de l'intestin postérieur de *Microdeutopus gryllotalpa* A. Costa (Amphipodes-*Aoridae*). Ann. Sci. Nat. Bot. 2: 625-634.
———. 1962. *Arundinula galatheae* n. sp., et *Taeniella galatheae* n. sp., Trichomycètes Eccrinacées, parasites de *Galathea strigosa* L. (Crustacés, Décapodes). Vie Milieu 13: 453-466.
Manier, J.-F., and A. Raibaut. 1970. Évolution des kystes de *Amoebidium parasiticum* Cienkowski, 1861 (Trichomycète, Amoebidiale). Bull. Soc. Zool. Fr. 95: 31-33.
Manier, J.-F., J.-A. Rioux, and B. Juminer. 1964. Présence en Tunisie de deux Trichomycètes parasites de larves de Culicides. Arch. Inst. Pasteur, Tunis 41: 147-152.
Manier, J.-F., J.-A. Rioux, and H. C. Whisler. 1961. *Rubetella inopinata* n. sp. et *Carouxella scalaris* n. g., n. sp., Trichomycètes parasites de *Dasyhelea lithotelmatica* Strenzke, 1951 (Diptera, Ceratopogonidae). Natur. Monspeliensia, Bot. 13: 25-38.
———. 1965. Validation du genre *Carouxella* et de l'espèce *Carouxella scalaris* Manier, Rioux et Whisler, 1961. Nat. Monspeliensia, Bot. 16: 87.
Manier, J.-F., and J. Théodoridés. 1965. A propos d'une Eccrinale parasite de Coléoptère Passalide du Laos. Ann. Parisitol. Hum. Comp. 40: 497-504.
Moss, S. T. 1970. Trichomycetes inhabiting the digestive tract of *Simulium equinum* larvae. Trans. Brit. Mycol. Soc. 54: 1-13.
———. 1972. Occurrence, cell structure and taxonomy of the Trichomycetes, with special reference to electron microscope studies of *Stachylina*. Ph.D. Thesis, Univ. of Reading, 340 pp.
———. 1974. A note on the nuclear cytology of *Stachylina grandispora* (Trichomycetes, Harpellales). Mycologia 66: 173-178.
———. 1975. Septal structure in the Trichomycetes with special reference to *Astreptonema gammari* (Eccrinales). Trans. Brit. Mycol. Soc. 65: 115-127.
———. 1976. Formation of the trichospore appendage in *Stachylina grandispora* (Trichomycetes). Pages 279-294 *in* R. Fuller and D. Lovelock, eds. Microbial ultrastructure. Annual technical series of the Society of Applied Bacteriology. Academic Press, London.
Moss, S. T., and R. W. Lichtwardt. 1976. Development of trichospores and their appendages in *Genistellospora homothallica* and other Harpellales and fine-structural evidence for the sporangial nature of trichospores. Canad. J. Bot. 54: 2346-2364.
———. 1977. Zygospores of the Harpellales: an ultrastructural study. Canad. J. Bot. 55: 3099-3110.
Moss, S. T., R. W. Lichtwardt, and J.-F. Manier. 1975. *Zygopolaris*, a new genus of Trichomycetes producing zygospores with polar attachment. Mycologia 67: 120-127.
Moss, S. T., and T. W. K. Young. 1978. Phyletic considerations of the Harpellales and Asellariales (Trichomycetes, Zygomycotina) and the Kickxellales (Zygomycetes, Zygomycotina). Mycologia 70: 944-963.
Poisson, R. 1927. Sur une Eccrinide nouvelle: *Taeniellopsis orchestiae* nov. gen., nov. sp., Protophyte parasite du rectum de l'*Orchestia bottae* M.-Edw. (Crust.Amphipode). Son cycle évolutif. C. R. Hebd. Séanc. Acad. Sci., Paris 185: 1328-1329.
———. 1928. *Eccrinopsis mercieri* n. sp., Eccrinide parasite du rectum de l'*Oniscus asellus*. Son cycle évolutif. C. R. Hebd. Séanc. Acad. Sci., Paris 186: 1765-1767.

———. 1929. Recherches sur quelques Eccrinides parasites de Crustacés Amphipodes et Isopodes. Arch. Zool. Exp. Gén. 69: 179–216.

———. 1931a. Recherches sur les Eccrinides. Deuxième contribution. Arch. Zool. Exp. Gén. 74: 53–68.

———. 1931b. A propos du cycle évolutif des *Amoebidium* (Eccrinideae Amoebidina). C. R. Séanc. Soc. Biol. 106: 354–357.

———. 1932a. *Asellaria caulleryii* n. g., n. sp., type nouveau d'Entophyte, parasite intestinal des Aselles (Crustacés Isopodes). Description des stades connus et d'une partie de son cycle évolutif. Bull. Biol. Fr. Belg. 66: 232–254.

———. 1932b. Sur deux Entophytes parasites intestinaux de larves de Diptères. Ann. Parisitol. Hum. Comp. 10: 435–443.

———. 1936. Sur un Endomycète nouveau: *Smittium arvernense* n. g., n. sp., parasite intestinal des larves de *Smittia* sp. (Diptères Chironomides) et description d'une nouvelle espèce *Stachylina* Lég. et Gauth., 1932. Vol. jubilaire du Pr. L. Daniel, Rennes: 75–86.

Rajagopalan, C. 1967. An *Enterobryus* (Trichomycetes, Eccrinales) in a milliped. Curr. Sci. 36: 20–22.

Reichle, R. E., and R. W. Lichtwardt. 1972. Fine structure of the Trichomycete *Harpella melusinae* from blackfly guts. Arch. Mikrobiol. 81: 103–125.

Robin, C. 1853. Histoire naturelle des végétaux parasites qui croissent sur l'homme et les animaux vivants. J. B. Baillère, Paris. 1–45 pp.

Sangar, V. K., and P. R. Dugan. 1973. Chemical composition of the cell wall of *Smittium culisetae* (Trichomycetes). Mycologia 65: 421–431.

Sangar, V. K., R. W. Lichtwardt, J. A. Kirsch, and R. N. Lester. 1972. Immunological studies on the fungal genus *Smittium* (Trichomycetes). Mycologia 64: 342–358.

Scheer, D. 1944. Ein neuer parasitärer Pilz aus dem darm der Wasserassel (*Asellus aquaticus* L.). Zentbl. Parasitenk. 13: 275–282.

———. 1972a. Über Pilze (Asellarien) aus dem darm von Wasserasseln des Subwassers. Z. Binnenfischerei DDR 19: 369–373.

———. 1972b. Eingliederung des Pilzes *Recticharella aselli* Scheer 1944 in die Asellariaceae (Eccrinales, Endomycetes). Arch. Protistenk. 114: 343–348.

———. 1976a. Der wahre wirt von *Astreptonema longispora* Hauptfleisch (Trichomycetes, Eccrinales) und die konsequenzen aus seiner ermittelung. Arch. Protistenk. 118: 11–17.

———. 1976b. *Parataeniella mercieri* (Poisson) (Trichomycetes, Eccrinales) und ihre wirte in der Deutschen Demokratischen Republik. Arch. Protistenk. 118: 202–208.

———. 1977. *Nodocrinella hylonisci* n. g., n. sp., eine neue Eccrinacee (Trichomycetes, Eccrinales) aus dem darm von *Hyloniscus riparius* (C. L. Koch) (Crustacea, Isopoda). Arch. Protistenk. 119: 163–177.

Starr, M. 1977. Sterol analysis of cultured Trichomycetes. M.A. Thesis, Univ. of Kansas. 92 pp.

Starr, M. P. 1975. A generalized scheme for classifying organismic associations. Pages 1–20 *in* D. H. Jennings and D. L. Lee, eds. Symbiosis. Symposia of the Society for Experimental Biology 29. Cambridge University Press, London and New York.

Svoboda, J. A., and W. E. Robbins. 1975. Insect steroid metabolism. Ann. Entomol. 20: 205–220.

Thaxter, R. 1920. Second note on certain peculiar fungus-parasites of living insects. Bot. Gaz. 69: 1–27.

Trotter, M. J., and H. C. Whisler. 1965. Chemical composition of the cell wall of *Amoebidium parasiticum*. Canad. J. Bot. 43: 869–876.

Tuzet, O., and J.-F. Manier. 1947a. *Palavascia philoscii* n. g., n. sp., Entophyte Eccriniforme parasite de *Philoscia couchii* Kin. C. R. Hebd. Séanc. Acad. Sci. Paris 224: 1854–1856.

———. *Orphella culici* n. sp., Entophyte parasite du rectum des larves de *Culex hortensis* Fclb. C. R. Hebd. Séanc. Acad. Sci. Paris 225: 264–266.

———. La sexualité et les spores durables des Eccrinides du genre *Enterobryus*. C. R. Hebd. Séanc. Acad. Sci. Paris 226: 1312–1314.

———. 1948b. La reproduction sexuée chez *Palavascia philoscii* Tuzet et Manier et chez *Palavascia sphaeromae*, nouvelle espèce de Palavasciées parasites de *Sphaeroma serratum* F. C. R. Hebd. Séanc. Acad. Sci. Paris 226: 2177–2178.

———. 1950a. Les Trichomycètes. Révision de leur diagnoses. Raisons qui nous font y joindre les Asellariées. Ann. Sci. Nat. Zool. 12: 15–23.

———. *Lajassiella aphodii* n. g., n. sp., Palavascide parasite d'une larve d'*Aphodius* (Coleoptère Scarabaeidae). Ann. Sci. Nat. Zool. 12: 465-470.
———. 1951a. Sur quelques Eccrinides du Brésil. Ann. Sci. Nat. Zool. 13: 145-147.
———. 1951b. Le cycle de l'*Amoebidium parasiticum* Cienk. Revision du genre *Amoebidium*. Ann. Sci. Nat. Zool. 13: 351-364.
———. 1953. Recherches sur quelques Trichomycètes rameux: *Asellaria armadillidii* n. sp., *Genistella choanifera* n. sp., *Genistella chironomi* n. sp., *Spartiella barbata* Tuzet et Manier. Ann. Sci. Nat. Zool. 15: 373-390.
———. 1954a. Importance des cultures de Trichomycètes pour l'étude du cycle et de la classification de ces organismes. C. R. Hebd. Séanc. Acad. Sci. Paris 238: 1904-1905.
———. 1954b. Trichomycètes commensaux de l'intestin postérieur des Myriapodes Diplopodes récoltés dans la forêt de la Mandraka (Madagascar). Mém. Inst. Scient. Madagascar 9: 1-13.
———. 1955a. Étude des Trichomycètes de l'intestin des larves de *Simulium equinum* Linné récoltés aux Eyzies (Dordogne). Ann. Sci. Nat. Zool. 17: 55-62.
———. 1955b. Sur deux nouvelles espèces de Génistellales: *Genistella rhitrogenae* n. sp., et *Genistella mailleti* n. sp., observées dans les larves de *Rhitrogena alpestris* Eat. et *Baetis bioculatus* L., récoltées aux Eyzies (Dordogne). Ann. Sci. Nat. Zool. 17: 67-71.
———. 1957a. Écologie parasitaire chez *Glomeris marginata* Villers. Vie Milieu 8: 58-71.
———. 1957b. Troisième contribution à la connaissance des *Eccrinida* commensaux de l'intestin postérieur de Myriapodes Diplopodes du Brésil. Révision des Eccrinida déja identifiés chez les Diplopodes. Arch. Zool. Exp. Gén. 94: 121-147.
———. 1962. *Enteromyces callianassae* Lichtwardt, Trichomycète Eccrinale commensal de l'estomac de *Uca pugilator* Latreille. Ann. Sci. Nat. Bot. 3: 615-617.
———. 1967. *Enterobryus oxidi* Lichtwardt, Trichomycète Eccrinale parasite du Myriapode Diplopode *Oxidus gracilis* (Kock). (Cycle, ultrastructure). Protistologica 3: 413-421.
Tuzet, O., J.-A. Rioux, and J.-F. Manier. 1961. *Rubetella culicis* (Tuzet et Manier, 1947), Trichomycète rameux parasite de l'ampoule rectale des larves de Culicides (morphologie et specificité). Vie Milieu 12: 167-187.
Van Beneden, P. J. 1876. Animal parasites and messmates. D. Appleton & Co., New York.
Whisler, H. C. 1960. Pure culture of the Trichomycete, *Amoebidium parasiticum*. Nature 186: 732-733.
———. 1962. Culture and nutrition of *Amoebidium parasiticum*. Amer. J. Bot. 49: 193-199.
———. 1963. Observations on some new and unusual enterophilous Phycomycetes. Canad. J. Bot. 41: 887-900.
———. 1965. Host-integrated development in the Amoebidiales. J. Protozool. 13: 183-188.
———. 1968. Developmental control of *Amoebidium parasiticum*. Develop. Biol. 17: 562-570.
Whisler, H. C., and M. S. Fuller. 1968. Preliminary observations on the holdfast of *Amoebidium parasiticum*. Mycologia 60: 1068-1079.
Williams, M. C., and R. W. Lichtwardt. 1971. A new *Pennella* (Trichomycetes) from *Simulium* larvae. Mycologia 63: 910-914.
———. 1972a. Infection of *Aedes aegypti* larvae by axenic cultures of the fungal genus *Smittium* (Trichomycetes) Amer. J. Bot. 59: 189-193.
———. 1972b. Physiological studies on the cultured Trichomycete, *Smittium culisetae*. Mycologia 64: 806-815.
Young, T. W. K. 1969. Ultrastructure of aerial hyphae in *Linderina pennispora*. Ann. Bot. 33: 211-216.
———. 1974. Ultrastructure of the sporangiospore of *Kickxella alabastrina* (Mucorales). Ann. Bot. 38: 873-876.
Zhuzhikov, D. P. 1970. Permeability of the peritrophic membrane in the larvae of *Aedes aegypti*. J. Insect. Physiol. 16: 1193-1202.

chapter 9

The Laboulbeniales and Their Arthropod Hosts

by ISABELLE I. TAVARES*

ABSTRACT

The relative thickness of the cuticular layers and the shape of the epidermal cells of the integument of the arthropod hosts of the Laboulbeniales vary with the species and the part of the body, as well as with the stage of development. The cuticle subtending the foot of *Laboulbenia borealis* Speg. is paler than the surrounding cuticle; there is no perpendicular haustorial tube beneath the foot such as that produced by the haustorium of *Hesperomyces virescens* Thaxt. The haustoria of *Herpomyces* may cause some damage to epidermal cells of cockroach antennae. In the Laboulbeniales, the ability to germinate on a particular host genus depends on its availability (determined by geographical distribution, habitat preference, and living habits) and without doubt also on the physical and chemical characters of its integument and hemolymph. A list of host groups parasitized by genera of the Laboulbeniales shows various patterns of parasitism. Genera of the Laboulbeniales may occur on many groups of hosts or they may be quite restricted; there are clusters of genera that have undoubtedly evolved with their hosts; occurrence of several different structural types on one host family reflects recurrent waves of infection. Taxonomically significant structural characteristics include the position and cell wall thickness of the basal and stalk cells of the perithecium, the number and relative height of perithecial outer wall cells, the number of receptacle cells, the presence of secondary receptacular axes, and the structure of the antheridia and primary appendage.

INTRODUCTION

Although the Laboulbeniales cannot accurately be called commensals, these obligate parasites do no appreciable harm to their hosts. The fact that

*Herbarium, Department of Botany, University of California, Berkeley, California 94720.

INSECT-FUNGUS SYMBIOSIS /Batra (ed.) / Allanheld, Osmun, Montclair, NJ

the insects live essentially normal lives even though infected by these fungi has caused some doubt about their actual relationships since the time of the discovery of the Laboulbeniales in 1849 by Auguste Rouget and Ferdinand Schmidt (See Rouget 1850; von Hauer 1850). Robin (1853) assumed that they might take nutrients from the air; Cavara (1899) suggested that the vesicular appendages of *Rickia wasmannii* Cavara might function as assimilatory organs, which derive nutrients from such materials as the glandular secretions of the ant hosts. The suggestion, unsupported by any evidence, was made by Rick (1903) that some kind of symbiosis might exist between *R. wasmannii* and the ants on which they live, the insects possibly deriving sugar from the fungi. As late as 1917, Spegazzini proposed that the nutrients might be assimilated from the external environment by means of pigments of the thallus.

It might be said that these fungi compete with the host itself for the nutrients of the hemocoel and that normally there is enough nourishment for both. Problems apparently arise when some structural damage is caused by haustorial growth in sensitive areas, but this is a subject that has not been investigated by detailed anatomic studies of the host. As early as 1869, Karsten reported damage to the muscle sheaths of flies infected by *Stigmatomyces,* although the activities of the hosts seemed unimpaired. The first convincing report of tissue damage caused by the Laboulbeniales was that of Chatton and Picard (1909), who observed that the haustorium of *Trenomyces* ramified through the fat body of the host. The fat disappeared as a result and the chromatin of the nuclei became condensed; eventually the chromatin was destroyed and the cells became confluent. Unfortunately, no illustrations of this process were published by these authors.

THE INTEGUMENT AND BODY CAVITY OF THE HOST

Because the thalli of the Laboulbeniales are attached to the integument of the host, a knowledge of this part of the insect body is essential to an understanding of the interrelationships between fungus and host. As Locke (1974) pointed out, the integument serves as an outer covering, a skeletal structure, and a reserve supply of food during times of starvation. This exoskeleton varies somewhat in different groups of insects and at different stages of development, but in general it consists of the following:

Epicuticle—a thin outer layer little more than 1 μm thick (cf. electron micrograph, p. 144, Locke, 1974, and Fig. 9-4); as Locke pointed out, it appears as a thin refractive line in sections viewed with the light microscope. This layer, which contains no chitin, is generally considered to consist of four layers:

OUTER CEMENT LAYER—(also called the tectocuticle— see Richards 1951)—secreted by the dermal glands, which are located in the epidermal cell layer; the cement is transported to the surface through the ducts of the glands (see three-dimensional diagram of Richards 1951, p. 4); its composition is not well understood, but functionally it can be compared with varnish; it apparently is deposited on the surface after the wax layer has started to form following ecdysis; it is not present in all cuticles (Locke 1974).

WAX LAYER—(also called the lipid epicuticle—see Richards 1951)—produced by the epidermal cells; Locke (1974) presumed that wax moves to the surface as liquid crystals in the homogeneous inner epicuticle (they are visible at the level at which the pore canals terminate; Locke suggested that the motive force driving lipids to the surface may be surface tension); the waxes are usually mixtures of hydrocarbons and esters of fatty acids and alcohols; this layer apparently is the first to be deposited on the surface after the molt and deposition may continue during the intermolt period in larvae (see Locke 1974, pp. 168, 170, diagrams).

CUTICULIN—(the equivalent of part of the polyphenol layer of Wigglesworth 1961)—a thin layer approximately the thickness of a cell membrane, which covers the entire surface of the insect including tracheae and gland ducts; it is produced at the surface of the epidermal cells at the beginning of the molting period; its composition is not known (Locke 1974).

DENSE HOMOGENEOUS INNER LAYER—(the equivalent of the cuticulin of Wigglesworth 1961)—a wider layer (shown in electron micrograph, Locke, 1974, p. 144, to be a little more than 1 μm in thickness), which is laid down by the epidermal cells below the cuticulin; its composition is not known, but according to Locke (1974), during some periods, at least, the outer part probably contains polyphenols; the wax canals pass through it. (Wigglesworth, 1961, believed that the pore canals traversed it—it should be noted that structural features are difficult to detect with the light microscope; in addition, early work about the layers relied on staining and chemical reactions.)

Procuticle—a thick, laminated inner layer, composed of the polysaccharide chitin and a number of proteins, which combine with chitin to form glycoproteins; some parts of the procuticle are rubber-like, because of the presence of an insoluble protein, resilin (Hackman 1974); the procuticle is typically traversed by pore canals. It consists of two layers:

EXOCUTICLE—the outer, stabilized layer, which is produced at the surface of the epidermal cells beneath the epicuticle before molting (Locke 1974); after ecdysis, this layer becomes hardened and usually darkened (tanned); this process takes place before the deposition of the wax layer, according to Locke (Richards, 1951, referred to the darkening process as sclerotization).

ENDOCUTICLE—the thick inner layer, which is secreted during and after molting; it is not hardened and consequently retains the original lamellate structure; pore canals are conspicuous in this layer (they are not always present, however); there may be a mesocuticle between the endocuticle and exocuticle (see Locke, 1974, for the structure of the endocuticle; see Lower, 1964, for mesocuticle characters; see Figs. 9.1–9.4).

Epidermal Cell Layer—also called hypodermis; the layer of cells covering the body beneath the cuticle; it is typically one cell thick, but there are periclinal divisions at the time of seta formation (Lawrence 1967), and oenocytes (large cells of ectodermal origin with homogeneous cytoplasm filled with endoplasmic reticulum, according to Smith, 1968), as well as nerves and tracheoles, may be present (see Smith, 1968, p. 14; Locke, 1974). The epidermal cells may be polypoid; the upper surface may form filaments that extend up into the pore canals (Locke 1974) (see Figs. 9.1–9.3).

Basement Membrane—a thin, amorphous, granular layer up to 0.5 μm thick, which separates the cuticle from the body cavity or hemocoel; it has been thought to be a layer of neutral mycopolysaccharide secreted by hemocytes (Locke 1974) (see Fig. 9.2).

The hemocoel consists of the internal body cavities (separated by diaphragms or fat bodies) in which the blood or hemolymph circulates; the volume of hemolymph varies from 0.9–45.4% of the body weight; most of it may be absorbed during starvation; several kinds of hemocytes may be found in the hemolymph; the circulation of the hemolymph is in a forward direction through the dorsal heart and in a backward direction laterally and ventrally; in the wings, the hemolymph enters along the anterior margin and returns along the posterior margin (Jones 1964). The blood contains little fermentable sugar and glycogen, but there are large amounts of

---

Figures 9.1–9.4. 9.1–9.3. Cockroach antennae. (X 250.) *9.1.* Longitudinal section, *Blattella germanica* L. bearing *Herpomyces ectobiae (HE);* seta (*se*) at left emerges from setal socket; wide endocuticle with pore canals contrasts with dark exocuticle (arrow, left); interannular membrane (arrow, right) is lamellate but lacks pore canals; in epidermal cell layer below fungus, cytoplasm is homogeneous and one dark-staining nucleus is visible (inner arrow). (Berkeley, California. Iron hematoxylin.) *9.2–9.3.* Transverse sections. (Collection of Dr. Barbara Stay, Harvard University.) (Iron hematoxylin, orange G, preparation by Miss Nel Rem). *9.2. Periplaneta americana* bearing *H. periplanetae* Thaxt. (above, shield visible); cytoplasmic strands extend from epidermal cell layer to setae; a row of epidermal cell nuclei just below the fungus is subtended by an area in which there are only a few scattered nuclei that seem to have lost their capacity to absorb stain; basement membrane (*bm*) visible at right; endocuticle absorbs stain, so is darker than exocuticle. *9.3. Blaberus craniifer* bearing *H. paranensis* Thaxt. (shield cells and perithecium, upper left); haustorial strands (*ha*) pass through setal sockets below the fungus; lamellae visible; size of setae is indicated by width of channels through cuticle. *9.4.* Prothorax of *Cycloneda sanguinea* bearing thallus of *Hesperomyces virescens;* thin surface line is epicuticle; lamellae are distinct, pore canals indistinct (indicated by row of lines); cytoplasm in foot curves downward, extending through haustorial tube and emerging to form narrow haustorial branches in hemocoel. (Preparation by Nel Rem; specimen from Dr. N. H. Hussey, Miss Barbara Gurney, Glasshouse Crops Research Institute, Rustington, Littlehampton, Sussex, England.) (Iron hematoxylin, orange G. X570.) (For abbreviations, see p. 244.)

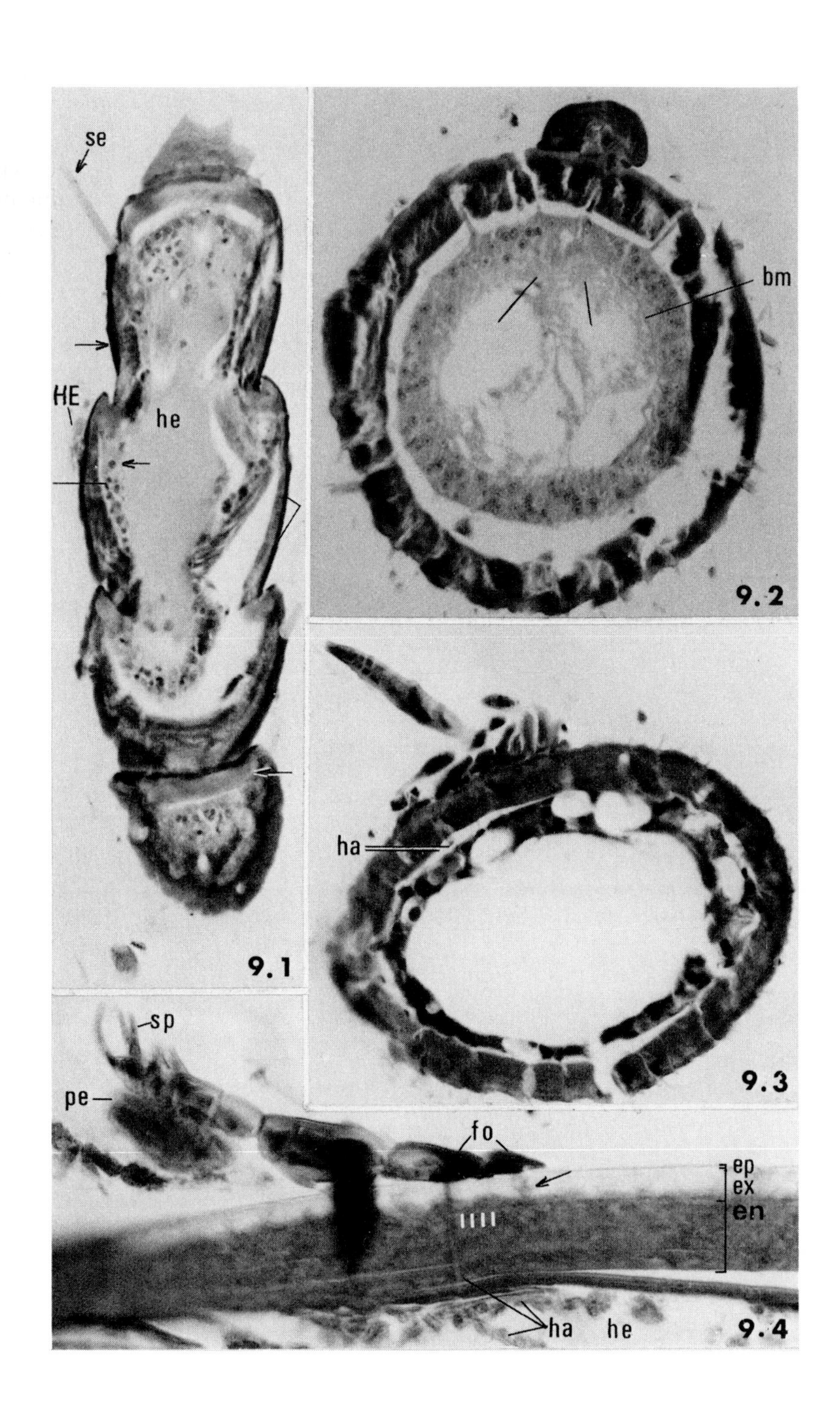
se
HE
he
9.1
bm
9.2
ha
9.3
sp
pe
fo
ep
ex
en
ha
he
9.4

glycerol and organic phosphates; the carbohydrate trehalose is characteristic of insects; there is a high concentration of amino acids similar to that found in the cells; quantities of uric acid, urea, and ammonia may be stored (see Florkin and Jeuniaux 1974).

In addition to the ventral nerve cord, the digestive tract, Malpighian tubules, nerves, striated muscles, tracheae, tracheoles, and reproductive organs, the body cavity contains the fat body, which consists of compact or scattered aggregations of fat cells; oenocytes may be present in the fat body; glycogen and protein reserves fill the large fat cells during the feeding periods of larvae, but starvation depletes the reserves (Kilby 1963).

The formation of new cuticle begins before each molt when the old endocuticle separates from the epidermal cells and cuticulin is formed in patches on the outer surface of the cells; these patches coalesce to form a continuous layer (see electron micrograph, Smith 1968, pl. 2). The beginnings of the surface sculpturing of the integument may appear before the cuticulin layer is complete (Noble-Nesbitt 1967). Molting fluid containing chitinase and protease is secreted, which digests the old endocuticle and causes the old exocuticle to separate from it; the cuticulin layer contains minute pores which presumably enable small molecules from the digested endocuticle to pass through, but it protects the newly developing cuticle because it is impenetrable to the molting fluid (Smith 1968). The products of digestion of the old endocuticle are incorporated into the newly formed endocuticle (Locke 1974).

After the cuticulin and the underlying homogeneous layer are formed, chitin microfibers are laid down at the surface of the epidermal cells, originating from precursors formed within the cells (Locke 1974); layers (lamellae) are produced. Each lamella is made up of many laminae of microfibers, arranged parallel with the surface of the cuticle; however, the direction changes slightly in successive laminae, so that in oblique sections a parabolic pattern appears (Smith 1968; see also diagrams of Locke 1974). It has been observed in *Schistocerca gregaria* Forsk. (the desert locust) that at night, several lamellae may be formed, whereas during the day, nonlamellate layers are produced in which the microfibers are oriented in a longitudinal direction and consequently these layers appear to be isotropic in transverse section (Neville 1967). Most of the protein molecules are attached to the chitin near the epidermal cell surface, as shown by experiments with labeled proteins, but some of the protein moves up through the integument and becomes incorporated in other parts of the cuticle (Noble-Nesbitt 1967).

According to Wigglesworth (1961), the endocuticular lamellae in elytra of Coleoptera may be made up of parallel or anastomosing strands, laterally compressed (called beams or Balken). They are embedded in a homogeneous matrix in successive layers and the strands run at an angle of about 60°. In the elytra of *Cycloneda sanguinea* (L.), crossed fibers can be seen faintly, together with shallow pits in the outer cuticle (Fig. 9.6, arrow) (cf. diagram of crossed fibrillar orientation, Neville 1967, p. 221; see section of elytron, p. 193, and also p. 244, Richards 1951).

Pore canals are not always present—for example, they are lacking in the interannular endocuticle between annulations (segments; according to Imms 1939, 1940, the flagellum of the antennae in Pterygota—winged insects—does not consist of true segments) of cockroach antennae (see Fig. 9.1; also see Haas 1955). Although pore canals are present in the cuticle of the prothorax of *Cycloneda sanguinea* (Coleoptera, Coccinellidae), they are less conspicuous than they are in the cockroaches (Blattaria) (cf. Figs. 9.1-9.4).

Pore canals are most conspicuous in the endocuticle; the tanning process either alters or obscures them in the exocuticle (Noble-Nesbitt 1967). Although pore canals have been thought to be helical, Locke (1974) has shown that this appearance results from their following the arrangement of fibers in the lamellae. He also pointed out that they are extracellular because the plasma membranes of the epidermal cells do not line the canal walls; however, the canals contain one or more filaments from the epidermal cells. Contents are later hardened by chitin or other material, but usually there is still a free passage through the canal (David 1967).

When the old exuvium (the old exocuticle and epicuticle) is cast off (ecdysis), the new exocuticle becomes hardened. It is believed that dihydric phenols pass to the surface for oxidation to quinones and then return to tan the outer chitinous layers of the cuticle (Noble-Nesbitt 1967). There is a pale spot of exocuticle to which the thallus of *Hesperomyces virescens* Thaxt. is attached in Fig. 9.4. Such a colorless hardened exocuticle is in contrast to the usual darkened layer and may be the result of low concentrations of *o*-quinone at the time that the tanning process takes place (see Hackman 1974).

After the exocuticle hardens, the cement and wax form a protective layer on the surface of the cuticulin. Cockroaches are covered with very low melting point waxes, and the cement probably serves chiefly to soak up mobile lipids, according to Locke (1974), who believed that a reserve of such wax could play an important role in rapidly sealing surface abrasions. Because abrasion can readily damage the epicuticular layers, they probably do little to hinder the penetration of haustoria of the Laboulbeniales, although they are of vital importance to the insects. As Locke (1974) pointed out, these layers determine the surface properties and most of the impermeability of the integument.

The relative thickness of the layers of the cuticle varies in different insects and in different parts of the body, such as the hard sclerites and the soft interconnecting membranes where the exocuticle may be lacking (see Dennell and Malek 1954; also Fig. 9.1). Variation may also occur at different times in the life cycle—for example, the exocuticle may be very thin in larvae, whereas it is often very thick, dark, and hard in adults (Locke 1974).

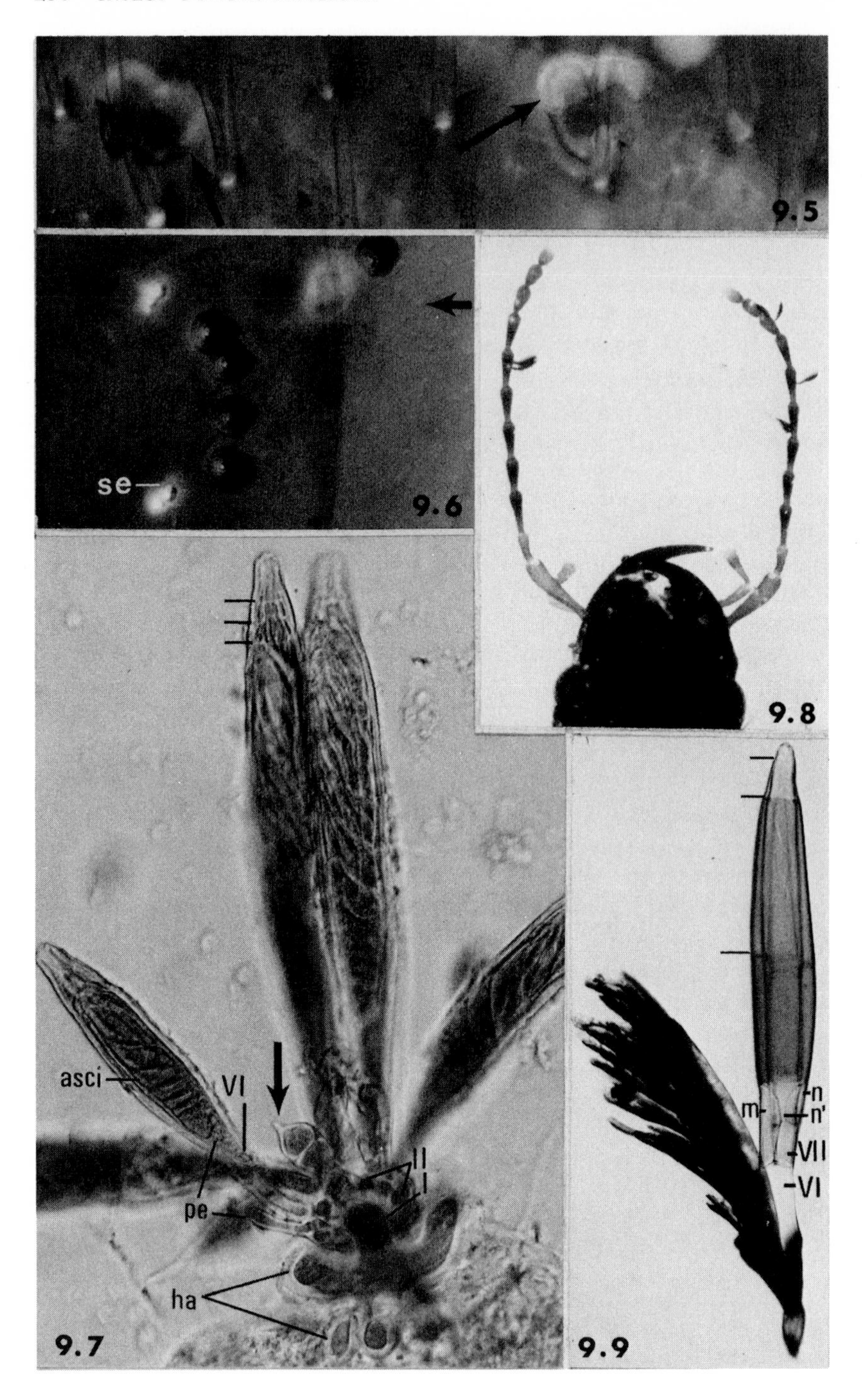
9.5
se
9.6
9.8
asci
VI
II
I
pe
ha
9.7
n
m
n'
VII
VI
9.9

The epidermis of the more highly evolved winged insects (Endopterygota) is typically a uniform layer of cuboid cells with relatively large nuclei situated near the basement membrane; the cell walls are well defined (Lower 1964). By contrast, in primitive wingless insects (Apterygota), the epidermal cells tend to be ovate with the long axis parallel to the surface of the body; their nuclei are variable in size and position, and the limits of the cells are usually not clear. Both types of layers are found in the Exopterygota (those insects, such as cockroaches, that do not undergo complete metamorphosis); in this group, the epidermis in the more advanced orders resembles that in the Endopterygota (see Lower 1964; Steinhaus and Tanada 1971). The epidermal cells may degenerate in adult insects after the last molt (Steinhaus and Tanada 1971).

In the cockroach antenna, the epidermal cell layer in the distal annulations (cf. Fig. 9.2) may be thicker than it is in the proximal part (Fig. 9.3). Whereas sections of antennae of *Blaberus craniifer* Burmeister of smaller diameter are very similar to those of *Periplaneta americana* (L.) (Fig. 9.2), at the level shown in Fig. 9.3, the hemocoel is empty and the epidermis is narrow. The nuclei are almost all dark-staining (there are a few normal nuclei at right similar to those in the section of *Periplaneta americana*). The large cells subtending the setae contain homogeneous cytoplasm (see cell at lower right that contains a dark nucleus) and resemble oenocytes. In the larger, empty cells, the cytoplasm is limited to the cell margins. In the prothorax of *Cycloneda sanguinea*, there is little evidence of epidermal cells on the dorsal side of the body, beneath the pronotum (Fig. 9.4); however, the ventral cells form a narrow but more or less continuous series.

The Laboulbeniales are sometimes found on Diplopoda and Acarina (Arachnida). According to Richards (1951), the integument is essentially the same in all arthropods (see his diagram of the integument, p. 4), although in some there are calcified layers. The hard cuticle of many millepedes is highly calcified (Lower 1964), although thin, elastic cuticle forms the intersegmental membranes. Despite this, Laboulbeniales have

---

Figures 9.5–9.9. Laboulbenia borealis feet on piece of slightly bleached integument of abdomen tip of *Gyrinus plicifer*, viewed from below over bright light; setae and surface sculpture visible. (Berkeley, California.) (× 570.) *9.6. Hesperomyces virescens* feet on elytron of *Cycloneda sanguinea*, viewed from below over bright light; setal sockets, crossed fibers, and surface characteristics visible. (Glasshouse Crops Research Institute.) (Cotton-blue stain. × 570.) *9.7. Trenomyces histophtorus* female with branched, thick-walled haustorium; dome-shaped primary appendage is visible (arrow); basal cells of perithecia indistinct. (Farlow Herbarium 706, on *Menopon*, isotype, south France, Chatton and Picard.) (× 480.) *9.8–9.9. Corethromyces cryptobii* Thaxt. (Ithaca, New York.) *9.8.* Habit on antennae of *Homaeotarsus bicolor* (Grav.). (× 18.) *9.9.* Mature thallus, showing differences in height of outer wall cells in perithecium. (× 225.) (Photo *9.8* by Victor Doran. Photo *9.9* by Alfred Blaker.) (For abbreviations, see p. 244.)

been reported to occur all over the millepede (Colla 1932; Scheloske 1969); the exact positions of the thalli were not indicated by these authors, although Colla published a drawing of a thallus of *Troglomyces* growing on a seta. In an adult tick, Richards (1951) showed that the gland and setal ducts are very conspicuous and the pore canals are readily visible. In Diplopoda, pore canals are present in the lamellate procuticle; apparently, the trichogen cells may be quite large in comparison to the other epidermal cells (Richards 1951, p. 155, Figs. 32, A–C).

Natural apertures reaching to the surface of the epidermis include tracheae (which penetrate into the body cavity), dermal gland ducts, and setal sockets. Of these, the setal sockets with the emerging setae are of greatest interest in a study of the relations of the Laboulbeniales to their hosts. These fungi often germinate on setae, which are covered by a cuticle much thinner than that of the integument. Each seta is formed by a trichogen cell in the epidermal cell layer. Accompanying this cell is a tormogen cell which forms the surrounding sheath through which the protoplast of the trichogen cell grows (see Smith 1968, pl. 4; Lawrence 1967; Snodgrass 1935). In accordance with the terminology of Snodgrass (1935), *seta* is being used here, rather than *hair, bristle,* or *microtrichia* (see Wigglesworth 1961; Lawrence 1967); in a study of the Laboulbeniales, the cuticle and socket are of more importance than the ontogeny and function of these structures. Snodgrass referred to setae with associated nerves as *innervated setae,* whereas they are customarily called *sensilla* (see Wigglesworth 1961).

The structure of only a few insects has been carefully studied. Consequently, the structure of the cuticles of most hosts of the Laboulbeniales is unknown. Observations should be made whenever possible, because the characteristics of the cuticle may have a pronounced effect on the fungus. The following characteristics should be noted: Are pore canals present and what are their dimensions? Is exocuticle present? Are there distinct lamellae? What are the characteristics of the epidermal cell layer (for example: thickness, presence of other cells)? The appearance of cells and particularly of nuclei should be observed in parts of the host near the fungi—this includes the muscles and fat body, as well as the epidermis (see Fig. 9.2).

When specific insects are found that seem to show a reaction to the Laboulbeniales (such as an apparent visual disturbance noticed by Peyritsch, 1871, in a male fly with an infection of *Stigmatomyces* on its head), sections should be made if possible. Portions of insects in suitable fixatives may be placed in agar blocks, which are then embedded in paraffin and sectioned (Richards and Smith 1956, used celloidin and paraffin, with Mallory's triple stain to provide a contrast between fungus and host). A thin solution of Parlodion coating the slides should reduce loss of sections. Heidenhain's iron haematoxylin is an excellent cytological stain.

As Steinhaus and Tanada pointed out (1971), it is desirable to compare an insect that appears to be suffering from a disease of the integument with a normal, healthy specimen. Even though one is familiar with the normal appearance of an insect, it may still be difficult to detect physical abnormalities related to the disease. These authors cautioned against interpretation of extremes of normal variations as the effect of disease.

THE RELATIONSHIP OF THE FUNGUS TO THE HOST

All of the three possible sources of nutrients have been reported for various species of the Laboulbeniales—the cuticle, the epidermal cells underlying the cuticle, and the body cavity of the host. At first Thaxter (1896) was convinced that absorption of food materials took place through the thin membrane by which the foot of a species such as *Laboulbenia hagenii* Thaxt. is attached to the host integument (see 1896, pl. 3, Fig. 4). He later (1914) expressed the belief that all species of external fungus parasites of insects, regardless of the type of attachment, obtain their food supply from the same source—the circulatory system of the host. However, he later reiterated his earlier opinion (1931, p. 115), remarking that absorption of nutrients seemed to take place through the thin membrane applied to the pore canals.

These last two statements are not necessarily contradictory, although his earliest proposal prompted Picard (1908) to suggest that the fungi must obtain their nourishment by hydrolyzing chitin, the damage being repaired by secretions of the epidermal cells. The endocuticle is actually subject to digestion by the insect itself at times of starvation (Locke 1974), and consequently if some nutrients are actually absorbed from the cuticle itself, the epidermis, if still active, is capable of producing more endocuticle. Satisfactory sections have not yet been made to demonstrate the effect on the cuticle of taxa that do not have visible haustoria.

If one compares the photographs of the feet of *Laboulbenia borealis* Speg. (1915) on *Gyrinus plicifer* LeConte (Fig. 9.5) with those of *Hesperomyces virescens* on *Cycloneda sanguinea* (Fig. 9.6), it is easy to see why there is confusion about the source of nutrients. Spots of bright light show through the feet of *H. virescens*, which indicate that haustorial tubes perpendicular to the surface of the elytra penetrate through the upper integument (the lower integument is very thin; cf. Fig. 9.4.). On the other hand, no bright spot is visible below the feet of *Laboulbenia borealis*; instead, the thin portion of the wall of the foot is visible. This area is paler than the surrounding integument, which indicates that the fungus has some effect on the cuticle (the feet are viewed from below on a portion of the abdominal integument, bleached slightly). Perhaps the pore canals under the entire surface of the so-called absorptive membrane have been altered. It should be remembered that Richards and Smith (1956) doubted that nourishment could be obtained through the cuticle without penetration unless the fungus alters the cuticle to destroy its barrier properties.

The absence of a bright spot does not mean that a haustorium is not present—it merely indicates that there is none penetrating to the epidermis perpendicular to the surface. It is possible that a short haustorium could progress laterally to a nearby setal socket and penetrate into the body cavity by this route (cf. Richards and Smith 1956, who showed that haustoria can enter through the setal socket from the exterior in *Herpomyces*).

Colla (1926) reported the presence of a long sinuous structure that seemed to be attached to the foot of *Laboulbenia vulgaris* Peyr.; it became horizontal after penetrating about 3.5 μm into the integument. Tonghini published drawings (1913) purporting to show haustoria of *Laboulbenia* penetrating the integument of *Aulonogyrus* (Gyrinidae). However, the oblique channels he showed might have been fractures (they could have indicated an unusual fragility, such as that reported by Richards and Smith 1956). It is very possible that in some taxa haustoria may occasionally grow laterally between laminae of the cuticle. The short haustorium shown by Colla (1926) penetrating a short distance into the integument would extend into the area of the pore canals. Pore canals undoubtedly do offer less resistance to haustoria than solid cuticle unless they are plugged with hard substances.

Haustorial penetration was first clearly demonstrated by Richards and Smith (1956), who published photographs of the haustoria of *Herpomyces stylopygae* Speg. (1917) passing through the integument into the epidermal cell layer of the host (Mallory's triple stain was used, which stained the fungus blue and the epidermal cells red). The larger haustoria enlarge into large bulbs within the epidermis and are not known to penetrate the basement membrane; they have been found also in the host of *Herpomyces ectobiae* Thaxt. (Tavares, unpublished). In *Stigmatomyces* Peyritsch had reported (1871) that haustoria pass through the integument and enlarge below its surface into small spherical knobs. Dainat and Manier (1974) published photographs showing similar subcuticular swellings in *Stigmatomyces scaptomyzae* Thaxt. A more dramatic demonstration of the occurrence of haustoria was presented by Kamburov, Nadel, and Kenneth (1967); they showed that the haustoria of *Hesperomyces virescens* form several narrow branches radiating out into the body cavity. For such elongate narrow haustoria, the term rhizomycelia suggested by Benjamin (1971) is very appropriate. Similar large haustoria may also occur in *Fanniomyces* (Tavares, unpublished), at least below large clumps of thalli; it has not been determined whether these haustoria represent a fusion of those of several thalli.

There are distinct cell walls on the narrow haustoria of *Hesperomyces virescens* (Fig. 9.4) and on the broad, lobed haustorium of *Trenomyces* (Fig. 9.7); on the contrary, a cell wall cannot be clearly seen on the haustorial bulbs of *Herpomyces stylopygae* (Richards and Smith 1956). Perhaps wall layers are deposited only on those haustoria that penetrate into the hemocoel. Several other genera have large haustoria—*Arthrorhyn-*

chus, Microsomyces, Rhizomyces, and *Rhizopodomyces*; in none of these genera has the relationship to the host been carefully studied.

Colla (1926) and Richards and Smith (1956) concluded from their studies that the haustoria must rely on enzymatic action, rather than on mechanical action. However, David (1967) suggested that some mechanical force is probably involved in penetration by various insect parasites. Small depressions apparently may be made in the epicuticle by *Metarrhizium anisopliae* (Metchn.) Sorok. Mercier and Poisson (1927) observed a slight thinning of the cuticle at the point of attachment of the foot of *Stigmatomyces*; a depression is just above the discolored area under *Hesperomyces* foot (arrow, Fig. 9.4). Although it would be assumed that lipases, proteases, and chitinases would be necessary to penetrate the integument, there is said to be no protease in *Metarrhizium* (David 1967); presumably penetration can be in part effected by the utilization of pore canals, possibly at abraded places.

The actual uptake of material by *Laboulbenia* from the hemolymph was demonstrated by Scheloske (1969), when he was able to achieve sufficient concentration of Nile blue sulfate dye by injecting it into an elytron of the host beetle.

In *Trenomyces histophtorus* attached to the surface of a sclerite, Chatton and Picard (1909) found the size to be small and the appearance abnormal when compared to thalli occurring on softer integument of an articulation. Species probably adapt to differences in the integument by varying the form of the haustorium to some extent. If a thallus cannot penetrate directly, possibly it will send a haustorium to the nearest place from which it can obtain nourishment.

THE DEVELOPMENT OF THE THALLUS AND TAXONOMIC CHARACTERISTICS

After the haustorium (or the equivalent absorptive mechanism) is formed, the cells of the germinating spore divide in the manner characteristic of the genus (Figs. 9.12, 9.18). In genera in which an extensive primary appendage system is formed, this part of the thallus often becomes well developed before there is any indication of the perithecial initial cell (Fig. 9.15).

The receptacle is formed from the lower cell of the two-celled germinating spore and the primary appendage develops from the upper cell of the spore (usually the smaller of the two cells). The primary appendage is usually highly developed in *Laboulbenia, Corethromyces* (Figs. 9.9, 9.15), and several related genera, although in *C. piesticola* Thaxt. (1931), it breaks off early and scarcely branches (Figs. 9.18, 9.19). In the genera closely related to *Corethromyces,* there are only three cells in the receptacles besides the cells that constitute the perithecium and its stalk.

The perithecium in most genera grows from the middle cell of the three cells of the young receptacle (Fig. 9.18). The perithecial initial extends

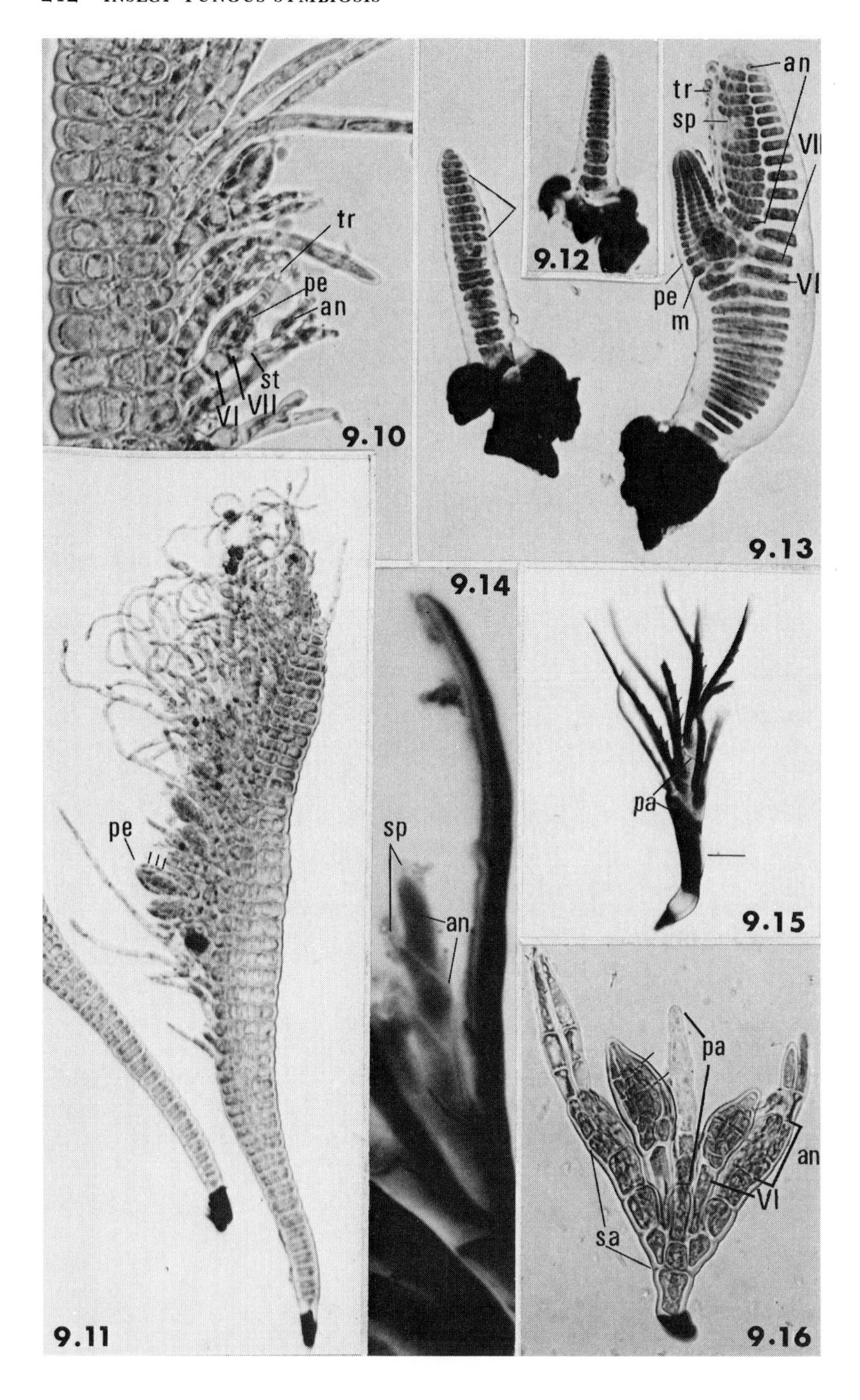
tr
pe
an
st
VI
VII
9.10
9.12
an
tr
sp
VII
VI
pe
m
9.13
pe
9.11
9.14
sp
an
pa
9.15
pa
an
VI
sa
9.16

on other orders is based primarily on Brues, Melander, and Carpenter (1954). * Indicates unpublished combinations.

Class Insecta—six-legged arthropods.
Order Coleoptera—forewings modified into rigid elytra; beetles.
Suborder Adephaga—beetles with notopleural sutures (lateral grooves on the underside of the prothorax); hind coxae dividing first visible abdominal segment.

Carabidae—antennae inserted at sides of head between bases of mandibles and eyes; predacious ground beetles.

1. *Cochliomyces*—uniseriate receptacle; primary appendage unbranched; many short equal outer wall cells in perithecium; unbranched stalk appendage; terminal phialide.
2. *Euzodiomyces* (fortuitous—Scheloske, 1969)—multiseriate; many short lateral appendages; perithecium—see 1.
3. *Pseudoecteinomyces*—like 1, but with rudimentary stalk appendage and lateral phialides. Rossi, 1977, Mycologia 69: 1075.
4. *Rhachomyces*—main axis is uniseriate secondary receptacle; short lateral appendages, few unequal cells in each vertical row of outer wall cells of perithecium; perithecial basal cells have well-developed cell walls.
5. *Laboulbenia*—usually five receptacle cells; usually dark insertion cell above receptacle, bearing an outer and an inner appendage; appendages often abundantly branched; phialides (as in 1–4); perithecium—see 4.
6. *Misgomyces* and an unpublished segregate—in *Misgomyces,* base of appendage is compound antheridium; in unpublished genus, appendage bears phialides; receptacle multiseriate or biseriate above; perithecium—see 4 (except unequal number of cells in different rows in segregate).
7. Unpublished segregate of *Misgomyces*—uniseriate multicellular receptacle; appendage (probably antheridial) on cell *III*.
8. *Apatomyces*—apparently dioecious; upper receptacle adnate to stalk cell of perithecium; branched appendage; three-celled receptacle; perithecium—see 4.
9. *Corethromyces*—monoecious; receptacle three-celled (cells *III* and *VI* not laterally adnate); branched appendage; perithecium—see 4.
10. Unpublished segregate of *Dioicomyces*—dioecious; three-celled receptacle; unbranched appendage; perithecium—see 4; cell *VI* free from upper receptacle.
11. *Peyritschiella*—short two-celled appendages with cells separated by thick black septum; multiseriate receptacle; compound antheridia sessile on receptacle; perithecium—see 4.
12. *Enarthromyces*—long uniseriate receptacle; compound antheridia and short lateral appendages; perithecium—see 4.
13. *Dimorphomyces*—dioecious; cell *I* extends laterally below row of receptacle cells; perithecial basal cells with indistinct walls; few unequal wall cells in each vertical row; compound antheridia.
14. *Dimeromyces*—dioecious; no lateral extension of receptacle; perithecium—see 13.

15. *Eucantharomyces*—monoecious; primary appendage is compound antheridium; receptacle three-celled; perithecium—see 4.

Cicindelidae—antennae arising between eyes; legs long; tiger beetles.

5. *Laboulbenia.*

Noteridae—undivided eyes; two slender claws on hind tarsi; streamlined body; burrowing water beetles.

16. *Chitonomyces*—appendages as in 11; few-celled receptacle; antheridia simple, innate just above perithecium; perithecium—see 4.

Dytiscidae—undivided eyes; streamlined body; scutellum shows centrally between bases of elytra; short maxillary palps; predacious diving beetles.

17. *Autoicomyces* (fortuitous?)—receptacle with few cells below perithecium; stalk and secondary stalk cells are intercalary cells of primary axis; perithecium has many short equal outer wall cells; cells in all four rows of perithecium similar in height.
16. *Chitonomyces.*

Gyrinidae—divided eyes; streamlined body; short maxillary palps; whirligig beetles.

16. *Chitonomyces.* 5. *Laboulbenia.*

Haliplidae—greatly expanded hind coxal plates; crawling water beetles.

16. *Chitonomyces*—perithecium laterally adnate to upper receptacle.
18. *Hydraeomyces*—perithecium free laterally; two or more short secondary appendages; perithecium—see 4 (but vertical rows differ).

Suborder Polyphaga—no notopleural sutures; first visible abdominal segment usually not interrupted.

Hydrophilidae—undivided eyes; streamlined body; long maxillary palps; water scavenger beetles; some occur in dung.

19. *Thaumasiomyces*—primary axis subdivided at level of perithecium into many branches; perithecium—see 17, but with long neck.
17. *Autoicomyces.*
20. *Plectomyces*—like 17 but with long lower receptacle.
21. *Rhynchophoromyces*—like 17 but with long perithecial neck; base of perithecium usually adnate laterally to primary axis.
22. *Ceratomyces*—like 17 but with cells of different lengths in alternate vertical wall cell rows of perithecium; two vertical rows narrow.
23. *Eusynaptomyces*—like 22 but with long lower receptacle.
24. *Phurmomyces*—like 23 but with outer wall cells approximately equal in width.
25. *Synaptomyces*—like 23 but perithecial apex with row of outgrowths.
26. *Zodiomyces*—massive thallus with many perithecia borne at apex.
27. *Hydrophilomyces*—long primary receptacle; perithecium similar to 4 but with tall accessory cell on outer side of lower wall cells.
28. *Chaetarthriomyces*—perithecium normal, like 4; unbranched primary appendage.
29. Segregate of *Misgomyces*—perithecium like 4, but with an extra cell at base or on inner side of lower wall cells; much-branched primary appendage.
30. *Limnaiomyces*—compound antheridium borne below perithecium; appendages on one side only, like 11; receptacle cells in tiers; perithecium—see 4.

31. *Rickia*—receptacle multiseriate, cells irregularly arranged; appendages like 11; perithecium—see 4.
32. *Cantharomyces*—like 15, but with appendage extending beyond antheridium.

Limnebiidae—antenna terminates in five-segmented pubescent club; minute moss beetles.

33. *Thripomyces*—long uniseriate receptacle; lower cells on side of perithecium bear minute branchlets; perithecium like 17 or 1; walls indistinct at maturity.
34. *Drepanomyces ?*—(host group uncertain) like 33 but no perithecial branchlets.
27. *Hydrophilomyces.*

Staphylinidae—slender, flexible body; short elytra exposing abdomen; rove beetles.

35. *Euceratomyces*—receptacle few-celled; branched appendages with phialides; perithecium—see 1.
2. *Euzodiomyces.*
36. *Kainomyces*—receptacle many celled; perithecium on long, uniseriate unbranched stalk; rostrate perithecium, like 4 but with many short equal outer wall cells.
37. *Compsomyces*—perithecia like 36 but not rostrate, borne on uniseriate secondary axes.
38. *Clematomyces*—like 37 but perithecium on multiseriate secondary axis or suprabasal cell complex.
39. *Balazucia*—multiseriate secondary axis on either side of primary; perithecium like 4.
40. *Clonophoromyces*—uniseriate secondary axis with terminal cluster of slender branchlets; perithecium like 4.
4. *Rhachomyces*—asymmetrical; row of short lateral appendages.
41. *Chaetomyces*—uniseriate primary axis with antheridial appendage below perithecium, which is like 4.
42. *Diplomyces*—broad, multicellular receptacle terminating in small branchlets and broad two- to three-celled upgrowths; perithecium like 4.

NOTE: ALL PERITHECIA LIKE 4 in nos. 43 TO 72 EXCEPT 49, 60–62.

43. *Idiomyces*—no broad upgrowths; broad primary appendage stump visible.
44. *Sandersoniomyces*—complex of suprabasal cells extends upward into paired secondary axes bearing branches on both sides.
45. *Symplectromyces*—suprabasal cell complex has many short appendages; antheridia are intercalary cells of secondary appendages.
46. *Teratomyces*—similar but secondary appendages bear sessile phialides.
47. *Smeringomyces*—narrow multicellular receptacle; short vertical appendages rising from near base of perithecium terminate in inconspicuous phialides.
48. *Skelophoromyces*—perithecia borne on lateral stalks bearing rhizoid-like downgrowths.
49. Unpublished segregate of *Misgomyces* (see 6) with short, branched primary appendage and perithecium like 27.

29. Segregate of *Misgomyces*. 5. *Laboulbenia*. 6. *Misgomyces*.
50. *Diclonomyces*—one or two secondary appendages produced by cell *II*; phialides few; diagonal or vertical septum between cells *II* and *III*.
51. *Stichomyces*—primary appendage elongate, symmetrically branched with corner cells bearing short branches with phialides; perithecia one or more.
52. *Rhadinomyces*—usually cell *III* bears cluster of branches with phialides.
53. *Sphaleromyces*—primary appendage bears row of inner branches with intercalary phialides; five cells in each vertical row of outer wall cells, rather than four as in *Corethromyces*.
9. *Corethromyces*—differs from 51 by asymmetrical branching; from 53 by number of wall cells; from 52 by lack of branch from *III* (Figs. 9.8, 9.9).
54. *Ilyomyces*—cells *I* and *II* side by side; paired phialides terminate short appendage; perithecium has four terminal lobes.
55. *Polyascomyces*—cells *II* and *VI* laterally adnate; perithecial apex has four terminal lobes; ascogenic cells numerous; basal perithecial cells normal.
56. *Acallomyces*—like 55 but ascogenic cells few; basals displaced laterally.
57. *Zeugandromyces*—cells *I-III* superposed; simple appendage with lateral row of phialides.
58. *Amorphomyces*—dioecious; female appears to have no appendage.
59. *Tetrandromyces*—dioecious; female with simple, short appendage; male with more than one apical phialide.
60. **Phaulomyces*—no appendages on cell *III* or lower receptacle; basal cells of perithecium with indistinct cell walls; extra protruding cell in one vertical row of outer wall cells.
61. *Meionomyces*—like 60 but with short, simple appendages on cell *III*, which is extremely small.
62. **Aporomyces*—usually dioecious; persistent terminal trichogyne; indistinct basal cells of perithecium.
63. *Mimeomyces*—cell *I* tall, obconic; *II* flat, broad; five cells in each vertical row of outer wall cells; primary appendage branches bear phialides and/or upper receptacle bears compound antheridia.
11. *Peyritschiella*. 31. *Rickia*.
64. *Diaphoromyces*—like 31 except with central row of vertically elongate cells.
65. *Diandromyces*—receptacle uniseriate, short; antheridia borne just below perithecia; no black septa.
13. *Dimorphomyces*. 14. *Dimeromyces*. 32. *Cantharomyces*.
66. *Camptomyces*—primary appendage is short, broad, compound antheridium with terminal discharge tube; antheridium has few cells.
67. *Euhaplomyces*—like 66 but antheridium long, narrow, gradually tapered.
68. *Haplomyces*—like 66 but antheridial cells many, irregularly arranged.
69. *Neohaplomyces*—like 68 but antheridial cells in diagonal rows.
70. *Kleidiomyces*—primary appendage undeveloped; compound antheridia intercalary on branches from cell *II*; perithecia borne directly on *II*.
71. *Monoicomyces*—like 70 but perithecia borne on antheridial branches.
72. *Eumonoicomyces*—like 71 but massive flask-shaped antheridia.

Pselaphidae—short elytra, wide, inflexible abdomen; short-winged mold beetles.

73. *Pselaphidomyces*—perithecium borne on multicellular uniseriate unbranched stalk; perithecium like 37; phialides on primary axis.
74. *Autophagomyces*—short appendage with terminal phialide; perithecium like 4.
75. *Apatelomyces*—short cell *I* bears long *II* and short, rounded *III*, which subtends dense cluster of appendages; perithecium—see 4.
76. *Cryptandromyces*—primary appendage long, simple or sparsely branched, bearing lateral phialides; perithecium—near 73.
77. *Peyerimhoffiella*—like 76 but short antheridial branch from *III*.
78. *Porophoromyces*—primary appendage is large compound antheridium.

Ptiliidae—feather-winged beetles; minute.

79. *Ecteinomyces*—uniseriate multicellular receptacle; phialides on branchlets above perithecium; perithecium like 4.
80. *Siemaszkoa*—similar receptacle; sessile phialides; perithecium like 60. Tavares and Majewski, 1976, Mycotaxon 3: 202.
31. *Rickia.*

Scaphidiidae—small, convex, long-legged, shiny; truncate elytra; shining fungus beetles.

81. *Scaphidiomyces*—perithecia like 4, borne near bases of uniseriate, branched secondary axes.
31. *Rickia.*

Scydmaenidae—elytra full length; ant-like stone beetles; often in ant nests.

82. **Acompsomyces*—row of sessile phialides on short unbranched primary appendage; perithecium like 4 but subapical outer wall cells flared outward and apex tapered to point.
83. *Acrogynomyces*—persistent trichogyne base at apex of perithecium, which is otherwise like 4; cells *I* and *II* parallel.
76. *Cryptandromyces.* 50. *Diclonomyces.* 31. *Rickia.*
14. *Dimeromyces* as *Jeanneliomyces*, on larva.

Leiodidae (Liodidae)—strongly convex, shiny, often roll up; round fungus beetles.

84. *Cucujomyces*—uniseriate secondary axes curve down on both sides of primary axis; perithecium like 4, often with dark rings.
85. *Euphoriomyces*—secondary appendages on receptacle below or opposite perithecium, which is like 60.

Leptodiridae (Catopidae, Colonidae)—head deflexed, thin integument; small carrion beetles.

86. *Colonomyces*—perithecium like 35.
87. *Columnomyces*—massive columnar receptacle terminating in crowded branches; perithecium like 4, borne near base of receptacle.
88. Unpublished segregate of *Corethromyces* with cells *II* and *VI* laterally adnate; cells *I* and *II* similar in height.
9. *Corethromyces*—cells *II* and *VI* superposed.
89. *Asaphomyces*—perithecia borne laterally on multicellular uniseriate to biseriate receptacle; antheridial appendage just above perithecium, which is like 60, but with no protruding apical cell.
63. **Mimeomyces*—phialides on primary appendage branches.

Histeridae—hard, compact, shiny, with clubbed geniculate antennae; slightly shortened elytra; hister beetles.

90. *Histeridomyces*—perithecia and phialides with spherical necks borne on multicellular uniseriate secondary axes; perithecium near 73.
91. *Rhipidiomyces*—receptacle wide, of vertically elongate cells; innate lateral phialides; innate perithecium.
92. *Homaromyces*—multiseriate receptacle bearing free perithecia, which are like 74.
5. *Laboulbenia.*

Elateridae—articulating prothorax; body elongate; click beetles.

5. *Laboulbenia.*
93. *Stemmatomyces*—cells *II* and *III* extend down on opposite sides of *I*; row of phialides on outer side of appendage.

Byrrhidae—convex, with deflexed head; pill beetles.

62. *Aporomyces.*

Limnichidae—oval, convex, with visible short antennae; minute marsh-loving beetles.

62. *Aporomyces.* 32. *Cantharomyces.*

Dryopidae—short, hidden antennae; long-toed water beetles.

94. *Helodiomyces*—perithecium like 17, but with lateral branches near base.
32. *Cantharomyces*

Heteroceridae—small body, short antennae, large mouth parts; variegated mud-loving beetles.

95. *Botryandromyces*—cluster of sessile phialides at top of receptacle; perithecium like segregate of 6. Tavares and Majewski, 1976, Mycotaxon 3: 195.

Passalidae—sublamellate antennal clubs; peg or bess beetles.

31. *Rickia.*

Scarabaeidae—lamellate-clubbed antennae; scarab beetles.

5. *Laboulbenia.* 31. *Rickia.*

Anthicidae—deflexed head, entire elytra, ant-like flower beetles.

74. *Autophagomyces*—monoecious; short, slender appendage (sometimes branched) bearing narrow terminal phialides.
96. *Dioicomyces*—dioecious; male with one phialide; female primary appendage has one or two cells.
59. **Tetrandromyces.* 14. *Dimeromyces.*

Tenebrionidae—hard body, mostly black; eyes usually transverse and pronotum wider than head; darkling beetles.

5. *Laboulbenia.*
97. *Synandromyces*—cells *II* and *III* extend downward on opposite sides of *I*; compact series of phialides on inner side of appendage; perithecium like 4.
60. *Phaulomyces.* 13. *Dimorphomyces.* 14. *Dimeromyces.* 64. *Diaphoromyces.*

?Tenebrionoidea.

98. *Sympodomyces*—perithecium like 4; secondary axes spreading laterally on both sides of primary axis, bearing branches above and below.

Chrysomelidae—often bright-colored; usually with bilobed third tarsal segment; leaf beetles.
beetles.
5. *Laboulbenia.* 14. *Dimeromyces.*

Cleridae—furry body, bright-colored; mostly clubbed antennae; checkered beetles.
14. *Dimeromyces.*

Biphyllidae—oval, somewhat convex, pubescent; false skin beetles.
84. *Cucujomyces.* 14. *Dimeromyces.*

Cisidae—cylindrical to oval, hairy; head deflexed; minute tree-fungus beetles.
60. *Phaulomyces.* 85. *Euphoriomyces.*

Coccinellidae—ladybird beetles, usually bright-colored, spotted, and convex.
99. *Hesperomyces*—perithecium like 4, with complex of apical lobes.

Colydiidae—hard, shiny, usually elongate; cylindrical bark beetles.
100. *Amphimyces*—massive multiseriate receptacle; simple antheridia on appendages below perithecium; no thick black septa; perithecium like 4.
101. *Dipodomyces*—primary axis undeveloped; secondary foot; secondary axis terminates in cluster of phialides; perithecium like 4, but with slender apical outgrowths.
96. *Dioicomyces.* 61. *Meionomyces.* 31. *Rickia.* 14. *Dimeromyces.*

Cryptophagidae—elongate, oval, pubescent; clubbed antennae; silken fungus beetles.
84. *Cucujomyces.* 74. *Autophagomyces.* 82. *Acompsomyces.* 97. *Synandromyces.* 60. *Phaulomyces.* 31. *Rickia.* 14. *Dimeromyces.*

Cucujidae—elongate, with parallel or oval sides; flat bark beetles.
84. Cucujomyces.
102. *Microsomyces*—large cell *I* corticated by small cells; phialides terminate small appendage; perithecium like 4; ascogenic cells many.
97. *Synandromyces.* 14. *Dimeromyces.*

Endomychidae—more or less oval; colorful; handsome fungus beetles.
38. **Clematomyces.*
103. *Trochoideomyces*—perithecium like 38, borne on uniseriate multicellular stalk from suprabasal cell complex.
104. *Kruphaiomyces*—lower receptacle two-celled; slender appendage with phialides arises from cell *III*; perithecium like 4 except beaked.
31. *Rickia.* 14. *Dimeromyces.*

Erotylidae—hemispherical to elongate, bright-colored; pleasing fungus beetles.
5. *Laboulbenia.* 31. *Rickia.*

Lathridiidae—elongate oval shape, three-segmented tarsi; minute brown scavenger beetles.
105. *Rickia* segregate—biseriate to triseriate receptacle of cubical cells; no black septa; perithecium like 4. 82. *Acompsomyces.*

Mycetophagidae—obovate, depressed body covered with short hairs, sometimes orangish or reddish markings; hairy fungus beetles.
84. *Cucujomyces.* 99. *Hesperomyces.*

Nitidulidae—bright-spotted; three-segmented antennal club; sap beetles.
85. *Euphoriomyces.*

106. *Carpophoromyces*—sessile phialides on branches of secondary axis; perithecia like 60.
97. *Synandromyces.* 31. *Rickia.*

Orthoperidae (Corylophidae)—head often covered by pronotum; minute fungus beetles.

105. *Rickia* segregate. 5. *Laboulbenia.* 74. *Autophagomyces.* 60. *Phaulomyces.* 31. *Rickia.*

Discolomidae—minute; broadly oval, convex; prothorax much widened behind.

60. *Phaulomyces.* 14. *Dimeromyces.*

Phalacridae—convex, oval, dark; shining flower beetles.

105. *Rickia* segregate. 74. *Autophagomyces.* 31. *Rickia.* 14. *Dimeromyces.*

Rhizophagidae—small, dull; two-segmented antennal club; root-eating beetles.

105. *Rickia* segregate. 31. *Rickia.*

Anthribidae—usually scaly; antennae not geniculate; beak broad; fungus weevils.

84. *Cucujomyces.*

Dermaptera—abdomen terminated by pair of forceps; earwigs.

107. *Filariomyces*—elongate uniseriate receptacle bears short branchlets and perithecia laterally; perithecium like 4; phialides on branchlets.
76. **Cryptandromyces.*
108. *Distolomyces*—phialides on small appendage bent toward perithecium; perithecium like 4 but with apical outgrowth.
109. *Nanomyces*—apparent male has terminal sterile cell; appendage of apparent female is short, cylindrical, not tapered; probably most are hermaphrodites.
110. *Dermapteromyces*—phialide sessile or on short appendage from cell *III*; primary appendage with short, broad lateral branchlets; perithecium like 89, but with two visible cells in base.
14. *Dimeromyces.*

Hymenoptera, Formicoidea—abdomen strongly constricted at base; ants.

5. *Laboulbenia.* 31. *Rickia.* *13.* **Dimorphomyces.*

Diptera—hind pair of wings replaced by halteres.

Pupipara—parasitic flies, external on birds, bats, and other mammals.

Nycteribiidae—parasites of Chiroptera (bats); wingless, head folds back; long legs—bat-tick flies.

111. *Arthrorhynchus*—conspicuous haustorium; normal foot lacking; unbranched appendage with crowded phialides; perithecium like 4.

Streblidae—mostly on bats, more or less normal in shape; bat flies.

112. *Gloeandromyces*—cells *II* and *VI* parallel; densely branched primary appendage; normal foot; perithecium like 4.
113. *Nycteromyces*—dioecious; cells *II* and *III* parallel; perithecium like 13.

Hippoboscidae—bird and mammal parasites (not on bats); more or less normal form, sometimes winged.

114. *Trenomyces*—dioecious; enlarged cell *I* and large haustorium; compound antheridia; perithecium like 13. Benjamin, 1967, Aliso 6 (3): 132.

Brachycera (various families of flies)—short antennae; free-living.

5. *Laboulbenia.*

115. *Ilytheomyces*—three-celled receptacle; pair of phialides near base of simple primary appendage; perithecium like 4.
116. *Rhizomyces*—like 115 but phialides at base of each branch of appendage; often with conspicuous haustorium.
117. *Stigmatomyces*—unbranched appendage bears lateral phialides; cells *III* and *VI* adnate laterally; perithecium like 4.
118. *Fanniomyces*—like *Stigmatomyces* but elongate, branched appendage. Majewski, 1972, Acta Mycol. 8: 229.
14. *Dimeromyces.*

Hemiptera, suborder Heteroptera—basal part of forewings toughened; terminal part membranous; bugs. See Benjamin, 1967, Aliso 6(3): 111–112, 132–135; 1970, 7(2): 165.

Cryptocerata—antennae short, usually hidden.

Corixidae—swim with dorsal side upward; water boatmen.

119. *Coreomyces*—perithecium develops inside upper part of receptacle (perithecial structure like 4); lateral appendages below perithecium.

Gymnocerata—longer, visible antennae.

Veliidae (broad-shouldered water striders), Mesoveliidae, Macroveliidae (semiaquatic bugs).

5. *Laboulbenia.* 74. *Autophagomyces.* 96. *Dioicomyces.*
120. *Prolixandromyces*—like 74 but with apical perithecial outgrowth; paired phialides with long necks. See Benjamin, 1970, Aliso 7(2): 174.

Hebridae—minute, broad, velvety; semiaquatic bugs.

121. *Rhizopodomyces*—dioecious; caducous small appendage on female; prominent haustorium.

Anthocoridae—minute pirate-bugs; flower bugs.

82. **Acompsomyces.*

Lygaeidae—chinch bugs; plant bugs.

9. *Corethromyces.*

Scutelleroidea (Pentatomoidea), Plataspididae, Pentatomidae—terrestrial bugs.

122. *Polyandromyces*—dioecious; male has terminal compound antheridium; female has diagonal cells in receptacle; perithecium like 13.

Thysanoptera—four narrow wings fringed with long bristles; thrips.

96. *Dioicomyces.* Balazuc, 1972, Bull. Soc. Entomol. France 76: 228–231.

Mallophaga—biting lice external on birds and mammals (infected lice on several orders of birds and on rodents); small, flat, wingless insects.

114. *Trenomyces.*

Isoptera—social insects, often pale in color; not narrow between prothorax and abdomen; termites.

5. *Laboulbenia.* 14. *Dimeromyces.*

Orthoptera, Saltatoria, Gryllioidea—hind legs usually adapted for jumping.

Gryllidae—long antennae; color usually dark; crickets.

5. *Laboulbenia.* 14. *Dimeromyces.*

Gryllotalpidae—forelegs enlarged for digging; mole crickets.

123. *Tettigomyces*—compound antheridium above perithecium, which is like 17.
31. *Rickia.*

Blattaria—head partly covered by pronotum; running insects; cockroaches.

124. *Herpomyces*—perithecia borne on secondary axes from undeveloped primary axis; vertical wall cell rows many-celled; internal perithecial structure unique.

5. *Laboulbenia.*

Acarina—four pairs of legs in adult; mites (no reports from ticks).

5. *Laboulbenia.* 31. *Rickia.* 13. *Dimorphomyces.* 14. *Dimeromyces*

Diplopoda—many pairs of legs, usually two pairs to a segment; millepedes.

Rhachomyces report, Balazuc, 1973, Inst. Spéol. "Emile Racovitza," Livre Cinquant., Colloque Natl. Spéol. Pp. 463–477 is incorrect (personal communication).

125. *Troglomyces*—narrow cells *II* and *III* adnate to perithecium; phialide borne at base of primary appendage.

31. *Rickia.*

LITERATURE CITED

Arnett, R. H. 1963. The Beetles of the United States (a manual for identification). Catholic University of America Press, Washington, D.C. 1112 pp.

Benjamin, R. K. 1971. Introduction and supplement to Roland Thaxter's contribution towards a monograph of the Laboulbeniaceae. Bibliotheca Mycol. 30: 1–155.

Brues, C. T., A. L. Melander, and F. M. Carpenter. 1954. Classification of insects, keys to the living and extinct families of insects, and to the living families of other terrestrial arthropods. Bull. Mus. Compar. Zool. Harvard: 108: 1–917.

Cavara, F. 1899. Di una nuova Laboulbeniacea *Rickia Wasmannii* nov. gen. e nov. spec. Malpighia 13: 173–188. Pl. VI.

Chatton, E., and F. Picard. 1909. Contribution a l'étude systématique et biologique des Laboulbéniacées: *Trenomyces histophtorus* Chatton et Picard, endoparasite des poux de la poule domestique. Bull. Soc. Mycol. Fr. 25: 147–170. Pls. VII, VIII.

Colla, S. 1926. Sull'organo d'assorbimento della specie del gen. "Laboulbenia" Rob. Atti Accad. Sci. Torino 61: 277–280.

———. 1932. "Troglomyces Manfredii" n. gen. et n. sp.: nuova Laboulbeniacea sopra un miriapode. Nuovo Giorn. Bot. Ital. (n. s.) 39: 450–453.

Dainat, H., and J.-F. Manier. 1974. Haustoria de *Stigmatomyces scaptomyzae* Thaxter (Laboulbéniale) parasite de *Scaptomyza graminum* Fallen (Diptère, Drosophilide). Bull. Soc. Mycol. Fr. 90: 217–221.

David, W. A. L. 1967. The physiology of the insect integument in relation to the invasion of pathogens. Pages 17–35 *in* J. W. L. Beament and J. E. Treherne, eds. Insects and physiology. Oliver and Boyd, Edinburgh, London.

Dennell, R. E., and S. R. A. Malek. 1954. The cuticle of the cockroach *Periplaneta americana*. I. The appearance and histological structure of the cuticle of the dorsal surface of the abdomen. Proc. Roy. Soc. London, Ser. B 143: 126–136.

Florkin, M., and C. Jeuniaux. 1974. Hemolymph: composition. Pages 255–307 *in* M. Rockstein, ed. The physiology of insecta. 2nd ed., Vol. 5. Academic Press, New York, London.

Hackman, R. H. 1974. Chemistry of the insect cuticle. Pages 215–270 *in* M. Rockstein, ed. The physiology of insecta. 2nd ed., Vol. 6. Academic Press, New York, London.

Haas, H. 1955. Untersuchungen zur Segmentbildung an der Antenne von *Periplaneta americana* L. Arch. Entwicklungsmech. 147: 434–473.

Hauer, F. von. 1850. Versammlungen von Freunden der Naturwissenschaften in Laybach. Ber. Mitt. Freunden Naturwiss. [Wien] 6: 174–184.

Imms, A. D. 1939. On the antennal musculature in insects and other arthropods. Quart J. Microscop. Sci. (n. s.) 81: 273–320.

——. 1940. On growth processes in the antennae of insects. Quart. J. Microscop. Sci. (n. s.) 81: 585-593.

Jones, J. C. 1964. The circulatory system of insects. Pages 1-107 *in* M. Rockstein, Ed. The physiology of insecta. Vol. 3. Academic Press, New York, London.

Kamburov, S. S., D. J. Nadel, and R. Kenneth. 1967. Observations on *Hesperomyces virescens* Thaxter (Laboulbeniales), a fungus associated with premature mortality of *Chilocorus bipustulatus* L. in Israel. Israel J. Agr. Res. 17: 131-134.

Karsten, H. 1869. Chemismus der Pflanzenzelle. Eine morphologisch-chemische Untersuchung der Hefe. W. Braumüller, Vienna. 90 pp.

Kilby, B. A. 1963. The biochemistry of the insect fat body. Pages 111-174 *in* J. W. L. Beament, J. E. Treherne, and V. B. Wigglesworth, eds. Advances in insect physiology. Academic Press, London, New York.

Lawrence, P. A. 1967. A simple model of the embryo. Pages 53-68 *in* J. W. L. Beament and J. E. Treherne, eds. Insects and physiology. Oliver and Boyd, Edinburgh, London.

Locke, M. 1974. The structure and formation of the integument in insects. Pages 123-213 *in* M. Rockstein, ed. The physiology of insecta. 2nd ed., Vol. 6. Academic Press, New York, London.

Lower, H. F. 1964. The arthropod integument. Pages 275-288 *in* M. Thiel, ed., Studium Generale. 17 (5). Springer-Verlag, Berlin, Göttingen, Heidelberg.

Mercier, L., and R. Poisson. 1927. Une Laboulbéniale, *Stigmatomyces ephydrae* n. sp., parasite d'*Ephydra riparia* Fall. (Dipt. *Ephydridae*). Bull. Soc. Zool. Fr. 52: 225-231.

Neville, A. C. 1967. Chitin orientation in cuticle and its control. Pages 213-286 *in* J. W. L. Beament, J. E. Treherne and V. B. Wigglesworth, eds. Advances in insect physiology. Academic Press, New York, London.

Noble-Nesbitt, J. 1967. Aspects of the structure, formation and function of some insect cuticles. Pages 3-16 *in* J. W. L. Beament and J. E. Treherne, eds. Insects and physiology. Oliver and Boyd, Edinburgh, London.

Peyritsch, J. 1871. Über einige Pilze aus der Familie der Laboulbenien. Sitzungsber. Kaiserl. Akad. Wiss., Math.-Naturwiss. Cl., Abt. 1 [Wien] 64: 441-458. Taf. I-II.

Picard, F. 1908. Sur une Laboulbéniacée marine (*Laboulbenia marina* n. sp.), parasite d'*Aepus Robini* Laboulbéne. C. R. Hebd. Séances Mém. Soc. Biol. [Paris] 65: 484-486.

Richards, A. G. 1951. The integument of arthropods, the chemical components and their properties, the anatomy and development, and the permeability. Univ. of Minnesota Press, Minneapolis. 411 pp.

Richards, A. G., and M. N. Smith. 1956. Infection of cockroaches with *Herpomyces* (Laboulbeniales). II. Histology and histopathology. Ann. Entomol. Soc. Amer. 49: 85-93.

Rick, J. 1903. Zur Pilzkunde Vorarlbergs. V. Oesterr. Bot. Z. 53: 159-164.

Robin, C. P. 1853. Histoire Naturelle des Végétaux Parasites qui Croissent sur l'Homme et sur les Animaux Vivants. Vol. 1, Atlas. J.-B. Baillière, Paris. 702 pp., Atlas 24 pp., Pls. I-XV.

Rouget, A. 1850. Notice sur une production parasite orservée [sic] sur le *Brachinus crepitans*. Ann. Soc. Entomol. Fr. (sér. 2) 8: 21-24.

Scheloske, H.-W. 1969. Beiträge zur Biologie, Okologie, und Systematik der Laboulbeniales (Ascomycetes) unter besonderer Berücksichtigung des Parasit-Wirt Verhältnisses. Parasitol. Schriftenreihe Heft 19: 1-176.

Smith, D. S. 1968. Insect cells, their structure and function. Oliver and Boyd, Edinburgh. 372 pp.

Snodgrass, R. E. 1935. Principles of insect morphology. McGraw-Hill. New York, London. 667 pp.

Spegazzini, C. 1915. Laboulbeniali ritrovate nelle collezioni di alcuni musei italiani. An. Mus. Nac. Hist.Nat. Buenos Aires 26: 451-511.

——. 1917. Revisión de las Laboulbeniales argentinas. An. Mus. Nac. Hist. Nat. Buenos Aires 29: 445-688.

Steinhaus, E. A., and Y. Tanada. 1971. Diseases of the insect integument. Pages 1-86 *in* T. C. Cheng, ed. Current topics in comparative pathobiology. Vol. 1. Academic Press, New York, London.

Tavares, I. I. 1965. Structure and development of *Herpomyces stylopygae* (Laboulbeniales). Amer. J. Bot. 53: 311-318.

Thaxter, R. 1896. Contribution towards a monograph of the Laboulbeniaceae. Mem. Amer. Acad. Arts. Sci. 12: 187–429. Pls. I–XXVI.

———. 1914. On certain peculiar fungus parasites of living insects. Bot. Gaz. [Crawfordsville] 58: 235–253. Pls. XVI–XIX.

———. 1931. Contribution towards a monograph of the Laboulbeniaceae. Part V. Mem. Amer. Acad. Arts Sci. 16: 1–435. Pls. I–LX.

Tonghini, C. C. 1913. Ulteriori ricerche morfologiche e biologiche sulle Laboulbeniacee. Malpighia 26: 329–344, 477–518. Pl. XII.

Wigglesworth, V. B. 1961. The principles of insect physiology. 5th ed. E. P. Dutton and Co., New York, 546 pp.

chapter 10

Symbiosis, Commensalism And Aposymbiosis—Conclusions

by L. R. BATRA

It is clear from the prefatory song and Whisler's lively chapter, "The Fungi Versus the Arthropods," that our studies are important. He casts the net wide and covers saprophytism, symbiosis, commensalism and parasitism (Table 1.1). It is also evident, particularly from concluding remarks by Weber (Chapter 5), that, apart from academic interest, most symbioses discussed here are of great economic concern. I am pleased, however, that all contributors have been temperate and careful to avoid exaggerated claims of importance, success or unwarranted optimism.

This chapter deals with the following three points: (1) a comparative summary of major conclusions reached by the contributors; (2) some theoretical and practical considerations that may relate to the economic importance of symbionts; and (3) direction of future research and some problems or obstacles in conducting research in this field.

The great majority of multicellular animals are Arthropoda. They colonize vast terrestrial and aquatic habitats and are highly variable anatomically and physiologically. It is also notable that symbionts (fungi or bacteria) are found in association with many Insecta that are economically important: Blattaria, Coleoptera, Diptera, Hemiptera, Homoptera, Hymenoptera, and Isoptera; on the other hand Siphonoptera and Lepidoptera have no symbionts.

Host specificity of symbionts and commensals. One of the distinctive characteristics of fungus symbionts and commensals is their host specificity. Certain species are restricted to certain hosts or their close relatives (Table 3.1, Chapters 6, 9). All *Termitomyces* are associated with

INSECT-FUNGUS SYMBIOSIS /Batra (ed.) / Allanheld, Osmun, Montclair, NJ

the Macrotermitinae, and certain species are restricted to one termite genus. Similarly, basidiocarps of fungi from diverse Attini (Table 5.1, cultures W6, W7, W17, W22) are *Lepiota* spp. "Species-specificity" may have evolved to outpace competitive saprophytes and gain advantage of any particular physical, physiological or ethological hospitality the arthropod host may offer. Laboulbeniales are an outstanding example of species-specificity, and this may be comparable to a similar situation among the rusts, smuts and the Peronosporales, which are highly specific pathogens of vascular plants. Benjamin (1965, in Chapter 1) gives several examples of extreme host-specificity. Certain species may even be restricted to certain locations on the insect and then only on one sex. Laboulbeniales are considered parasites by some (Chapter 1) and commensals by others (Chapter 2). The vegetative mycelium is minimal, apparently metabolically efficient and effective, and it is soon transformed into a cellular fruiting body or "perithecium."

Ectosymbiosis. Ectosymbiosis between termites and fungi was first noted in 1779 by König (Chapter 6); 70 years later Leydig observed yeasts in aphid mycetocytes—a term incidentally coined by DeBary in 1887 along with "symbiosis"! For termites, ants and beetles, it is relatively easy to assign a possible dietary role to fungi that regularly occur in the vicinity of the brood, but the complex role of endosymbionts in the life of insects was not elucidated until recently and after many years of comparative histological investigations by Büchner, Koch and others (see references cited in Chapter 6). Speculations about the role of commensals also began very early (Chapter 8, 9).

The physical, physiological and behavioral characteristics of insects such as the ambrosia beetles (Chapters 2, 3), attine ants (Chapter 5) and macrotermitid termites (Chapter 6) so modify their microhabitat that the equally versatile, plastic, and pleomorphic fungi can readily colonize it. By their externally secreted enzymes (Abo-Khatwa, 1978; Chapters 5, 6) the fungi then mobilize nutrients such as cellulose. They abstract and concentrate the very low levels of nitrogen present in wood, they recycle nutrients from excreta (Chapters 5, 6) and in turn they are directly or indirectly significant in providing for the dietary needs of the insects.

Low nitrogen is generally considered to be a limiting factor in the decomposition of wood cellulose and lignin by fungi. The addition of organic nitrogen invariably stimulates, sometimes by over 60 percent, the decay rate by several non-symbiotic Basidiomycetes. The nitrogen that is recycled and concentrated from the excreta of the Attini and Macrotermitinae may well assist in the rapid decomposition of the vegetal matter newly incorporated into their gardens. As judged by mycelium dry weight, this is apparently the case with some Xyleborini (see Batra 1963, 1966 in Kok, Chapter 2) and Macrotermitinae. Data compiled by Swift (1977) from several sources reveal that the most usual ratio between mycelium and wood nitrogen in non-symbiotic fungi would be about 6:1, but it may range from 1:1 to over 14:1.

Vitamins, sterols and other growth factors. In certain symbioses, studies have progressed to a point where specific substances provided by the mycosymbiont have been identified. This has obviously been relatively easier in systems where the fungal substrate, the fungus and the insects are readily isolated. Thus Norris and his students (Chapter 2, 3) show that the fungi provide certain sterols required by insects for their normal growth and reproduction. Juvenile hormones are attributed to *"Xylaria"* from the termite fungus combs (see Sannasi 1969, in Chapter 6), and impairment of metabolism and pigmentation suggests endocrine deficiencies in the aposymbiotic insects (Büchner 1965, in Chapter 6). Jurzitza's meticulous and painstaking work demonstrates that the endosymbiotic yeasts of *Lasioderma* provide certain vitamins, essential amino acids, choline—a sterol, and at the same time recycle uric acid, a waste product (Chapter 4).

Endosymbiosis. In endosymbiosis the fungus cell comes in closest contact with the host protoplast. Perhaps for this reason the greatest challenge to us even today is that we have been unable to obtain axenic cultures of most of the endosymbionts and such commensals as the Trichomycetes and the Laboulbeniales. Having such cultures on hand should facilitate further experimental research on the interactions between the two symbionts and elucidate the probable significance of one to the other, even though we can never be sure of the sterility of aposymbiotic animals. Nutritional requirements of the mycosymbiont may be complex or they may be simple but very exacting, such as a certain level of anaerobiosis in their environment. Having axenic cultures of fungi of course does not solve all our problems. We must also be sure that they perform, as they ought to, when they meet their macrosymbiont and establish specific symbioses. And they must form some sort of distinctive reproductive structures to be classified among their respectable peers and not dumped among the "Mycelia Sterilia."

With respect to fungal endosymbiosis, it may be worth remembering that rapid progress is being achieved in the study of characteristics of nitrogen-fixing bacterial symbionts of plants. Here I realize that there are certain basic differences between these and the fungus-arthropod systems. However I see some possibilities, particularly with respect to techniques, where the exact contribution of fungal symbionts in the life of macrosymbionts may be ascertained by using macrosymbiont single cell cultures, biochemical mutants of the microsymbiont, radioactively labeled nutrients or growth factors, or by transducing phages. It would be unimaginative not to borrow technology from sister fields (cf.

"aposymbiosis" below) even though the prospects of success may be remote.

Coevolution. In nature the symbionts and commensals discussed in this work do not live apart from their arthropod hosts, including the primary ambrosia fungi which are often intermixed with or followed by auxiliary species. Except for the Laboulbeniales, where no experimental data are forthcoming in this regard, most terrestrial mycosymbionts receive protection from desiccation and suitable metabolic substrata for colonization or overwintering.

Larvae of ambrosia beetles are usually mycetophagous, and in this respect they are apparently more advanced than their bark- or wood-feeding relatives. Earlier I stated ". . . that fungi that may have been fortuitously carried at one time into the tunnels of bark- and wood-inhabiting beetles today live symbiotically with ambrosia beetles in a truly mutualistic relation. Initially the fungi established themselves beneath the bark, ramifying into the frasspacked tunnels. The fungi were occasionally consumed along with bark, as with *Ips, Tomicus,* and *Dendroctonus* spp. During the course of evolution some fungi perhaps invaded wood and made it possible for the beetles to penetrate weakened xylem while feeding on fungi. Eventually some beetles abandoned the eating of bark and wood, became wholly adapted to mycetophagy, and developed mutualism with ambrosia fungi." (Batra 1966, in Chapter 2). Among the termites, similar coevolutionary trends may have led to the use of the ectosymbiont *Termitomyces* and wood decomposed by fungi rather than by the protozoan endosymbionts. Hand in hand with the change in dietary habit, the development of certain structures such as mycangia, used for transport, storage and for the multiplication of starter inoculum, by ambrosia beetles, Attini and Siricidae, is a spectacular coevolutionary phenomenon. Likewise, adaptation to cavernicolous mode of life or growth in insect tunnels by the mycosymbionts is equally notable.

As pointed out by Moss (Chapter 8), for the perpetuation of the commensalism the Trichomycetes must provide for the close ". . . coordination of the vegetative growth and reproductive development of the commensal with metabolism, ecdysis and death of the host." This is also true of all mycosymbionts discussed here. The Trichomycetes have evolved fast-developing, effective holdfasts that secure the fungus to cuticular lining of the gut. Some of these fungi have also evolved special types of spores ". . . for the trip outside of the animal." (Whisler, Chapter 1). Additional adaptive characteristics useful in the perpetuation of Trichomycete commensalism are ". . . limited vegetative growth, conversion of the entire or majority of the thallus cytoplasm into infestive propagules, sequential maturation of asexual spores in the Eccrinales and Harpellales and spore appendages in genera of the

Eccrinales and Harpellales infesting aquatic hosts" (Moss, Chapter 8). A most notable ecological adaptation to the life cycle of the host is the type of spore production in *Amoebidium parasiticum*. During the intermoult period it produces rigid, elongate sporangiospores that may infest other hosts or, more commonly, attach to the same host. "Immediately prior to moulting, or upon injury or death of the host there is a switch from rigid to amoeboid sporangiospore production." (Fig. 8).

Systematics of mycosymbionts and commensals. Pioneering work on symbiotic and commensalistic fungi was undertaken by specialists in the Arthropoda, and it is notable that about half of the contributors to this symposium are professors of entomology. In his monumental compendium, Büchner (1965, in Chapter 6) repeatedly chided microbiologists for their lack of cooperation and regretted their apathy towards experimental work on endosymbionts, but fortunately the situation since then has somewhat improved. In the absence of interdisciplinary teamwork, meaningful data in our field are hard to obtain, and efforts are often unrewarding and frustrating.

As stated in the preface, symbiosis is an ecological phenomenon, with various levels of intimacy and nutritional interdependence between the symbionts. Just as parasitism often provides useful taxonomic and coevolutionary information, mutualism too can provide similar characteristics of value. Different degrees of pleomorphism in bacterial and fungal endosymbionts and in some ectosymbionts, such as the fungi of ambrosia beetles, ants and termites, may be important characteristics (Chapters 3, 4, 5, 6). The ambrosia or 'kohlrabi' phase of the fungus is attributed to the influence of the insect, for the morphologically similar yeastlike cells from mycetocytes, mycangia or the 'garden' give rise to taxa belonging to such diverse groups as the Hemiascomycetes, the Euascomycetes, and the Fungi Imperfecti (Batra and Batra, 1967, in Chapter 6). Each taxonomic group varies only within certain morphological limits, however.

Symbiotic fungi in general, and pleomorphic species in particular, present some technical problems:

First, when freshly isolated, many of these fungi are extremely slow growing, even on the most complex media. In this respect they resemble the mycobionts of the lichens and some of the mycorrhizal species.

Second, their natural pleomorphism makes it extremely difficult and laborious to recover them from the insect symbiont, particularly the intracellular yeasts which often assume the appearance of cell organelles. During their primary isolation one must be extremely meticulous and continuously monitor microscopically the development of fungus propagules from the insect or its vicinity. Only then can one be sure of their legitimacy as true symbionts.

Third, Koch's postulates must be fulfilled. Here we face the problem of obtaining aposymbiotic or "sterile" hosts. Acceptance/rejection tests with fungi and ants by Weber (Chapter 5), similar tests by Norris with the Xyleborini beetles (Chapter 3), and more refined tests by Jurzitza with the anobiid beetles (Chapter 4) fulfill such postulates well.

Fourth, pleomorphism may be a direct outcome of *insect behavior* towards the fungus or *metabolic products,* at least in the ectosymbionts, or cytoplasmic *stimuli* in the intracellular species. Ethological observations on the Attini (Chapter 5) and the Macrotermitinae (Chapter 6) reveal that the fungus staphylae and the spherules on the comb, respectively, are tended rather similarly to the insects' brood. The questions of whether or not these fungus structures mimic elements of the brood, and the chemotactic or thigmotactic interactions with the insects, are worth further exploration.

Aposymbiosis. Since mutualistic insects cannot survive in nature for long without their mycosymbionts, it is evident that controlling the fungus may also control a pest. Jurzitza demonstrated that aposymbiotic *Lasioderma* showed signs of malnutrition and did not survive long (Chapter 4). Other workers cited by him similarly concluded that in the absence of the microsymbiont, the host size, vigor and fecundity of diverse pests were severely impaired and larval mortality was higher than the controls. In the laboratory, aposymbiosis has been achieved, mostly with the endosymbionts, by surgical removal of mycetocytes, surface disinfestation, thermal inactivation, or by the incorporation into their diet of antibiotics or other chemicals such as sulphanomides, gentian violet, etc. Strategies to synergistically predispose target pests should be carefully considered in any program.

At present there are no practical ways to control insect pests by aposymbiosis, and any chemicals to control symbionts must not select resistant forms. It is surprising that hardly any basic research is being conducted on the role of aposymbiosis as a tool to restrain arthropod pests, within the overall strategy of integrated pest management, including control by microbial pathogens which are well reviewed (Burgess and Hussey, 1971; Cantwell, 1975; DeBach, 1964; Evlakhova, 1974; Franz and Krieg, 1972; Koval, 1974; Müller-Kögler, 1965; Steinhaus, 1967; Weiser, 1966). Aposymbiosis may debilitate our enemy, and anything that debilitates our enemy must be looked into —may it even be a chariot wheel!

LITERATURE CITED

Abo-Khatwa, N. 1978. Cellulase of fungus-growing termites. A new hypothesis on its origin. Experientia 34: 559–60.
Burgess, H. D., and N. W. Hussey. 1971. Microbial control of insects and mites. Academic Press, New York. 861 pp.
Cantwell, G. E. 1975. Insect diseases. Marcel Dekker, New York. 567 pp.
DeBach, P. 1964. Biological control of insect pests and weeds. Chapman and Hall, London. 844 pp.
Evlakhova, A. A. 1974. The entomopathogenic fungi: systematics, biology and practical importance. Naukha, Leningrad. 260 pp.
Franz, J. M., and A. Krieg. 1972. Biologische Schädlingsbekämpfung. P. Parey, Berlin. 208 pp.
Koval, E. Z. 1974. Guidebook to entomophilic fungi of USSR. Naukowa Dumka, Kiev. 260 pp.
Müller-Kögler, E. 1965. Pilzkrankheiten bei Insekten. P. Parey, Berlin. 444 pp.
Steinhaus, E. A. 1967. Insect Pathology: An advanced treatise. Academic Press, New York. Vol. 1, 661 pp; Vol. 2, 687 pp.
Swift, M. J. 1977. The ecology of wood decomposition. Sci. Prog. Oxford 64: 175–99.
Weiser, J. 1966. Neomci Hymzu. Narladatelstve Ceskoslv. Acad. Ved. (Prague). 554 pp.

Index of Authors*

*Full citations appear on page numbers given in italics.

Subject Index